ATOMIC PROCESSES IN PLASMAS

Published Proceedings in the Series of Conferences on Atomic Processes in Plasmas

	Year	Held in	Publisher	ISBN
15th	2007	Gaithersburg, Maryland	AIP Conf. Proceedings vol. 926	978-0-7354-0436-6
14th	2004	Santa Fe, New Mexico	AIP Conf. Proceedings vol. 730	0-7354-0211-6
13th	2002	Gatlinburg, Tennessee	AIP Conf. Proceedings vol. 635	0-7354-0090-3
12th	2000	Reno, Nevada	AIP Conf. Proceedings vol. 547	1-56396-976-9
11th	1998	Auburn, Alabama	AIP Conf. Proceedings vol. 443	1-56396-802-9
10th	1996	San Francisco, California	AIP Conf. Proceedings vol. 381	1-56396-552-6
9th	1993	San Antonio, Texas	AIP Conf. Proceedings vol. 322	1-56396-411-2
8th	1991	Portland, Maine	AIP Conf. Proceedings vol. 257	0-88318-939-9
7th	1989	Gaithersburg, Maryland	AIP Conf. Proceedings vol. 206	0-88318-769-8
6th	1987	Santa Fe, New Mexico	AIP Conf. Proceedings vol. 168	0-88318-368-4

To learn more about these titles, or the AIP Conference Proceedings Series, please visit the webpage
http://proceedings.aip.org/proceedings

ATOMIC PROCESSES IN PLASMAS

15th International Conference on
Atomic Processes in Plasmas

Gaithersburg, Maryland   19 – 22 March 2007

EDITORS

John D. Gillaspy
John J. Curry
Wolfgang L. Wiese
National Institute of Standards and Technology
Gaithersburg, Maryland

All papers have been peer reviewed.

SPONSORING ORGANIZATION
U. S. Department of Energy, Fusion Energy Sciences
Lawrence Livermore National Laboratory
Los Alamos National Laboratory
National Institute of Standards and Technology

CD-ROM INCLUDED

Melville, New York, 2007
AIP CONFERENCE PROCEEDINGS ■ 926

Editors

John D. Gillaspy
John J. Curry
Wolfgang L. Wiese

National Institute of Standards and Technology
Gaithersburg, Maryland
20899

E-mail: john.gillaspy@nist.gov
         jjcurry@nist.gov
         wolfgang.wiese@nist.gov

L.C. Catalog Card No. 2007931859
ISBN 978-0-7354-0436-6
ISSN 0094-243X

Printed in the United States of America

CONTENTS

HIGH ENERGY DENSITY PHYSICS

FUSION PLASMAS

*Italicized names indicate authors who presented the papers.

SMALL LASER PLASMAS

FUNDAMENTAL DATA AND MODELING

PLASMA DIAGNOSTICS

INDUSTRIAL PLASMAS

The 15th International Conference on Atomic Processes in Plasmas and these Proceedings are dedicated to the memory of

Yong-Ki Kim
(1932 - 2006)

in recognition of his many important contributions to this field and his more than two decades of service to this conference.

Preface

The 15th International Conference on Atomic Processes in Plasmas (APiP) was held March 19-22, 2007 at the National Institute of Standards and Technology in Gaithersburg, Maryland. The Executive Committee dedicated the 2007 conference to Yong-Ki Kim, for his many scientific contributions to this field, and for several decades of service to this conference series. Yong-Ki effectively served as co-chair of the present conference until his untimely death in an automobile accident in September, 2006. A memorial lecture was delivered by Philip Stone in the Fundamental Data and Modeling Session, and a related article is contained in this volume.

In anticipation of the release of an Interagency Task Force Report on the topic of "High Energy Density Physics (HEDP)", and the growing interest in this topic in general, this year we began the conference with several sessions on HEDP, and had a plenary lecture by Philip Bucksbaum about the Linac Coherent Light Source at the Stanford Linear Accelerator Center.

This year we also began referring to the conference as "International" rather than "APS Topical" in order to better reflect the true nature of the conference. This change was enacted by a majority vote of the executive committee, after the American Physical Society expressed their desire to cease using the term "APS Topical" for conferences that now appear to be continuing indefinitely into the future.

The number of registered participants was 158, which after 14 cancellations yielded the 144 participants listed in an appendix at the end of this volume. There were 37 invited speakers, and approximately 80 contributed posters. The coauthors of the posters represented 18 different countries and half the states in the U.S. The fraction of participants (and speakers) from outside of the U.S. was 22%. The fraction of participants from government, university, industry, or other were 53%, 40%, 6%, and 1%, respectively. The institutions with the largest number of participants were NIST (23), Livermore (11), Los Alamos (9), and NRL/Artep (7).

This volume contains contributions from all the invited speakers. Most of these contributions are peer-reviewed manuscripts, and the rest are abstracts (sometimes revised and extended, and/or with references added). We have limited the use of color in the printed versions of the manuscripts, but have provided full color versions in a CD included in this volume, as well as online. We note that the last-minute replacement of one speaker (Biedermann) by another (Ralchenko) was caused by severe airline delays due to inclement weather on the East Coast, but the general topic (and even title) of this talk was unchanged.

The poster abstracts, photographs of the invited speakers, details of the excursion to the National Air and Space Museum Udvar-Hazy Center, and other information is available on the conference web site at http://physics.nist.gov/Meetings/APIP.

Financial support for this conference was provided by the U.S. Department of Energy (Office of Fusion Energy Sciences), the Lawrence Livermore National Laboratory, the Los Alamos National Laboratory, and the National Institute of Standards and Technology.

We would like to also acknowledge here the session Chairs and people who gave introductory remarks: James Hill, William Ott, Francis Thio, Chris Keane, Dan Stutman, Dieter Schneider, Randall Smith, Todd Ditmire, Ron McKnight, Wolfgang Wiese, Joe Abdallah, John Seely, and James Whetstone.

The Executive Committee selected the Lawrence Livermore National Laboratory as the host of the next APiP conference, under the Chairmanship of Kevin Fournier.

John D. Gillaspy, Chair
John J. Curry
Wolfgang L. Wiese

June 2007
Gaithersburg, Maryland

Executive and Program Committee

John Gillaspy, Chair (National Institute of Standards and Technology)
James Babb (Harvard-Smithsonian Center for Astrophysics)
James Bailey (Sandia National Laboratories)
Jacques Bauche (University of Paris XI)
James Cohen (Los Alamos National Laboratory)
Gary Ferland (University of Kentucky)
Kevin Fournier (Lawrence Livermore National Laboratory)
Marcel Klapisch (Naval Research Laboratory)
Richard Lee (Lawrence Livermore National Laboratory)
Roberto Mancini (University of Nevada, Reno)
Stephane Mazevet (Commissariat à Énergie, France)
Fred Meyer (Oak Ridge National Laboratory)
John Rice (Massachusetts Institute of Technology)
Jorge Rocca (Colorado State University)
William Rowan (University of Texas)
David Schultz (Oak Ridge National Laboratory)
Phillip Stancil (University of Georgia)

Local Organizing Committee

John Gillaspy, Chair (National Institute of Standards and Technology)
Paul Bergstrom (National Institute of Standards and Technology)
Michael Crisp (U.S. Department of Energy)
John Curry (National Institute of Standards and Technology)
Uri Feldman (Naval Research Laboratory)
Michael Finkenthal (The Johns Hopkins University)
Jeffrey Fuhr (National Institute of Standards and Technology)
Larry Hudson (National Institute of Standards and Technology)
Terrence Jach (National Institute of Standards and Technology)
Martin Laming (Naval Research Laboratory)
Howard Milchberg (University of Maryland)
Joshua Pomeroy (National Institute of Standards and Technology)
Yuri Ralchenko (National Institute of Standards and Technology)
Joseph Reader (National Institute of Standards and Technology)
Steve Rolston (University of Maryland)
Edward Saloman (National Institute of Standards and Technology)
Barry Schneider (U.S. National Science Foundation)
John Seely (Naval Research Laboratory)
Eric Silver (Harvard-Smithsonian Astrophysical Observatory)
Randall Smith (U.S. National Aeronautic and Space Administration)
Philip Stone (National Institute of Standards and Technology)
Endre Takacs, (University of Debrecen, Hungary)
Joseph Tan (National Institute of Standards and Technology)
Wolfgang Wiese (National Institute of Standards and Technology)

Sponsoring Organizations

U.S. Department of Energy, Fusion Energy Sciences
Lawrence Livermore National Laboratory
Los Alamos National Laboratory
National Institute of Standards and Technology

Meetings in This Series

1. Knoxville, Tennessee   February 16-18, 1977   *not published*
2. Boulder, Colorado   January 17-19, 1979   *not published*
3. Baton Rouge, Louisiana   February 25-27, 1981   *not published*
4. Princeton, New Jersey   April 13-15, 1983   *not published*
5. Pacific Grove, California   February 25-28, 1985   *not published*
6. Santa Fe, New Mexico   September 28-October 2, 1987   AIP vol. 168
7. Gaithersburg, Maryland   October 2-5, 1989   AIP vol. 206
8. Portland, Maine   August 25-29, 1991   AIP vol. 257
9. San Antonio, Texas   September 19-23, 1993   AIP vol. 322
10. San Francisco, California   January 14-18, 1996   AIP vol. 381
11. Auburn, Alabama   March 23-26, 1998   AIP vol. 443
12. Reno, Nevada   March 19-23, 2000   AIP vol. 547
13. Gatlinburg, Tennessee   April 22-25, 2002   AIP vol. 635
14. Santa Fe, New Mexico   April 19-22, 2004   AIP vol. 730
15. Gaithersburg, Maryland   March 19-22, 2007   AIP vol. 926

HIGH ENERGY DENSITY PHYSICS

Ultrafast X-ray Science at SLAC: Preparing for LCLS

Philip H. Bucksbaum

PULSE Center, Stanford Linear Accelerator Center, Stanford University
2575 Sand Hill Road, Menlo Park, CA 94025

Abstract. Ultrafast lasers (t less than 1 ps) can capture the quantum dynamics of single vibration in a crystal lattice or in a molecule, and they have also been used to view the transient molecular-scale transformations of chemical reactions. Hard x-rays (E greater than 1 keV) can probe the structure of matter on the length scale of a chemical bond. Until recently, only relatively weak sources based on laser-induced plasma radiation were capable of capturing these ultrafast dynamics and also viewing them on the scale of a single chemical bond. The recent Sub-Picosecond Pulse Source experiment at SLAC was the first instrument based on synchrotron radiation from an undulator that could do both. During its two-year run, its 8 keV, 80 fs x-ray pulses were the brightest ultrafast x-rays ever produced. The planned X-ray free electron laser at SLAC (LCLS) will be far brighter, generating focused x-ray fields as strong as atomic binding fields, comparable to today's highest intensity lasers. These new tools are creating some special opportunities for new science, and also some challenges. I will discuss these, and present recent progress in ultrafast x-ray sources and science. Additional information on the topic of my talk can be found in a recent publication [1].

REFERENCES

1. A. L. Cavalieri et al., *Phys. Rev. Lett.*, **94**, 114801 (2005).

CP926, *Atomic Processes in Plasmas—15th International Conference on Atomic Processes in Plasmas*
edited by J. D. Gillaspy, J. J. Curry, and W. L. Wiese
© 2007 American Institute of Physics 978-0-7354-0436-6/07/$23.00

Spectroscopic Investigation of the Particle Density and Motion in an Imploding z-Pinch Plasma

E. Kroupp, L. Gregorian[1], G. Davara[2], A. Starobinets, E. Stambulchik, and Y. Maron

Faculty of Physics, Weizmann Institute of Science, Rehovot, 76100, Israel

Yu. Ralchenko

Atomic Physics Division, NIST, Gaithersburg, MD 20899, USA

S. Alexiou

University of Crete, 71409 Heraklion, TK2208, Greece

Abstract. Experimental investigations of the ion density and flow as a function of time and space in z-pinch plasmas are of key importance for improving the understanding of z-pinch dynamics [1-5]. For such studies, measurements of emission- line shapes can be highly useful [6].

In the present experiment line emission of oxygen ions is used to investigate the ion density and motion in the imploding plasma in a 0.6-μs, 220-kA z-pinch experiment. For the time period studied here (220 – 85 ns before the stagnation on-axis), the plasma properties have been extensively characterized previously, employing various spectroscopic methods to determine the time-dependent radial distributions of the ion velocities [7], the magnetic field [8], the charge-state composition [9], the electron temperature [10], and the particle densities [11]. In particular, the electron density was determined from the absolute intensities of spectral lines [9], from the ionization times in the plasma [11], and from momentum-balance considerations, based on the previously measured time-dependent magnetic field radial distribution. The electron density determined was also found to be consistent with energy-balance considerations, as described in Ref. [10].

Using the values of the electron density and temperature as a function of the radial coordinate and time [10-11], the Stark widths of all emission lines observed (of O II – O VI) were calculated (we

---

[1]   Present address: Philips Medical Systems, Haifa, Israel
[2]   Present address: Orbotech, Yavne, Israel

CP926, *Atomic Processes in Plasmas—15th International Conference on Atomic Processes in Plasmas*
edited by J. D. Gillaspy, J. J. Curry, and W. L. Wiese
© 2007 American Institute of Physics 978-0-7354-0436-6/07/$23.00

note that all lines used are isolated, i.e, their Stark shapes are Lorentzian). For the Stark broadening computations we employed two independent methods, namely, a quantum-mechanical method based on the Baranger formula [6] and a non-perturbative semi-classical method [12-13]. For the quantum-mechanical calculations of the line widths performed, we use electron-collision cross sections calculated using two approaches: the Coulomb-Born-Exchange method [14] and the convergent close-coupling (CCC) method [15]. The latter has been successfully applied to many atomic-collision experiments, for example for the analysis of spectral-line broadenings in Li- and Be-like ions [16-18]. The calculation results of the quantum-mechanical and of the semi-classical methods were found to be similar for the purpose of present discussion.

The total widths predicted for each line were then determined by convolving the calculated Stark widths with the Doppler broadening (assuming equal ion and electron temperatures; i.e., $T_i = T_e$) and with the measured instrumental broadening. It was found that these widths are significantly smaller than the observed widths. For example for the 3144.7-Å line of OV, the calculated Stark width is found to be 0.20 ± 0.08 Å, the Doppler broadening (due to $T_i = T_e = 13$ eV) is 0.42 ± 0.09 Å, and the instrumental broadening is 0.21 Å. The width resulting form the convolution of these three contributions is 0.51±0.13Å. This value is much smaller than the observed width, 0.98 ± 0.03 Å. Similar results were obtained for the other O III – O VI lines.

Since the uncertainties in the Stark-broadening calculations are believed to be significantly smaller than the difference between the computed and experimental line widths, an additional Doppler broadening is suggested. Energy balance considerations, based on the radial distributions of the electron temperature, electron density, charge state, and magnetic field previously determined (Refs. [8-11]) allow for demonstrating that an ion temperature much higher than T_e is unlikely. For explaining the extra broadening we thus suggest the presence of turbulent ion motion at the outer plasma boundary, which develops in the plasma during the implosion. The spatial scale of the turbulence is believed to be smaller than the spatial resolution of the measurements, which is ≈ 0.5 mm.

Based on this explanation, it is inferred that the ion kinetic energy associated with the turbulence can be up to 70% of the radially-directed kinetic energy. It should be emphasized that these non-thermal ion velocities are inferred for the imploding plasma that was observed to be with no geometrical disruptions, i.e, the small-scale hydrodynamic turbulence here considered results in no spatial distortion of the plasma, to within the 0.5-mm spatial resolution of the measurements.

5

Keywords: Z-pinch dynamics, spectral line broadening, turbulence.

PACS: 52.58.Lq, 52.30.-q, 52.35.Ra, 52.25.Xz, 32.70.Jz

ACKNOWLEDGMENTS

It is a pleasure to thank H.R Griem, A. Fisher, C. Deeney, and A Velikovich for invaluable critical comments. This work was supported in part by Minerva (Germany), Israel Science Foundation (ISF), and Sandia National Laboratories (USA).

REFERENCES

1. D. D. Ryutov, M. S. Derzon, and M. K. Matzen, *Rev. Mod. Phys.* **72**, 167 (2000).

2. M. A. Liberman, J. S. De Groot, A. Toor, and R. B. Spielman, *Physics of High-Density Z-Pinch Plasmas* (Springer-Verlag, New York, 1999).

3. D. Mosher, B. V. Weber, B. Moosman et al., *Laser Part. Beams* 19, 579 (2001).

4. C. Deeny, P.D. LePell, B.H. Failor et al., *Phys.* Rev. E **51**, 4823 (1995).

5. A.L Velikovich, J. Davis, J.W. Thornhill et al., *Phys. Plasmas* **7**, 3265 (2000).

6. H.R. Griem, *Spectral Line Broadening* by *Plasmas* (Academic, New York, 1974).

7. M.E. Foord, Y. Maron, G. Davara, L. Gregorian, and A. Fisher, *Phys. Rev. Lett.* **72**, 3827 (1994).

8. G. Davara, L. Gregorian, E. Kroupp, and Y. Maron, Phys. *Plasmas* **5**, 1068 (1998).

9. L. Gregorian, V.A. Bernshtam, E. Kroupp, G. Davara, and Y. Maron, *Phys. Rev. E* **67**, 016404 *(2003)*.

10. L. Gregorian et al, *Phys. Rev. E* **71**, 056402 (2005).

11. L. Gregorian et al, *Phys. Plasmas* **12**, 092704 (2005).

12. S. Alexiou, *Phys. Rev. Lett.* **75**, 3406 (1995).

13. S. Alexiou and R.W. Lee, *J. Quant. Spectrosc. Radiat. Transf.* **99**, 10 (2006).

14. V.P. Shevelko and L.A. Vainshtein, *Atomic Physics for Hot Plasmas* (IOP Publishing, Bristol, 1993).

15. I. Bray, D.V. Fursa, A.S. Kheifets, and A.T. Stelbovics, *J. Phys.* B **35**, R117 (2002).

16. H.R. Griem, Yu.V. Ralchenko, and I. Bray, *Phys. Rev. E* **56**, 7186 (1997).

17. Yu.V. Ralchenko, H.R. Griem, and I. Bray, *J. Quant. Spectrosc. Radiat. Transf.* **81**, 371 (2003).

18. Yu.V. Ralchenko, H.R. Griem, I. Bray, and D. V. Fursa, *Phys. Rev.* A **59**, 1890 (1999)..

X-ray Thomson Scattering from Dense Plasmas

S. H. Glenzer

Lawrence Livermore National Laboratory, L-399, University of California, P.O. Box 808, Livermore, CA 94551, USA

Abstract. X-ray Thomson scattering has been developed for accurate measurements of densities and temperatures in dense plasmas. Experiments with laser-produced x-ray sources have demonstrated Compton scattering and plasmon scattering from isochorically-heated solid-density beryllium plasmas. In these studies, the Ly-alpha or He-alpha radiation from nanosecond laser plasmas has been applied at moderate x-ray energies of E = 3 - 9 keV sufficient to penetrate through the dense plasma and to avoid intense bremsstrahlung radiation at lower energies. In backscattering geometry, the experiments have accessed the non-collective Compton scattering regime where the spectrum reflects the electron velocity distribution of the plasma, thus providing an accurate measurement of the temperature. In addition to the inelastic Compton scattering feature, the spectra also show elastic (Rayleigh) scattering from tightly bound electrons. The intensity ratio of these features yields the ionization state that has been applied to infer the electron density in isochorically-heated matter. Forward scattering in these conditions have observed plasmons that allow direct and accurate measurements of the electron density from the frequency shift of the plasmon peak from the incident probe energy. The back and forward scattering data are in mutual agreement indicating an electron density of n_e = 3 x 10^{23} cm^{-3}, which is also consistent with results from radiation hydrodynamic simulations. These findings indicate that x-ray Thomson scattering provides accurate characterization in the previously unexplored regime of high-energy density matter. Future work will explore applications to measure compressibility, collisions, and electronic properties of dense matter.

Keywords: Plasma Diagnostic, Dense Plasmas, Thomson scattering.
PACS: 52.25.Os, 52.35.Fp, 52.50.Jm

INTRODUCTION

Accurate characterization techniques of dense plasmas for measurements of temperature, density, and ionization state are important for understanding and modeling high-energy density science experiments.[1,2] Examples of important applications include the measurement of the electron temperature of inertial confinement fusion capsule implosions where the ratio of electron to Fermi temperature will characterize the fuel adiabat.[3] Moreover, x-ray scattering may be applied as a tool to resolve fundamental physics questions such as the equation of state in dense matter[4], structure factors in two-component plasmas[5], limits of the validity of the random phase approximation[6], and the role of collisions.[7]

Significant advances in the physics of low-density plasmas has been enabled by the development of optical lasers and the ability to characterize plasma conditions with Thomson scattering.[8-10] Ultraviolet lasers[11,12] and soft x-ray lasers[13] have been subsequently applied to characterize high-density laser-produced plasmas by Thomson scattering[14-17] and interferometry[18,19], respectively. However, to access dense matter

CP926, *Atomic Processes in Plasmas—15th International Conference on Atomic Processes in Plasmas*
edited by J. D. Gillaspy, J. J. Curry, and W. L. Wiese
© 2007 American Institute of Physics 978-0-7354-0436-6/07/$23.00

with densities of solid and above it has been necessary to develop powerful x-ray sources[20] that penetrate through dense or compressed materials[21] and that fulfill the stringent requirements[22] on photon numbers and bandwidth for spectrally-resolved x-ray Thomson scattering measurements in single shot experiments.[23]

In this work, laser-produced He-alpha and Ly-alpha x-ray sources have been applied to perform x-ray Thomson scattering measurements in isochorically-heated solid-density beryllium.[24,25] Measurements in backscatter geometry have accessed the Compton scattering regime where the scattering process is non-collective and the spectrum shows the Compton down-shifted line that is broadened by the thermal motion of the electrons, thus providing the temperature with high accuracy. In addition to the Compton scattering feature from inelastic scattering by free and weakly bound electrons, the non-collective scattering spectrum exhibit the unshifted Rayleigh scattering component from elastic scattering by tightly bound electrons. The latter occupy quantum states with ionization energy larger than the Compton energy deep in the Fermi sea that cannot be excited due to the Pauli exclusion principle. These electrons become available for inelastic scattering by x rays and contribute to the Compton feature by thermal excitation and ionization. The intensity ratio of the inelastic Compton to the elastic Rayleigh scattering component can hence be a sensitive measure of the ionization state.[24,26,27] For isochorically heated matter where the ion density is known *a priori*, the ionization state also provides a measurement of the electron density.

In the forward scattering regime, collective plasmon oscillations have been observed.[25] The spectrum provides directly the local electron density from the frequency shift of the plasmon peak from the incident probe x-ray energy. The experiments have been performed in a well-characterized, isochorically-heated, solid-density mm-scale beryllium cylinder. Forward scattering of the narrow-band chlorine Ly-alpha x-ray line at 2.96 keV accesses the collective scattering regime and measures the characteristic plasmon peak associated with the collective plasma (Langmuir) oscillations.[28] The results show that this technique provides densities in agreement with the Compton scattering method that is also consistent with radiation hydrodynamic calculations using the code LASNEX.[29] In addition, forward scattering is not dependent on knowledge of the ion density and is thus directly applicable to characterizing compressed matter.

THEORY OF X-RAY THOMSON SCATTERING

For conditions where the energy of the scattered radiation is close to the incident x-ray probe energy, E_0, i.e., for small momentum transfers, the scattering geometry and the probe energy determine the scattering vector $\mathbf{k}$ through the relation

$$k = |\mathbf{k}| = 4\pi \frac{E_0}{hc} \sin\left(\frac{\theta}{2}\right). \tag{1}$$

In collective scattering the probe samples plasma scale lengths larger than the plasma screening length, λ_S, which can be approximated by the Thomas-Fermi length, λ_{TF}, for a degenerate system. For classical plasmas, on the other hand, the usual Debye screening length, $\lambda_D = (\varepsilon_0 k_B T_e / n_e e^2)^{1/2}$, should be applied. Here, ε_0 is the permittivity of free space, k_B is the Boltzmann constant, T_e is the electron temperature, and n_e is the electron density. For weakly degenerate plasmas, calculating the Debye length at an effective temperature will result in a smooth interpolation between the degenerate and classical plasma limits.[30,31]

In the collective scattering regime, the scattering parameter $\alpha > 1$ with α being defined as

$$\alpha = \frac{1}{k\lambda_S}. \tag{2}$$

For example, to access collective scattering in a weakly degenerate solid-density beryllium plasma with electron temperature of the order of the Fermi temperature, $T_e = T_F = 15\,eV$, requires forward scattering with $\theta = 40°$ and x-ray probe energies or order $E_0 = 3\,keV$ ($\lambda \approx 4$ Å). In these conditions, the scattering is predominantly probing k-vectors with $k = 1$ Å^{-1}. Calculating the screening length at the effective temperature results in $\alpha = 1.6$. In this regime, the scattered light spectrum shows collective effects corresponding to scattering resonances off ion acoustic waves and off electron plasma waves, i.e. plasmons. The plasmon frequency shift from E_0 is determined by the plasmon dispersion relation that can be approximated for small values of k by[32]

$$\omega^2_{plasmon} = \omega^2_p + 3k^2 v^2_{th}(1 + 0.088 n_e \Lambda^3_e) + \left(\frac{\hbar k^2}{2m_e}\right)^2. \tag{3}$$

where $\omega_p^2 = (n_e e^2/\varepsilon_0 m_e)^{1/2}$ is the plasma frequency, $v_{th} = (k_B T/m_e)^{1/2}$ is the thermal velocity and $\Lambda_e = h/(2\pi m_e k_B T)^{1/2}$ is the thermal wave length. In Eq. (3), the first term is a result of electron oscillations in the plasma, the second term represents the effect on propagation of the oscillation from thermal pressure. The third term includes degeneracy effects from Fermi pressure, and the last term is the quantum shift.[33] The quantum shift and the electron oscillation terms are temperature independent, and the thermal pressure is weakly dependent on T_e for the partially degenerate plasmas of interest here. Therefore, the plasmon energy provides a sensitive measure of the plasma electron density.

In backscattering geometry and at high x-ray probe radiation energies, the momentum transfer during the scattering process results in a significant frequency shift to the scattered radiation, i.e. the Compton effect.[34] During the scattering process, the incident photons transfer the momentum $\hbar k$ and the energy $\hbar\omega = \hbar^2 k^2/2m_e$ to the electrons. Momentum and energy can only be transferred to the free electrons and weakly bound electrons with binding energy less than the Compton energy. The free electrons carry the information on the temperature of the plasma resulting in a spectral broadening, $\omega = \mathbf{kv}$. The spectrum directly reflects the distribution function showing a parabolic function in case of a degenerate plasma with the width reflecting the Fermi temperature, T_F, and in case of a classical plasma a Gaussian spectral profile indicates a Maxwell-Boltzmann distribution function with the width providing the electron temperature, T_e.

On the other hand, bound electrons with ionization energies larger than the Compton energy cannot be excited, and no energy can be transferred during the scattering process. The corresponding spectral feature is an un-shifted elastic scattering component. Clearly, with increasing ionization tightly bound electron

10

become available for inelastic Compton scattering, and the ratio of elastic to inelastic scattering is a measure of the ionization state. In isochoric heated matter, the electron density can be deduced, from $n_e = 6 \times 10^{23}\ Z/A\ \rho\ cm^{-3}$ with $\rho = 1.85$ for Be and for $Z = 2.5$, results in $n_e = 3 \times 10^{23}\ cm^{-3}$.

The scattering cross section is described in terms of the dynamic structure factor of all the electrons in the plasma[30,35,36]

$$\frac{d^2\sigma}{d\Omega d\omega} = \sigma_{Th}\frac{k_S}{k_0}S(\mathbf{k},\omega)\qquad(4)$$

where σ_{Th} is the usual Thomson cross section and $S(\mathbf{k},\omega)$ is the total dynamic structure factor defined as

$$S(\mathbf{k},\omega) = \left|f_I(k) + q(k)\right|^2 S_{ii}(\mathbf{k},\omega) + Z_f S_{ee}^0(\mathbf{k},\omega) + Z_C \int S_{ce}(\mathbf{k},\omega - \omega')S_s(\mathbf{k},\omega')d\omega'.\qquad(5)$$

The first term in Eq. (5) accounts for the density correlations of electrons that dynamically follow the ion motion. This includes both the core electrons, represented by the ion form factor $f_I(k)$ and the screening cloud of free (and valence) electrons that surround the ion, represented by $q(k) = \sqrt{Z}\,S_{ei}/S_{ii}$. $S_{ei}(\mathbf{k},\omega)$ and $S_{ii}(\mathbf{k},\omega)$ are the electron-ion and the ion-ion structure factors, respectively. This term accounts for the scattering peak around the incident x-ray probe energy E_0. Its spectral features cannot be resolved in present experiments that use laser-produced x-ray sources and is conveniently labeled ion feature or elastic scattering feature.

The k=0, $T_e = T_i$ limit is indeed known $S_{ii}(k = 0,\omega) = \left(\gamma_i + Z_f\right)^{-1}$ with γ_i being the compressibility ratio ($\gamma_i = 1$ for an ideal gas) and Z_f is the ionization state. For the experiments in isochorically heated beryllium, applying this limit overestimates the intensity of the ion feature relative to the intensity of the plasmon. This finding is also observed for more detailed analytical calculations.[37] Thus, detailed hypernetted chain calculations with quantum interaction potentials have been developed.[38] Several quantum interaction potentials may be applied to provide accurate estimates of the intensity of the ion feature.[39-42]

The second term in Eq. (5) calculates the contribution to the scattering spectrum from the free electrons that do not follow the ion motion. Here, $S_{ee}(\mathbf{k},\omega)$ is the high frequency part of the electron-electron correlation function[43] and it reduces to the usual electron feature[44] in the case of an optical probe. Under the assumption that inter particle interactions are weak, so that the nonlinear interaction between different density fluctuations is negligible, the dielectric function may be derived in the random phase approximation (RPA)[45,46] that includes the Compton shift but does not include corrections to describe scattering from strongly coupled plasmas. More recent improvements apply Mermin theory[47,48] including the effect of collisions. Several approximation are available for the dynamic collision frequency[7,49] that primarily affect the damping, i.e., the spectral width of plasmons in collective scattering. These models may be tested by simultaneous measurements of plasmons and the electron temperature. Compton backscatter measurements or detailed balance that affects the intensity ratio of down- and up-shifted plasmons may be applied for this purpose.[50]

The last term of Eq. (5) includes inelastic scattering by core electrons, which arises from Raman transitions to the continuum of core electrons within an ion,

$S_{ce}(\mathbf{k},\omega)$ modulated by the self-motion of the ions, represented by $S_s(\mathbf{k},\omega)$. The corresponding spectrum of the scattered radiation is that of a Raman-type band. For experiments with isochorically-heated beryllium, we find that this contribution is small compared to the free electron dynamic structure.

X-RAY THOMSON SCATTERING EXPERIMENTS

Solid-density matter has been characterized by spectrally-resolved x-ray scattering more than 70 years ago.[51] These data have observed the Fermi distribution for cold solids predicted by Chandrashekar.[52] For the characterization of warm dense matter, the integrated total scattering has been measured as function of scattering angle[21] before the first x-ray Thomson scattering have been performed on the Omega laser facility yielding spectrally resolved x-ray scattering spectra in the high-energy density physics regime.[24,25]

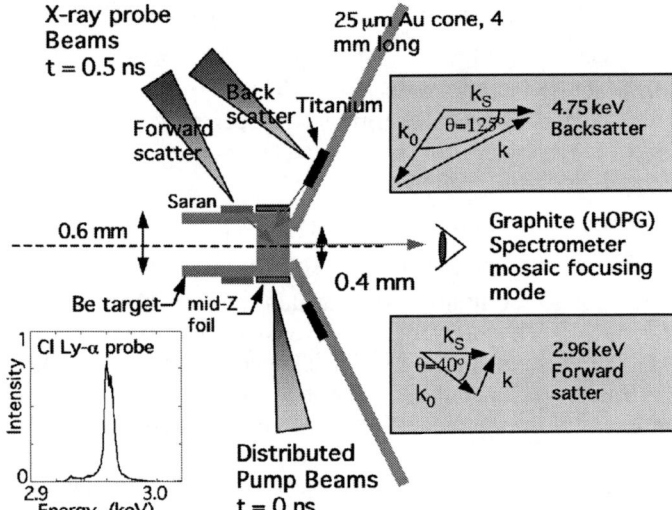

FIGURE 1. Schematic of scattering experiments employing the titanium He-alpha spectral line in backscattering at $\theta = 125°$ or the chlorine Ly-alpha line in forward scattering at $\theta = 40°$. The inset shows a Ly-alpha source spectrum. The k-vector diagrams show the averaged scattering vectors in both geometries yielding non-collective or collective scattering. Up to 15kJ of laser energy has been used for the broadband L-shell heating radiation and up to 7 kJ to produce the narrowband x-ray probe radiation.

Figure 1 shows a schematic of backward and forward scattering experiments on isochorically-heated solid-density beryllium. The beryllium is homogeneously and isochorically heated by L-shell x-rays from a mid-Z foil (Rh or Ag wrapped around the Be) and the dense plasma is probed after t = 0.5 ns with the narrowband x-ray probe radiation from titanium or chlorine. The high laser energy for producing the narrow band x-ray probe and the broadband heating radiation provides sufficient photons for producing homogeneous warm dense states of matter and probing in single shot experiments.

The scattered x rays have been observed through a 400-micron diameter diagnostic hole cut in the center of a 4 mm large Au shield. A gated Bragg crystal spectrometer has been employed in the mosaic focusing mode only observing x-rays scattered from the central homogeneously heated beryllium. In this configuration[26], the distance between the source and the crystal equals the distance between the crystal and the detector plane resulting in a focused image with a spectral resolution of 0.1-0.3%. A mosaic highly oriented pyrolytic graphite (HOPG) crystal has been chosen with a high reflectivity of order 1 mrad. The Au shield prevented a direct view of the source of the probe x rays or scattered radiation from shock waves. For a conversion efficiency of laser energy into resonance line radiation of 4×10^{-3}, a solid angle of beryllium plasmas that is probed of 10^{-2}, and 7kJ of laser energy, a total of 0.1J of probe x-rays or 10^{14} titanium He-alpha x-ray photons are provided at the beryllium plasmas for the duration of the detector gating time of 0.3 ns. The scattering fraction is $\sigma_{Th}n_eL = 3 \times 10^{-3}$ for a 0.4 mm long plasmas viewed by the spectrometer and the collection fraction is 0.1 rad x 1 mrad x 1% / $4\pi \approx 10^{-7}$. Thus, 30,000 photons are detected in a single shot.

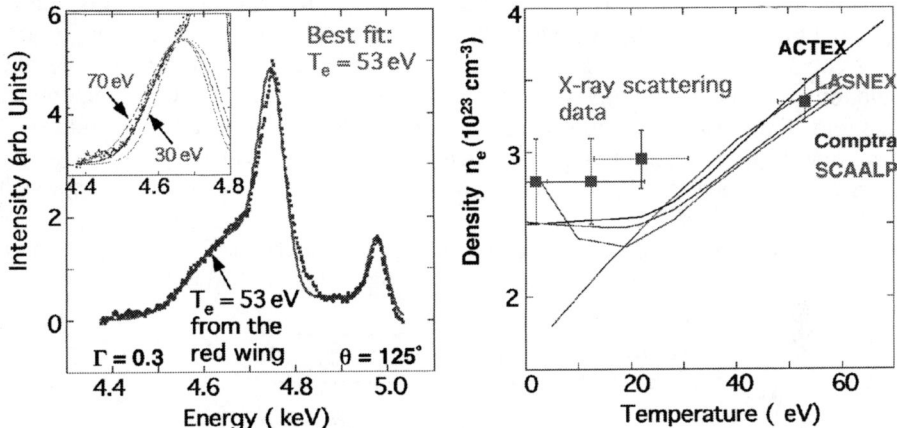

FIGURE 2. Scattering spectrum (blue dots) from isochorically heated beryllium shows elastic and inelastic (Compton) scattering components from titanium He-alpha and Ly-alpha probe x rays (left). The spectrum is fit well with the dynamic structure factor (red line) proving a temperature of $T_e = 53$ eV and density of $n_e = 3.3 \times 10^{23}$ cm^{-3}. These results have tested ionization balance calculations in the solid density plasma regime (right).

Figure 2 shows an example of a back-scattering spectrum from heated beryllium. Elastic scattering from both the titanium He-alpha radiation at 4.75 keV and Ly-alpha radiation at 4.96 keV is observed together with the downshifted Compton scattering feature. The spectrum is fit by the dynamic structure factor, Eq. (5), providing temperatures and densities from the broadening of the Compton downshifted line and from the intensity ratio of elastic and inelastic scattering components. The inset shows the sensitivity of the shape of the red Compton scattering wing to the electron temperature; T_e is inferred from these data with an accuracy of 10%.

A similar analysis has been performed when inferring the ionization state and density from the intensity ratio.[24,35] In conditions without weakly bound electrons the intensity of the inelastic scattering component is solely determined by the free electrons yielding the density with accuracy of about 10%. For lower temperature conditions, however, where bound electrons scatter non-elastically, the contributions from free and weakly bound electrons blend. Nonetheless, the bound-free spectrum is well known and can be distinguished from the Compton scattering spectrum of the free electrons yielding temperature and density information with larger error bars of about 30-50%.

In the experiment, the temperature of the beryllium plasma has been varied by using different mid-Z foils, i.e. Mo, Rh, and Ag, to convert the laser energy to L-shell heating x rays. In addition, on some shots only half of the laser energy per beam was applied probing a temperature phase space of up to T_e = 53 eV. At the highest temperatures, we observe that the measured electron density agrees with various ionization balance models, i.e., the codes ACTEX[53], Comptra[54], LASNEX[29] and SCAALP.[55] On the other hand, at lower electron temperatures, the role of delocalized electrons requires the calculation of all possible interactions between the plasma constituents including the screening of the bound states.[53] For large densities, the classical Debye-Hückel (Yukawa) potential needs to be replaced by quantum interaction potentials that approach the thermal de Broglie wavelength. This allows the calculation of the number of electrons that are no longer bound to a single ion. These electrons are free or weakly bound like the conduction electrons in a metal. Calculations that use interpolation functions between the zero and high temperature conductivity limits show deviations from the measured data at small temperatures.[55] This regime requires additional characterization of the physical properties, e.g., by using measurements at smaller scattering angles and lower x-ray energy to approach the collective regime, cf. Fig.

Figure 3 shows experimental scattering spectra from isochorically heated beryllium measured in forward scattering of the chlorine Ly-alpha line at 2.96 keV.

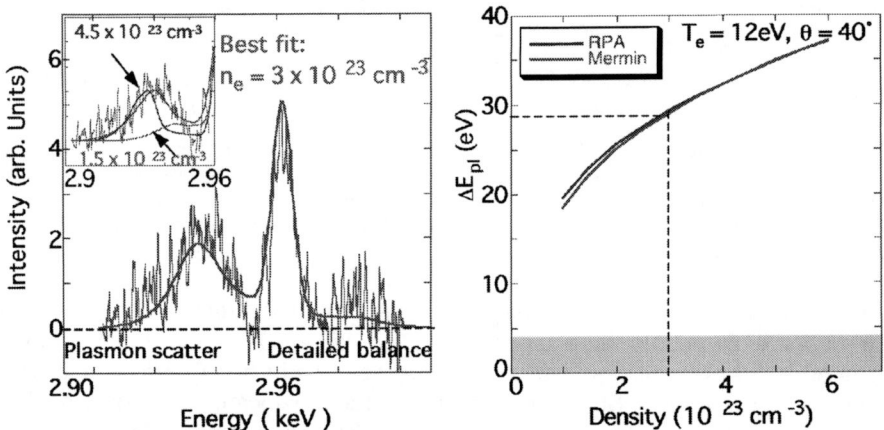

FIGURE 3. Forward scattering spectrum (black line) from isochorically heated beryllium shows elastic and inelastic (plasmon) scattering from chlorine Ly-alpha probe x rays (left). The spectrum is fit well with the dynamic structure factor (red line) proving a temperature of T_e = 12 eV and density of n_e = 3×10^{23} cm^{-3}. The dispersion relation for plasmons determines the frequency shift of the plasmon from the incident x-ray probe radiation yielding the electron density (right).

To resolve the plasmon frequency shift and damping in forward scattering, the x-ray bandwidth has to be smaller compared to the Compton scattering measurements. In Figure 3, the plasmon shift of ΔE = 28 eV was resolved by the chlorine Ly-alpha probe radiation with an effective bandwidth of 7.7 eV and no significant dielectronic satellite radiation on the red wing of the Ly-alpha doublet, see Fig. 1 (inset). Also shown are synthetic scattering profiles that represent a convolution of the theoretical form factor $S(\mathbf{k},\omega)$, calculated for the range of $\mathbf{k}$-vectors of the experiment, with the spectral resolution of 7.7 eV. The ion feature is observed as an elastic scattering peak at E_0 that is not resolved in this experiment. On the lower-energy wing of the ion feature we observe a strong plasmon resonance. On the higher-energy wing with nearly the same frequency shift, the data show a weak up-shifted plasmon signal. Compared to the intensity of down-shifted plasmon, the intensity is reduced by the Bose function $e^{-\hbar\omega/k_B T_e}$ reflecting the principle of detail balance. The intensity ratio of these plasmon features is thus sensitive to the temperature. In the present experiment, the signal to noise ratio only allows us to deduce an upper limit of $T_e < 25$ eV.

In $S(\mathbf{k},\omega)$, collision effects on plasmons are accounted for with a Mermin ansatz thus accounting for weak degeneracy effects. While the frequency shift is only marginally affected we observe that collisional damping cannot be neglected. The fit provides a temperature of 12 eV and density of $n_e = 3 \times 10^{23}$ cm^{-3}. Figure 3 (right) indicates that the density is accurately determined by the shift of the plasmon. For the small $\mathbf{k}$-vectors probed in the present experiment, significant deviations from the random phase approximation are not expected yielding an error bar in density that is solely determined by the signal to noise ratio and the quality of the fit. The inset in Fig. (3) shows calculations for densities of 1.5×10^{23} cm^{-3} and 4.5×10^{23} cm^{-3} indicating that the error bar in density is of order 20% in the present experiment; a value that may be improved with better signal to noise ratio.

Inferring the temperature from the collective scattering spectrum requires accurate measurements of the up-shifted plasmon and the application of detailed balance or alternatively the width of the plasmon may be used. The latter is determined by Landau damping and collisional damping.

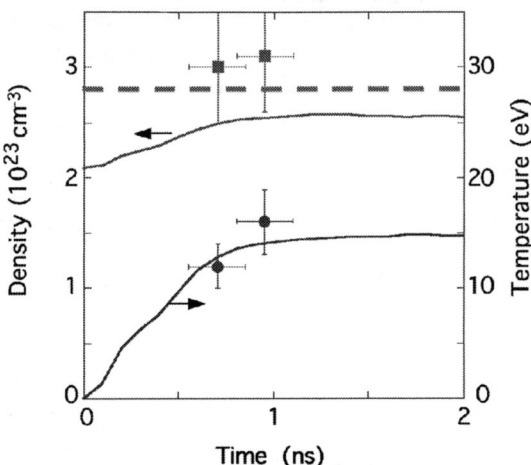

FIGURE 4. Plasma parameters from collective x-ray scattering and radiation hydrodynamics calculations are shown. The dashed curve shows the electron density for solid-density beryllium for Z = 2.3 measured in backscattering. The calculated density agrees marginally with the data from forward and backward scattering due to approximations in the calculation of Z in hydrodynamic modeling.

For our conditions, collisions account for an additional broadening of approximately 5 eV. The difference can be as large as a factor of 1.5 for scattering spectra calculated at a single scattering angle. Clearly, scattering experiments in a regime where damping by collisions dominates the broadening of plasmons are of interest to test models of the collisionality in dense plasmas. For the experiments discussed here, collisions provide a correction to the measured width. A good fit of the plasmon spectrum is obtained for a temperature of 12 eV and with the dynamic collision frequency calculated in Born approximation.

Figure 4 shows the experimental results from the forward scattering experiments. The temperature is lower in these data because 1) the detector was gated earlier during the heating of the plasma, 2) fewer laser beams have been used to illuminate the L-shell converter, and 3) a higher-Z element was used, i.e. Ag. Calculations show 20-30% lower conversion efficiency than for Rh or Mo. We find that the experimental temperature data agree well with radiation hydrodynamic simulations. In addition, the density is in good agreement with the expectations from the backward scattering experiments in the low temperature range, cf. Fig. (2) indicating that the plasmon spectrum provides a robust density diagnostic.

CONCLUSIONS AND OUTLOOK

X-ray Thomson scattering experiments on isochorically heated solid-density beryllium have shown accurate characterization of warm dense matter. Results from forward and backward scattering measurements have been shown to yield mutually consistent results. In particular, the electron density inferred from the frequency shift of plasmons agrees with the ionization balance measurements from the intensity ratio of the Compton scattering to elastic scattering component. Moreover, the electron temperature is inferred from the spectrum of the Compton downshifted line that is directly reflecting the electron velocity distribution providing temperatures from first principles. These results show that the x-ray scattering technique can be applied to measure the compressibility of dense matter from the plasmon frequency shift, or in case of a degenerate system, from the Compton scattering feature that will reflect the Fermi temperature. Future applications will include measurements of the conductivity from the collisional damping of plasmons and investigations of the validity of the random phase approximations at large values of the scattering vector k.

ACKNOWLEDGMENTS

This work was performed under the auspices of the U.S. Department of Energy by University of California Lawrence Livermore National Laboratory under contract No. W-7405-Eng-48. Supported by LDRD 05-ERI-003, SFB 652, and the Alexander-von-Humboldt society.

REFERENCES

1. J. D. Lindl, P. Amendt, R. L. Berger et al. *Phys. Plasmas* **11**, 339 (2004).
2. B. A. Remington, P. R. Drake, D. D. Ryutov., *Rev. Mod. Physics* **78**, 755 (2006).
3. M A. Kritcher, P. Neumayer, M. K. Urry et al., *High Energy Density Physics* (2007), in print.
4. L. B. DaSilva et al. *Phys. Rev. Lett.* **78**, 483 (1997); M. D. Knudson et al., *ibid* **90**, 035505 (2003).
5. S. Ichimaru and S. Tanaka, *Phys. Rev. A* **32**, 1790 (1985).
6. G. Gregori, S. H. Glenzer, and O. L. Landen, *J. Phys. A* **36**, 5971 (2003).

7. H. Reinholz, R. Redmer, G. Röpke, and A. Wierling, *Phys. Rev. E* **62**, 5648 (2000).
8. H.-J. Kunze E. Fünfer, B. Kronast, and W. H. Kegel, *Phys. Lett.* **11**, 42 (1964).
9. A. A. Offenberger W. Blyth, A. E. Dangor et al., *Phys. Rev. Letters* **71**, 3983 (1993).
10. A W. DeSilva et al., *Phys. Fluids B* **4**, 458 (1992).
11. S. H. Glenzer T. L. Weiland, J. Bower et al., *Rev. Sci. Instrum.* **70**, 1089 (1999).
12. A. J. Mackinnon S. Shiromizu, G. Antonini et al., *Rev. Sci. Intrum* **75**, 3906 (2004).
13. J. Dunn, Y. Li, A. L. Osterheld et al., *Phys. Rev. Letters* **84**, 4834 (2000).
14. S. H. Glenzer, C. A. Back, K. G. Estabrook et al., *Phys. Rev. Letters* **77**, 1496 (1996).
15. S. H. Glenzer, W. Rozmus, B. J. MacGowan et al., *Phys. Rev. Letters* **82**, 97 (1999).
16. S. H. Glenzer, K. B. Fournier, B. G. Wilson et al., *Phys. Rev. Letters* **87**, 045002 (2001).
17. D. H. Froula, P. Davis, L. Divol et al., *Phys. Rev. Letters* **95**, 195005 (2005).
18. J. Filevich, J. J. Rocca, M. C. Marconi et al., *Applied Optics* **43**, 3938 (2004).
19. R. F. Smith, J. Dunn, J. Nilsen et al., *Phys. Rev. Letters* **89**, 065004 (2002).
20. O. L. Landen et al., *Rev. Sci. Instrum.* **72**, 627 (2001).
21. D. Riley et al., *Phys. Rev. Letters* **84**, 1704 (2000).
22. M. Urry, G. Gregori, O.L. Landen et al., *J. Quan. Spectr. Rad. Transfer* **99**, 636 (2006).
23. O. L. Landen, S.H. Glenzer, M.J. Edwards et al., *J. Quan. Spectr. Rad. Transfer* **71**, 465 (2001).
24. S. H. Glenzer, G. Gregori, R. W. Lee et al., *Phys. Rev. Letters* **90**, 175002 (2003).
25. S. H. Glenzer, O. L. Landen, P. Neumayer et al., *Phys. Rev. Letters* **98**, 065002 (2007).
26. S. H. Glenzer, G. Gregori, F. J. Rogers et al., *Phys. Plasmas* **10**, 2433 (2003).
27. G. Gregori, S. H. Glenzer, F. J. Rogers et al., *Phys. Plasmas* **11**, 2754 (2004).
28. L. Tonks and I. Langmuir, *Phys. Rev.* **33**, 195 (1929).
29. G. Zimmerman and W. Kruer, *Comments Plasma Phys. Controlled Fusion* **2**, 85 (1975).
30. F. Perrot and M. W. C. Dharma-Wadarna, *Phys. Rev. B* **62**, 15636 (2000).
31. M. W. C. Dharma-Wadarna and F Perrot, *Phys. Rev. Letters* **84**, 959 (2000).
32. R. Zimmerman, *Many-Particle Theory of Highly Excited Semiconductors* (Teubner, Leipzig, 1987).
33. F. Haas, G. Manfredi, and M. Feix, *Phys. Rev. E* **62**, 2763 (2000).
34. A. H. Compton *Phys. Rev* **21**, 483 (1923).
35. G. Gregori, S. H. Glenzer, W. Rozmus, R. W. Lee, O. L. Landen, *Phys. Rev. E* **67**, 026412 (2003).
36. J. Chihara, *J. Phys.: Condens. Matter* **12**, 231 (2000).
37. G. Gregori, S. H. Glenzer and O. L. Landen, *Phys. Rev. E* **74**, 026402 (2006).
38. V. Schwarz, Th. Bornath, W. D. Kraeft et al., *Contr. Plas. Phys.* (2007), in print.
39. Yu. L. Klimontovich and W. D. Kraeft, *High. Temp. Physics(USSR)* **12** , 212 (1974).
40. G. Kelbg, *Ann. Phys. (Leipzig)* **12**, 219 (1963); **13**, 354 (1963); **14**, 394 (1964).
41. C. Deutsch, *Phys. Lett.* **60**A, 317 (1977).
42. A. V. Filinov, V. O. Golubnychiy, M. Bonitz et al., *Physical Review E* **70**, 046411 (2004).
43. S. Ichimaru, *Basic Principles of Plasma Physics* (Addison, Reading, MA, 1973).
44. E. E. Salpeter, *Phys. Rev.* **120**, 1528 (1960).
45. D. Pines and D. Bohm, Phys. Rev. **85**, 338 (1952). D. Pines and P. Nozieres, *The Theory of Quantum Fluids* (Addison-Wesley, Redwood, CA, 1990).
46. D. Kremp. M. Sclanges, W.-D. Kraeft and T. Bornath, *Quantum Statistics of Nonideal Plasmas* (Springer, Berlin 2005).
47. A. Höll, R. Redmer, G. Röpke, and H. Reinholz, Eur. Phys. J. D **29**, 159 (2004).
48. R. Redmer, H. Reinholz, G. Röpke et al., *IEEE Trans. Plas. Sci.,* **33**, 77 (2005).
49. R. Thiele et al., *J. Phys.. A* **39**, 4365 (2006)
50. A. Höll, Th. Bornath, L. Cao et al., *High Energy Density Physics* (2007), in print.
51. J. W. M. DuMond, *Phys. Rev.* **29**, 643 (1929). J. W. DuMond and H. A. Kirkpatrick, *ibid* **37**, 136 (1931).
52. S. Chandrasekhar, *Proceedings of the Royal Society* A **125**, 231 (1929).
53. F. Rogers, *Phys. Plasmas* 7, 51 (2000). . F. Rogers and D. A. Young, *Phys. Rev. E* **56**, 5876 (1997).
54. R. Redmer, *Phys. Rev. E* **59**, 1073 (1999). S. Kuhlbrodt and R. Redmer, *ibid* **62**, 7191 (2000).
55. P. Renaudin, C. Blancard, G. Faussurier, and P. Noiret, *Phys. Rev. Letters* **88**, 215001 (2002).

Using Laser-driven Shocks to Study the Phase Diagrams Of Low-Z Materials at Mbar Pressures and eV Temperatures*

P.M. Celliers[1], J.H. Eggert[1], D.G. Hicks[1], T.R. Boehly[2], J.E. Miller[2],
S. Brygoo[1,3], P. Loubeyre[3], D.K. Bradley[1], R.S. McWilliams[4],
R. Jeanloz[4] and G.W. Collins[1]

[1] Larence Livermore National Laboratory, Livermore, CA, 94550
[2] Laboratory for Laser Energetics, University of Rochester, Rochester, NY 14623
[3] Département de Physique Théorique et Applications, CEA, Commissariat a` l'Energie Atomique,
91680 Bruyères-le-Châtel, France
[4] Department of Earth and Planetary Science, University of California, Berkeley, CA 94720

Abstract. Accurate phase diagrams for simple molecular fluids and solids (H_2, He, H_2O, SiO_2, and C) and their constituent elements at eV temperatures and pressures up to tens of Mbar are integral to planetary models of the gas giant planets (Jupiter, Saturn, Uranus and Neptune), and the rocky planets. Laboratory experiments at high pressure have, until recently, been limited to around 1 Mbar. These pressures are usually achieved dynamically with explosives and two-stage light-gas guns, or statically with diamond anvil cells. Current and future high energy laser and pulsed power facilities will be able to produce tens of Mbar pressures in these light element materials. This presentation will describe the capabilities available at current high energy laser facilities to achieve these extreme conditions, and focus on several examples including water, silica, diamond-phase-carbon, helium and hydrogen. Under strong shock compression all of these materials become electronic conductors, and are transformed eventually to dense plasmas. The experiments reveal some details of the nature of this transition. To obtain high pressure data closer to planetary isentropes advanced compression techniques are required. We are developing a promising technique to achieve higher density states: precompression of samples in a static diamond anvil cell followed by laser driven shock compression. This technique and results from the first experiments with it will be described. Details about this topic can be found in some of our previous publications [1-2].

*This work was performed under the auspices of the U.S. Department of Energy by LLNL under contract number W-7405-ENG-48.

REFERENCES

1. R. Jeanloz, P. M. Celliers, G. W. Collins, J. H. Eggert, K. K. M. Lee, R. S. McWilliams, S. Brygoo, and P. Loubeyre, Proc. Natl. Acad. Sci. U. S. A., **104**, 9172 (2007).
2. K. K. M. Lee et al., J. Chem. Phys. **125**, 014701 (2006).

Study of Non-LTE Spectra Dependence on Target Mass in Short Pulse Laser Experiments

C.A. Back[1], P. Audbert[2], S.D. Baton[2], S. Bastiani-Ceccotti[2], P. Guillou[2], L. Lecherbourg[3], B. Barbrel[2], E. Gauci[2], M. Koenig[2], L. Gremillet[2], C. Rousseaux[3], E. Giraldez[1], S. Phommarine[1]

[1]*General Atomics, 3550 General Atomics Court, San Diego, CA 92121*
[2]*Laboratoire pour L'Utilisation des Lasers Intenses (LULI), UMR7605, CNRS – CEA Université Paris VI - Ecole Polytechnique, 91128 Palaiseau Cedex, FRANCE*
[3]*Commissariat à l'Energie Atomique, 91680 Bruyères-le-Châtel, FRANCE*

Abstract. Backlight sources created from short pulse lasers are useful probes of high energy density plasmas because of their short duration and brightness. Recent work has shown that the production of Kα radiation can be manipulated by the size and geometry of the targets. Empirical relationships suggest that the electron reflux in the target plays an important role in the heating of these targets to create x-ray backlight sources.

Keywords: X-ray backlight, Kα spectroscopy, fast electron transport.
PACS: 32.30.Rj, 52.50.Dg, 52.50.Jm

1. SHORT PULSE X-RAY SOURCES

Bright, high energy photon energy backlight sources are needed for future high-energy density plasma studies. For instance, photon energies of 20-40 keV will be needed to probe plasmas, such as those created from mid- to high-Z implosion experiments at the National Ignition Facility. X-ray sources in laser-produced plasma experiments are often created by using some of the multiple laser beams available to irradiate a secondary target, which itself becomes a plasma. In these sources, the plasma formation is dependent on the laser absorption, and the hydrodynamic evolution can be significant [1]. Unfortunately, backlights formed from "thermal" sources are not efficient enough to produce bright high energy backlights for energies greater than ~13 keV.

An alternative method for creating hot plasmas that can serve as x-ray backlights is to take advantage of the generation of fast particles in short pulse laser-solid interactions [2-7]. In this case, the plasma formation is dependent on the laser field. The electrostatic potential causes hot electrons to stay confined around the target, and the energy is transferred by electron-electron collisions. These kind of plasmas, in which collisional absorption starts to turn off for laser intensities $>10^{15}$ W/cm^2, are often characterized by a hot electron temperature. Recently, embedded silver micro-wire targets have successfully produced high-resolution two-dimensional radiographs using 22 keV photons that achieve a Kα conversion efficiency of $\sim 10^{-5}$[8].

CP926, *Atomic Processes in Plasmas—15th International Conference on Atomic Processes in Plasmas*
edited by J. D. Gillaspy, J. J. Curry, and W. L. Wiese
© 2007 American Institute of Physics 978-0-7354-0436-6/07/$23.00

2. EXPERIMENTS TO VARY TARGET MASS

In order to better understand fast electron transport, experiments on thin multilayer targets were performed at the LULI 100 TW laser facility. This facility uses a Nd:glass laser to provide up to 100 J before compression. After chirped pulse amplification, this laser can provide up to 30 J in 300 fs, a maximum of 5×10^{19} W/cm^2. In short-pulse experiments, higher conversion efficiency of laser energy to Kα emission is observed in smaller targets. Because electron refluxing plays a predominant role in the bulk heating of the target, for constant laser energy, a smaller target is expected to achieve higher temperatures. In these experiments, the laser energy and focal spot were kept constant and the spectra were recorded for targets of varying mass.

The multilayer targets were composed of V/Cu/Al and varied from 300 to 50 µm in diameter. The V layer was 0.2 µm thick and acted as a converter layer for producing electrons. The Cu layer was varied in thickness in order to study the electron heating through 5, 10, and 20 µm thicknesses Cu layers. The Al was 5 µm thick and originally designed to act as a diagnostic layer to monitor the heating of the back surface by the relative intensities of the 1s-2p transitions of Al which were expected to be sensitive to the electron temperature [9-10].

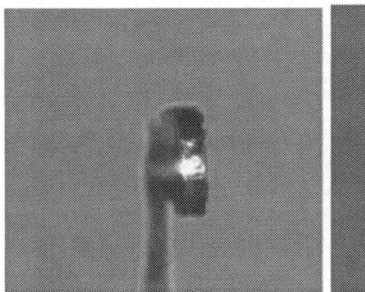

FIGURE 1. The smallest, lowest mass targets were 50 µm diameter disks mounted on a 6 µm dia. carbon fiber.

The targets were isochorically heated by a 20 J, 300 fs laser pulse that delivered $1{\sim}2\times10^{19}$ W/cm^2 to form a warm dense plasma. The target surface irradiated by the laser was circular or square. By using targets 50, 200, 300 µm in diameter, and also 100x100 µm and 200x200 µm square, in combination with the different Cu thicknesses, the mass of the target was varied by from 1 to 36 M, where M is defined as the mass of the smallest and lowest mass target, 0.13 µg.

Emission from the rear, unilluminated Al side was recorded simultaneously by three diagnostics. Two time-integrated diagnostics provided 2D imaging and spectra of the Cu. The 2D imager was fielded with a spherical quartz crystal at 1.54 Å to measure the Cu Kα at 8X magnification. The spectrometer used a conical TlAP crystal and covered a range of 7.6-8.4 Å to record both the Al emission in 1st order, and the Cu emission in 5th order. In addition, a sub-ps streak camera was fielded with another TlAP crystal spectrometer to obtain time-resolved spectra of either the Al Kα or Cu Kα emission.

3. EXPERIMENTAL RESULTS

The data from targets of different sizes and/or Cu layer thickness are compared and analyzed to better understand the heating of the target and temperature of the plasma. The Cu Kα imager shows that the emission decreases with target mass quite dramatically when the mass drops below 0.39 µg. The 50 µm diameter targets, which span a range of 1-3 M, are 40% or less than targets having a mass of 8 M.

Spectra including the Al-Kα, Al He-α, and Cu-Kα emission from the unirradiated Al side of the target show significant qualitative changes as a function of total mass. In general, a stronger continuum appears for the lower mass targets. But more evident, shown in Fig. 2 below, is the shift of the line emission to shorter wavelength as the plasma becomes more ionized for lower mass targets. When the target mass is decreased, the cold Kα of Al decreases, while the He-α from the ionized Al increases. The Kα disappears for targets <300 µm in diameter. This shift is expected, however, the more surprising result is that the cold Kα of Cu almost entirely disappears for the 50 µm diameter targets.

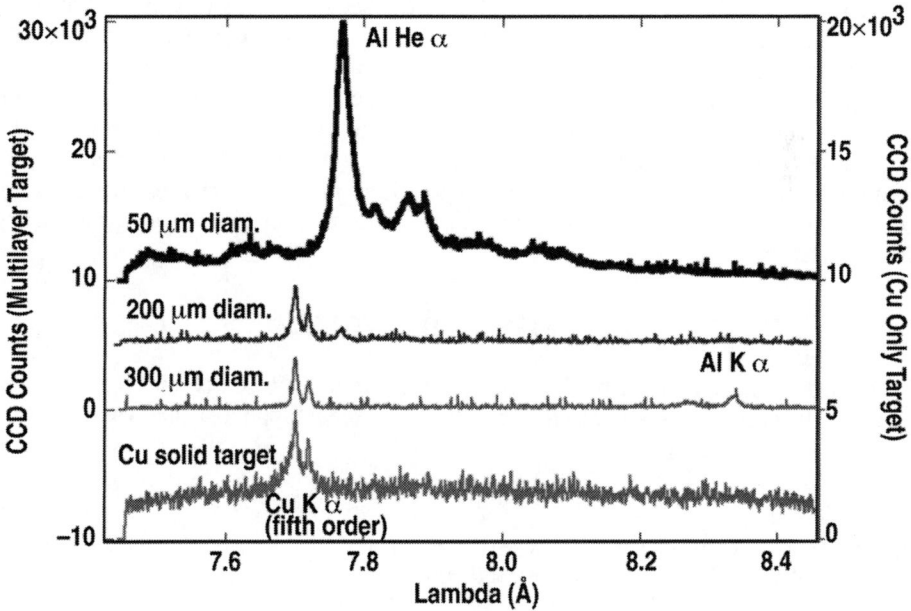

FIGURE 2. Spectra from a multilayer targets 50, 200, and 300 µm in diameter. All targets were composed of 0.2 µm V, 5 µm Cu, and 5 µm Al, and the laser irradiated the V side. The solid Cu target is shown to provide a reference spectra for the cold Cu Kα.

Calculations for the Cu spectra show that the thermal electron temperature effectively determines the charge state distribution for isochorically-heated targets. The effect of the hot electrons is to pump up the excited state populations so that the emission can be observed. The Kα lines for the L-shell ions at these densities and temperatures quickly become optically thick for plasmas even with the thinnest 5 μm thick Cu targets. Analysis of the Cu-Kα line at 8.064 keV shows that an electron temperature of 240 eV is most consistent with the measured spectra.

In these dense targets, the Al He-α line is expected to be opacity broadened. However, on the streak camera, this emission line does not appear to be reabsorbed. Instead, it appears optically thin, and has a relatively long duration of 5-15 ps. Preliminary modeling shows that the plasma expansion may be significant enough that the spectra of the 1s-2p transitions is not an effective electron temperature monitor of the dense plasma conditions in the Cu [11-12].

4. CONCLUSIONS

In conclusion, the spectra from low mass targets show trends that are consistent with the creation of a hot plasma. As the Cu is heated, the ionization balance changes and the cold Kα shifts and broadens. This has also been recently observed in experiments with chlorinated targets [13]. Furthermore, for the lowest mass targets in these experiments, the Kα nearly disappears.

Unfortunately, the 2-ps temporal resolution of the streak camera was not sufficient to record the Al spectra emitted during the time-of-interest when the plasma was hot and dense. Based on modeling, we now believe the hydrodynamic expansion compromised the use of the Al layer as a diagnostic of the electron temperature because it caused decompression of the Al plasma in the picosecond timeframe of the observation. Since the hot Al plasma takes a long time to cool due to low recombination rates, the lower density "expanding" plasma dominates the measured spectra. Future experiments are planned to continue refining this spectroscopic diagnostic.

ACKNOWLEDGMENTS

This work was supported by General Atomics Internal Research and Development funding. The authors thank the LULI laser crew for their support of the experiments.

REFERENCES

1. C. A. Back, *et al.* "Multi-keV X-ray Conversion Efficiency in Laser-Produced Plasmas," *Phys. Plasmas* **10**, 2047-2055 (2003) and references therein.
2. W. Theobald, *et al.,* "Hot Surface Ionic Line Emission and Cold K-inner Shell Emission From Petawatt-Laser-Irradiated Cu Foil Targets," *Phys. Plasmas* **13**, 043102 (2006).
3. T. Kawamura, *et al.,* "Population Kinetics on Kα Lines of Partially Ionized Cl Atoms," *Phys. Rev. E* **66**, 016402/1-016402/8 (2002).
4. D. J. Hoarty, *et al., J. Quant. Spectrosc. and Radiat. Transfer* **3**, 115-119 (2007).
5. K. U. Akli, *et al.,* "Temperature Sensitivity of Cu Kα Imaging Efficiency Using a Spherical Bragg Reflecting Crystal," *Phys. Plasmas* **14**, 023102 (2007).
6. G. Gregori, *et al.,* "Experimental Characterization of a Strongly Coupled Solid Density Plasma Generated in a Short-Pulse Laser Target Interaction," *Contrib. Plasma Phys.* **45**, 284-292 (2005).
7. R. B. Stephens, *et al.,* "Kα Fluorescence Measurement of Relativistic Electron Transport in the Context of Fast Ignition," *Phys. Rev. E* **69**, 066414 (2004).

8. H. S. Park, *et al.,* "High-Energy Kα Radiography Using High-Intensity, Short-Pulse Lasers," *Phys. Plasmas* **13**, 056309 (2006).

9. S. B. Hansen, *et al.,* "Temperature Determination Using Kα Spectra From M-Shell Ti Ions," *Phys. Rev. E* **72**, 036408 (2005).

10. P. Audebert, *et al.,* "Picosecond Time-Resolved X-ray Absorption Spectroscopy of Ultrafast Aliminum Plasmas," *Phys. Rev. Lett.* **94**, 025004 (2005).

11. H.-K. Chung, M. H Chen and R. W. Lee, "Extension of Atomic Configuration Sets of the Non-LTE Model in the Application to the Kα Diagnostics of Hot Dense Matter," *J. Quant. Spectrosc. and Radiat. Transfer* **85**, 57-64 (2007).

12. R. W. Lee and J. T. Larsen, "A Time-Dependent Model for Plasma Spectroscopy of K-shell Emitters," *J. Quant. Spectrosc. Radiat. Transfer* **56**, 535-556 (1996).

13. P. Neumayer (private communication, 2007).

X-Ray Absorption Spectroscopy Of Thin Foils Irradiated By An Ultra-short Laser Pulse

P. Renaudin[1], L. Lecherbourg[2-3], C. Blancard[1], P. Cossé[1], G. Faussurier[1], P. Audebert[2], S. Bastiani-Ceccotti[2], J.-P. Geindre[2], and R. Shepherd[4]

1 Département de Physique Théorique et Appliquée,
Centre DAM Ile-de-France, BP12, 91680 Bruyères-le-Châtel Cedex, France ;
2 Laboratoire pour l'Utilisation des Lasers Intenses, UMR 7605,
CNRS-CEA-Université Paris VI-Ecole Polytechnique, 91128 Palaiseau, France ;
3 Université du Québec, INRS Energie et Matériaux, Varennes, Québec, Canada ;
4 Lawrence Livermore National Laboratory, University of California, Livermore, CA 94550, USA

Abstract. Point-projection K-shell absorption spectroscopy has been used to measure absorption spectra of transient plasma created by an ultra-short laser pulse. The $1s$-$2p$ and $1s$-$3p$ absorption lines of weakly ionized aluminum and the $2p$-$3d$ absorption lines of bromine were measured over an extended range of densities in a low-temperature regime. Independent plasma characterization was obtained using frequency domain interferometry diagnostic (FDI) that allows the interpretation of the absorption spectra in terms of spectral opacities. Assuming local thermodynamic equilibrium, spectral opacity calculations have been performed using the density and temperature inferred from the FDI diagnostic to compare to the measured absorption spectra. A good agreement is obtained when non-equilibrium effects due to non-stationary atomic physics are negligible at the x-ray probe time.

Keywords: Ultra-short laser, dense plasma, x-ray absorption.
PACS: 52.50.Jm, 32.30.Rj, 52.25.Os

INTRODUCTION

The radiative opacities of dense plasmas are important for inertial confinement fusion plasmas [1] and stellar interior physics. It has been shown that updated CNO composition that suppresses the anomalous position of the Sun in the known galactic enrichment, leads to discrepancies between the standard model and solar seismic observations [2]. Such discrepancies may be due to the determination of the opacity coefficients in partially ionized elements for density greater than 0.1 g/cm^3.

Experimental investigation of plasma photoabsorption coefficients has been an active field of research for many years [3]. The method based on volumetrically radiatively heated foils provides uniform heating to temperatures of 80 eV and for densities from 0.001 to a few 0.01 g/cm^3 in conditions near local thermodynamic equilibrium (LTE) [4].

CP926, *Atomic Processes in Plasmas—15th International Conference on Atomic Processes in Plasmas*
edited by J. D. Gillaspy, J. J. Curry, and W. L. Wiese
© 2007 American Institute of Physics 978-0-7354-0436-6/07/$23.00

Higher density plasmas with minimized gradients can be created by intense laser irradiation of solids. When a sub-picosecond laser pulse irradiates a very thin foil (~tens of nm) with intensities in the range of 10^{14}-10^{16} W/cm^2, impulse heating followed by a rapid heat conduction produces a high-density, moderate-temperature plasma before any significant hydrodynamic expansion occurs.

We present quantitative data of transmission spectra in the warm dense matter regime. The $1s$-$2p$ and $1s$-$3p$ absorption lines of Al and $2p$-$3d$ absorption lines of Br were measured for densities up to 0.4 g/cm^3 and temperatures varying from 4 to 40 eV. Our experiment allows to investigate a higher-density plasma regime than the one commonly accessible using volumetrically radiatively heated foils. In the warm dense regime studied, the plasmas are weakly ionized and density effects are not negligible. The continuum lowering modifies the ionization balance and pressure ionization tends to delocalize atomic orbital. For instance, in the spectral range studied, many $1s$-np absorption lines of Al merge in the K-shell photoionization. Therefore, our experimental data allow to test the accuracy of some of the approximations which have been proposed to account for density effects in spectral opacity calculations.

EXPERIMENTAL SETUP

The experiment was performed at the 100 TW LULI facility, using two 300-fs laser beams. The experimental setup description has been detailed previously [5]. A 0.3-J (1.06-µm) heated a 45-nm thick Al or KBr layer, deposited on a self-standing 25-nm silicon nitride (Si$_3$N$_4$) substrate and cover with a 20-nm thick carbon layer. The heating beam was focused on the Si$_3$N$_4$ side of the target, limiting the longitudinal gradients inside the plasma. The laser beam was diaphragmed to provide an elliptic focal spot, 450-µm in the plane of incidence and 150-µm FWHM in the perpendicular direction. This shape was chosen to optimize the Bragg spectroscopy diagnostic as described below. The measured prepulse intensity contrast ratio was 10^{-7} at the fundamental frequency. A 8-J energy, (0.53-µm) beam focused on a samarium sample with an off-axis parabola to a spot size of 20-µm, produced the backlighter. The time-delay between the x-ray pulse and the heating laser at the target was measured before each shot with an uncertainty of 1 ps. Diagnostics viewed the opposite side of the irradiated layer side. A single-shot FDI diagnostic system [6] monitored the phase of a chirped probe laser beam (1.06-µm) of 35-ps duration, reflected at a 47° incidence angle on the rear surface of the plasma. X-ray transmission spectra in the 1470-1610 eV spectral region were recorded with a conically-bent potassium-hydrogen-phthalate (KAP) crystal spectrometer coupled to a cooled 1024×1024 16-bits CCD camera. This kind of spectrometer has the advantage of allowing easy magnification adjustments by moving the detector away from the focal line of the imaging crystal, without changing the wavelength range. The spectra were recorded with a magnification of 13 by setting the CCD camera at L=39-mm beyond the focus of the crystal and the backlighter source at d=3-mm from the plasma. The spatial resolution was limited by the x-ray source dimension, i.e., 20 µm. The spectral resolution was estimated from the width of the various $1s$-$2p$ lines of Al measured more than 50 ps after the heating pulse, when the instrumental width dominates the other broadening processes. The resolution was determined to be 2±0.2 eV. Due to the Bragg diffraction properties, different wavelengths cross the plasma at different positions. This motivated the choice of an elliptical focal spot, in order to cover the chosen spectral range in a homogeneous part of the plasma. For the spectral region studied here, the Sm plasma provided a 4-ps duration quasi-continuum backlighter spectrum [7].

Fig. 1 shows an example of an Al space-resolved absorption spectrum measured 12 ps after the heating laser at a laser intensity of 1.5×10^{15} W/cm^2. The spectra covered the 1s-2p absorption lines of Al^{4+} and Al^{5+}, the K-edge position at 1560 eV and the Al^{4+} and Al^{3+} 1s-3p absorption lines. The cold Al K-edge is visible outside the focal spot. The initial areal density of the Al layer was measured at each shot by comparing the tabulated cold Al and Si$_3$N$_4$ transmission near the K-edge with the measured transmission. We found that the areal density of the aluminum layer is equal to 12 ± 4 μg/cm^2.

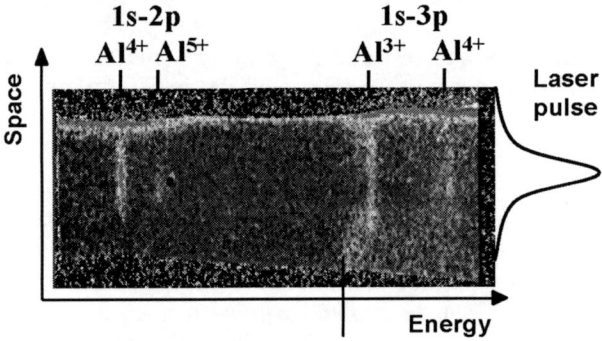

FIGURE 1. Space-resolved absorption spectrum obtained 12 ps after the heating laser at a laser intensity of 1.5×10^{15} W/cm^2. The solid vertical line gives the position of the Al K-edge at 1560 eV. The spatial laser intensity profile is represented on the right-hand side of the image

INTERPRETATION OF THE ABSORPTION SPECTRA IN TERMS OF SPECTRAL OPACITIES

To interpret the absorption spectrum in terms of spectral opacities, it is necessary to (i) infer the density and temperature of the layer during the absorption measurement, (ii) take into account the temporal and longitudinal gradients in the analysis, and (iii) check that the non-equilibrium effects due to transient atomic physics are negligible at the x-ray probe time.

Plasma Characterization

Since the energy is transferred to the target in a time short compared to any hydrodynamic expansion, the density and temperature during the plasma expansion depend mainly of the maximum temperature achieved at the end of the laser pulse. To infer the plasma parameters at the time of the absorption measurements we have used the one-dimensional hydrodynamic MULTI-FS code [8]. The uncertainty in the laser interaction parameter measurements leads to a non-negligible uncertainty in the laser intensity and, therefore, on the density and temperature calculated with the hydrodynamic code. Nevertheless a precise determination of the plasma parameters is necessary to interpret the absorption spectra in terms of spectral opacities. To overcome this problem, we used the FDI diagnostic system to infer the density and temperature of the expanding plasma during the absorption measurements. Indeed, the phase shift of the reflected light provided by the FDI diagnostic depends on the velocity of the critical surface, which is related to the laser intensity [9].

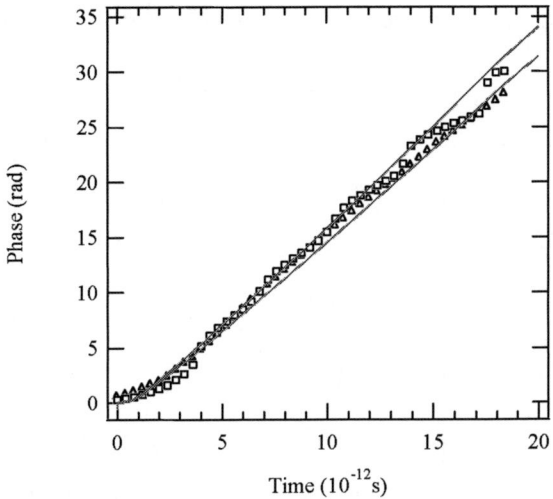

FIGURE 2. Temporal evolution of the phase of the probe laser beam reflected at the rear critical surface of the plasma, showing the best agreement between hydrodynamic simulations performed with MULTI-FS (solid lines obtained for two laser intensities 8×10^{14} and 10^{15} W/cm^2) and experimental data (markers) for a laser intensity of 9×10^{14} W/cm^2.

The Helmholtz equation is solved to calculate the propagation of the probe beam inside the subcritical plasma and its reflection near the critical surface. These calculations provide the phase variation of the probe beam as a function of the laser intensity and were used as a postprocessor for the MULTI-FS code. Fig. 2 shows the temporal evolution of the phase of the chirped probe laser beam reflected at the rear critical surface of the plasma. The best agreement between theoretical calculations and experimental data is obtained by adjusting the laser intensity in MULTI-FS. An input laser intensity of I±10% is determined from the best fit of the measured phase shift. The obtained laser intensity and the nominal laser intensities are equal within 30%. Thus, the accuracy of the FDI analysis allows us to reduce the uncertainty in the plasma parameters. An input laser intensity of ±10% leads to a 1 eV-temperature variation at the time of the absorption measurements, i.e., a lower variation than the one due to the temporal integration during the backlighter emission. For each experimental absorption spectrum, FDI analysis provides an independent characterization of the plasma expansion.

Plasma parameters at the probe time

MULTI-FS simulations were performed with laser intensity corresponding to the FDI measurement. The Si_3N_4 substrate does not contribute to the spectral transmission and allows one to reduce the gradients in the probe layer. Simulations give the plasma parameters during the 4-ps of the x-ray probe. The temperature and the density of the plasma at the time of the absorption measurements are mainly driven by the laser intensity and the time-delay between the heating pulse and the backlighter beam.

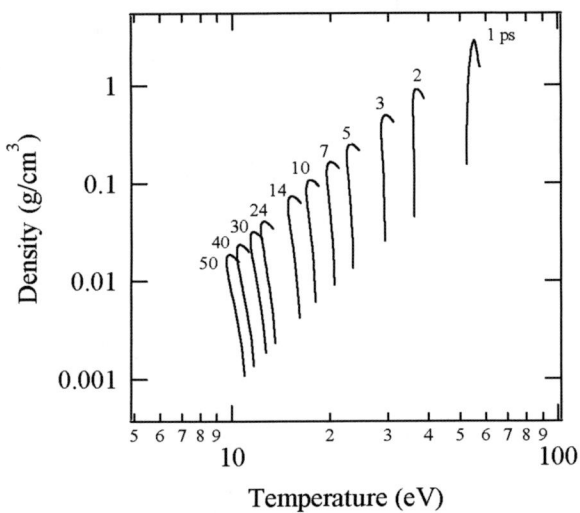

FIGURE 3. Thermodynamic path of the Al plasma at different times after the heating pulse. The laser intensity is equal to 9×10^{14} W/cm². MULTI-FS simulations were performed with 45-nm-thick Al layer deposited on 25-nm thick Si_3N_4 substrate irradiated with laser intensity corresponding to the FDI measurement (see Fig. 2). The Si_3N_4 substrate does not contribute to the spectral transmission and are not shown.

As our opacity calculations assume steady-state ions population, we have calculated the influence of non-equilibrium effects on the hydrodynamic evolution of the Al plasma. The temporal evolution of the average ionization $<Z>$ and the detailed populations of the plasma were calculated with the collisional-radiative code TRANSPEC [10] using the superconfiguration code AVERROES [10] as a postprocessor of the MULTI-FS code. Non-stationary effects become non-negligible when the plasma density decreases, i.e., after 15 ps. At this time, the stationary assumption leads to an underestimate of the ionic fractions. In a previous paper, we have shown that despite the rapid temporal evolution of the temperature, $<Z>$ given by AVERROES-TRANSPEC is in good agreement with the measured $<Z>$, suggesting that the plasma is close to LTE [11].

THEORETICAL CALCULATIONS

The opacity code OPAS [8] is used to interpret the transmission spectra assuming LTE conditions. The calculations are performed following three steps.

For given density and temperature, we use a non-relativistic average-atom model to self-consistently estimate the average ionization and the statistical sub-shell occupations [12]. From these quantities, a set of super-configurations (SC) is selected, spanning the relevant range of ionization stages. From these SC a large set of non-relativistic configurations (NRC) is generated and used in the initial phase of the next step.

The second step consists in calculating the occupation probability of each NRC by solving Saha-Boltzmann equations including a continuum lowering [13]. The configuration average energies are evaluated using one-electron wave-functions optimized per each SC. The chemical potential used in the Saha-Boltzmann equations is adjusted to match the average ionization with the one from the first step. In order to optimize the set of the selected NRC, new statistical sub-shell occupations are evaluated and compared to those issuing from the average-atom calculation. For an optimized NRC set, the chemical potential adjustment is generally limited to a few percents at maximum.

The third step is devoted to the opacity calculation, which is expressed as a sum including bound-bound, bound-free, free-free, and scattering contributions. The non-relativistic bound-bound (bound-free) oscillator strengths are evaluated using one-electron wave-functions optimized from the statistical sub-shell occupations coming from the previous step. Such a procedure preserves the one-electron oscillator strength sum rule. The length form of the electric dipole operator is used. The bound-free component of the opacity is evaluated from a detailed configuration accounting (DCA) approach using the above mentioned one-electron oscillator strengths. For each pair of NRC's involved in the transition, a pair of initial and final statistical relativistic configurations is generated. The energy threshold is then computed as a difference of the statistical relativistic configuration average energies. Such a procedure does not include the orbital relaxation effects (woOR). Assuming that these effects only modify the energy thresholds, they can be taken into account by performing a one-configuration Dirac-Fock calculation for each initial and final statistical relativistic configurations (wOR). An edge broadening is considered. It depends both on the thermal Doppler broadening and on the initial and final NRC energy variances [14].

The bound-bound component of the opacity is also evaluated using the above mentioned one-electron oscillator strengths. A DCA and a detailed line accounting (DLA) approaches are available. In the case of DCA calculations, for each pair of NRC's involved in the transition, a statistical relativistic splitting is introduced by generating two or three pairs of initial and final statistical relativistic configurations. A relativistic unresolved transition array (RUTA) approach is used to represent the overall lines connecting a pair of initial and final statistical relativistic configurations [15]. At this stage, the orbital relaxation effects are omitted. Each RUTA is assumed to be a gaussian profile, which average energy and variance can be exactly calculated in the framework of the retained approximations. Following the above mentioned procedure, the orbital relaxation effects can be taken into account by performing a one-configuration Dirac-Fock calculation for each initial and final statistical relativistic configurations. Each line shape is represented by a Voigt profile. The gaussian width depends on thermal Doppler and RUTA broadenings. The lorentzian width depends on natural (radiative and autoionization rates) and electron impact broadenings [16]. In the case of DLA calculations, a multi-configuration Dirac-Fock (MCDF) approach [17] is used to detail the lines connecting a pair of NRCs. The oscillator strengths are computed from the Babushkin gauge and the resulting total oscillator strength is reset to the one-electron oscillator strength. A Voigt profile is used for each line shape. The Gaussian width only depends on the thermal Doppler broadening. The lorentzian width depends on natural (radiative and autoionization rates) and electron impact broadenings. DCA and DLA approaches can be automatically mixed according to a coalescence criterion. This coalescence criterion is a function of the statistical number of lines connecting two NRC [18], on the individual line widths, and on the average line spacing. In the present work, the free-free component of the opacity is evaluated from the Kramers formula using the average ionization and a Gaunt factor equal to one. Bound-bound, bound-free, and free-free opacities are corrected for induced emission. The scattering contribution to opacity is taken into account using the Thomson scattering cross section.

Bromine absorption spectra were calculated with the DCA approach, whereas Al absorption spectra were calculated with the DLA approach.

RESULTS

The OPAS code is used with the density and temperature profiles given at the time of the absorption measurements derived from the simulations that reproduces the FDI measurements. The Al and KBr areal densities used to calculate the transmission spectra are equal to 8.5 $\mu g/cm^2$, in the lower limit of the measured areal density. The calculated transmission spectra are convolved with the spectral resolution, i.e., 2 eV and the spectral profiles in the center of the focal spot were extracted from all the space-resolved spectra, taking into account a 20-μm spatial width compatible with the Sm source diameter.

Fig. 4b and 4d shows lineout measured for two different time-delays between the heating pulse and the backlighter beam in the case of the Al layer. The delays are lower than 15 ps, allowing to probe the plasma when non-stationary effects are negligible. Fig. 4a and 4c shows the corresponding thermodynamic paths inferred from the FDI analysis.

The measured spectra exhibit $1s$-$2p$ absorption structures of Al^{4+} and Al^{5+} and extended structures in the Al^{3+} and Al^{4+} $1s$-$3p$ absorption. Concerning the $1s$-$3p$ absorption structures, their broadenings are primarily due to electron impact broadening which is known to be important in high density and moderate temperature plasmas [19]. At 7 ps and 12 ps, good agreement is obtained between the experimental spectra and the theoretical transmission. Calculations exhibit $1s$-$2p$ absorption structures of Al^{6+} which are not observed in the measured spectra. A red wing asymmetry can be observed for Al^{3+} $1s$-$3p$ absorption structure at 12 ps. The experimentally unresolved broad resonance and satellite lines of Al^{2+} are responsible of this asymmetry.

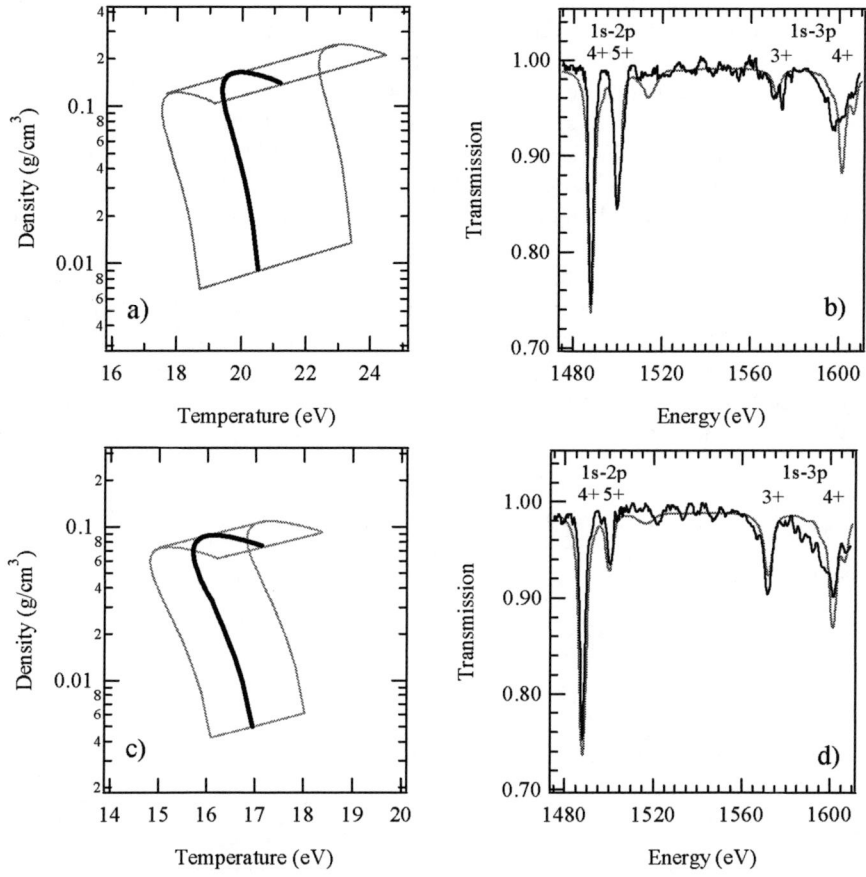

FIGURE 4. Thermodynamic path of the Al plasma a) 7 ps and c) 12 ps after the heating pulse. The laser intensity is equal to 9×10^{14} W/cm². MULTI-FS simulations were performed with 45-nm-thick Al layer deposited on 25-nm thick Si_3N_4 substrate irradiated with laser intensity corresponding to the FDI measurement. The Si_3N_4 substrate does not contribute to the spectral transmission and are not shown. The corresponding absorption spectra measured b) 7 ps and d) 12 ps after the heating pulse (black lines) are compared to the OPAS longitudinal-integrated absorption spectra (red lines).

Fig. 5a shows the thermodynamic path inferred from the FDI analysis in the case of the KBr layer. The laser intensity and the time-delay between the heating pulse and the backlighter beam are equals to 2×10^{15} W/cm² and 6 ps, respectively. Fig. 5b shows lineout measured in the center of the laser focal spot.

The measured spectrum exhibit extended structures due to $2p$-$3d$ absorption. Assuming that the areal density of KBr was the same as the one of Al, OPAS calculations are in agreement with the experimental spectrum when the density and temperature in the calculations are equal to the mean density and temperature in the layer, i.e., 0.4 g/cm³ and 38 eV. The density of the plasma is higher than the one of the

NaBr plasma previously measured with a z-pinch in Ref. [20]. In this experiment, the spectral transmissions of Br were measured at densities of about 1/100 of solid density, i.e. ten times lower than in the present study.

We checked the sensitivity of the opacity calculations by varying the density and the temperature around the values taken from the radiative-hydrodynamic simulations, i.e., 0.4 g/cm3 and 38 eV. We found that the theoretical spectrum is a weak function of density, and that a 10% variation in temperature leads to noticeable change in the theoretical spectrum. This change is due to the sensitivity of the ionization balance with respect to temperature.

The orbital relaxation treatment is crucial in this thermodynamic regime to reproduce the experimental transmission data. It allows to correctly predict the energies and the shapes of the main absorption structures. We have checked that differences between DLA and RUTA (wOR) calculations are non-perceptible at this density. Therefore, these experimental results can be used to test both atomic structure and statistical calculations.

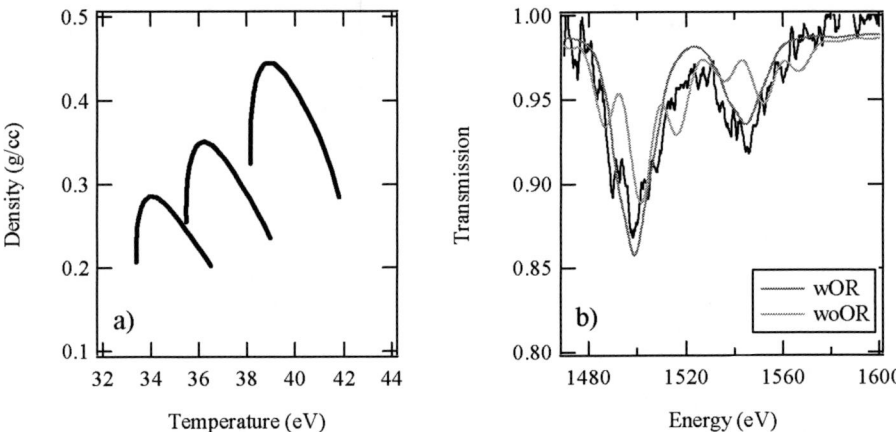

FIGURE 5. a) Thermodynamic path of the KBr plasma 6 ps after the heating pulse. The laser intensity is equal to 2×10^{15} W/cm^2. MULTI-FS simulations was performed with 45-nm-thick KBr layer deposited on 25-nm thick Si$_3$N$_4$ substrate irradiated with laser intensity corresponding to the FDI measurement. b) Measured absorption spectrum (black line) compared to the OPAS longitudinal-integrated absorption spectra (red and green lines) for 0.4 g/cm^3 and 38 eV.

CONCLUSIONS

Point-projection spectroscopy technique has been used to measure absorption spectra of LTE Al and KBr plasmas with minimized density and temperature gradients for densities up to 0.3 g/cm^3 and temperatures varying from 12 to 25 eV. The same setup has been used to estimate the precision of the hydrodynamic simulations (Al plasmas), by using K-shell opacity measurements, and L-shell spectral opacities of Br plasmas. Our experimental data are in good agreement with results of our LTE opacity code using a detailed line accounting approach using the results of the hydrodynamic simulations.

Independent characterization of the plasma parameters allowed an interpretation of the Al and Br absorption spectra in terms of spectral opacities. This agreement shows that the treatment of the density effects occurring in the experimentally accessed thermodynamic regime is well taken into account. Br opacity calculations including orbital relaxations agree well with the experimental transmission data. No significant differences were found for Br between detailed-line accounting calculations and orbital relaxation treatment coupled to a relativistic UTA formalism.

ACKNOWLEDGMENTS

The authors would like to thank O. Peyrusse for his help in the use of the TRANSPEC/AVERROES code and M. Millerioux (CEA/DAM-Ile-de-France) for making the deposited layer of aluminum. We acknowledge the invaluable support of the LULI laser operations staff. LLNL (Livermore, California) provided the various targets. Partial support for Ludovic Lecherbourg was provided by NSERC.

REFERENCES

1. J. D. Lindl, Phys. Plasma. **11**, 339 (2004).
2. S. Turck-Chièze *et al.*, Phys. Rev. Lett. **93**, 211102 (2004).
3. T. S. Perry *et al.* Phys. Rev. E **54**, 5617 (1996).
4. P. Renaudin *et al.*, J. Quant. Spectrosc. Radiat. Transfer **99**, 511 (2006) and references therein.
5. L. Lecherbourg *et al*, High Energy Density Phys., **3**, 175 (2007).
6. J.-P. Geindre *et al*, Optics Lett. **26**, 1612 (2001).
7. C. Chenais-Popovics *et al*,, SPIE Proceedings, vol. **5196**, 2005, SPIE, Bellingham, WA (2004).
8 K. Eidmann *et al*, Phys. Rev. E **62**, 1202, (2000).
9. R. Shepherd *et al.*, J. Quant. Spectrosc. Radiat. Transfer **71**, 711 (2001).
10. O. Peyrusse, J. Phys. B: At. Mol. Opt. Phys. **32** , 683 (1999); **33** 4303 (2000).
11.P. Audebert *et al.*, Phys. Rev. Lett. **94**, 025004 (2005).
12. C. Blancard *et al*, Phys. Rev. **69**, 16409 (2004).
13. J. C. Stewart and K. D. Pyatt, Astrophys. J. **144**, 1203 (1966).
14. J. Bauche *et al*, Phys. Rev. A **25**, 2641 (1982).
15. S. A. Moszkowski, Prog. Theoret. Phys. **28**, 1 (1962).
16.M. S. Dimitrijevic *et al*, A&A **172**, 345 (1987).
17.J. Bruneau, J. Phys. B **17**, 3009 (1984).
18.J. Bauche *et al*, J. Phys. B **20**, 1659 (1987).
19. D. Hoarty *et al.*, J. Quant. Spectrosc. Radiat. Transfer **99**, 283 (2006).
20. J. Bailey *et al.*, J. Quant. Spectrosc. Rad. Transfer **81**, 31 (2003).

Diagnostic Spectrometers for High Energy Density X-Ray Sources

L. T. Hudson*, A. Henins*, J. F. Seely[†] and G. E. Holland[†]

*National Institute of Standards & Technology, Gaithersburg, MD 20899 USA
[†]Naval Research Laboratory, Space Science Division, Washington DC 20375 USA

Abstract. A new generation of advanced laser, accelerator, and plasma confinement devices are emerging that are producing extreme states of light and matter that are unprecedented for laboratory study. Examples of such sources that will produce laboratory x-ray emissions with unprecedented characteristics include megajoule-class and ultrafast, ultraintense petawatt laser-produced plasmas; tabletop high-harmonic-generation x-ray sources; high-brightness zeta-pinch and magnetically confined plasma sources; and coherent x-ray free electron lasers and compact inverse-Compton x-ray sources. Characterizing the spectra, time structure, and intensity of x rays emitted by these and other novel sources is critical to assessing system performance and progress as well as pursuing the new and unpredictable physical interactions of interest to basic and applied high-energy-density (HED) science. As these technologies mature, increased emphasis will need to be placed on advanced diagnostic instrumentation and metrology, standard reference data, absolute calibrations and traceability of results.

We are actively designing, fabricating, and fielding wavelength-calibrated x-ray spectrometers that have been employed to register spectra from a variety of exotic x-ray sources (electron beam ion trap, electron cyclotron resonance ion source, terawatt pulsed-power-driven accelerator, laser-produced plasmas). These instruments employ a variety of curved-crystal optics, detector technologies, and data acquisition strategies. In anticipation of the trends mentioned above, this paper will focus primarily on optical designs that can accommodate the high background signals produced in HED experiments while also registering their high-energy spectral emissions. In particular, we review the results of recent laboratory testing that explores off-Rowland circle imaging in an effort to reclaim the instrumental resolving power that is increasingly elusive at higher energies when using wavelength-dispersive techniques. These efforts inform the optimization of diagnostic designs that will permit acquisition of high-resolution, hard x-ray spectra in the HED environment.

Keywords: x-ray spectroscopy, crystal spectrometer, Rowland circle focusing, resolving power, Cauchois geometry, hard x-ray sources
PACS: 07.85.Nc; 52.59.Px; 52.70.La, 32.30.Rj; 39.30+w

INTRODUCTION

One of the dominant trends in plasma physics today is the production of extreme states of matter that produce ever-more energetic and intense particle and radiation fields. For example, lasers for inertial confinement fusion are moving into the magajoule regime, while those for fast ignition or backlighter studies are gauged in units of femtoseconds and petawatts. The evolving performance of these and other

CP926, *Atomic Processes in Plasmas—15th International Conference on Atomic Processes in Plasmas*
edited by J. D. Gillaspy, J. J. Curry, and W. L. Wiese
2007 American Institute of Physics 978-0-7354-0436-6/07/$23.00

plasma confinement sources and novel high-peak-power x-ray sources will require parallel evolutions in design and implementation of hardened diagnostics of all types. In the case of x-ray diagnostics, the expected high incident flux recommends wavelength- rather than energy-dispersive methods, even as the smaller wavelengths challenge instrumental resolving power. Higher-energy photons also are more penetrating and can be a source of unwanted secondary fluorescence and diffuse background (scattering). This paper presents some instrumental strategies, trends, and laboratory results, using the crystal diffraction technique, that have been utilized for HED applications.

Previously we have reported on the design of bent crystal spectrometers that employ the so-called Cauchois[1] geometry, that is, x rays are incident on the convex face of a cylindrically-bent crystal lamella and detected on the concave side (via Laue diffraction).[2] We have also demonstrated the advantages of a particular implementation, illustrated in figure 1, for the intense environment of laser produced plasmas.[3] In this case, the center axis of the instrument coincides with a point-like x-ray source producing two, mirror-symmetric spectra with respect to this center axis. Alignment is gauged by pinhole imaging of the x-ray source along the instrument axis and a simple analytic expression relates wavelength to lateral distance from the center axis in the detector plane. The appropriateness of this design to HED applications derives from the ability to capture the diffracted spectrum at small diffraction angles (high photon energies) while shielding the detector from direct line-of-sight by means of a central septum (that holds the pinhole) and a narrow polychromatic crossover slit. We have employed crystals with relatively small radii of curvature resulting in both increased reflecting power and bandwidth as well as compact instrument packages covering 10 keV to 200 keV. The following section outlines design modifications and laboratory demonstrations that seek to anticipate the emerging diagnostic needs related to registering spectra from both high-intensity and high-energy x-ray sources.

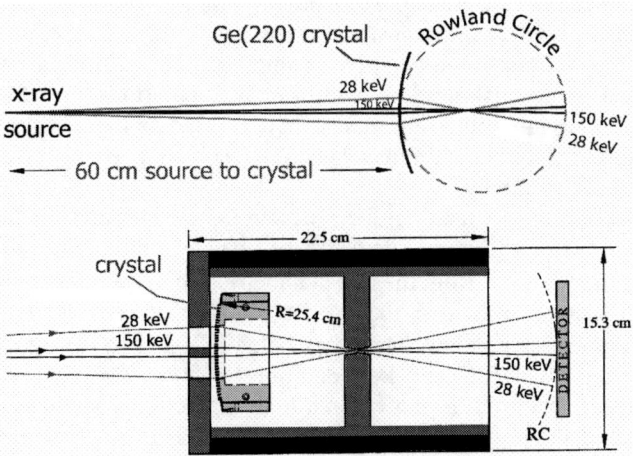

FIGURE 1. Example of a Cauchois-geometry crystal spectrometer that registers x rays from 28 keV to 150 keV from a point-like source. Below, a to-scale drawing of the instrument package shown with the detector placed near the Rowland circle (RC), that is, a distance behind the crystal equal to the crystal radius of curvature.

EXPERIMENTAL DESIGNS & MEASUREMENTS

The challenge of high-energy wavelength-dispersive spectroscopy is the nonlinear plate function that results in decreasing instrumental resolving power at high photon energies. For the spectrometer design shown in Figure 1, x rays of different energies are focused by the crystal to unique positions along the Rowland circle, which is shown tangent to the crystal and with a diameter equal to the radius of curvature of the crystal. While the focusing is not exact, conventional practice has been to locate the detector near the Rowland circle in order to capture a less-blurred spectrum. With the advent of large-area, inexpensive detectors such as image plates, the use of small crystal radii of curvature for high-bandwidth applications, and the interest in high-energy spectroscopy at good resolving power, we have investigated the effect of off-Rowland circle spectroscopy with both point-like and extended x-ray sources.

In cases where the x-ray source subtends a relatively small angle when viewed from the spectrometer, the extent along the convexly-curved crystal that satisfies the diffraction condition for a single energy is relatively small in extent and determined primarily by the width of the bent-crystal rocking curve. For tightly-bent crystal geometries, the contribution of dispersion to instrumental resolving power can exceed that of defocusing; in some cases this is true for arbitrarily distant detector positions.

A demonstration of this trend is shown in Figure 2. Here a quartz-crystal spectrometer, similar in format to that of Figure 1, was placed 0.5 m from a microfocus Mo x-ray tube with a nominal spot size of between 0.15 mm and 0.45 mm. The tube was operated at 1 mA and 50 kV and the Qz(10-11) crystal was bent to the form of a cylinder of radius 112 mm. An image plate was positioned on the Rowland circle (112 mm behind the crystal) and at locations noted on the figure that were -2 cm to +96 cm with respect to this nominal focusing distance. Column sums of spectral images acquired at six detector positions are shown in Figure 2 showing the region of the spectrum containing K-shell emission features of molybdenum (around 17 keV to 20 keV). The image plates were exposed from 0.5 min to 3 min depending on distance and were scanned giving two-dimensional images with 42 μm pixel spacing. Spectra were derived by column summing pixel values and are displayed with the same relative abscissa scale making line widths and separations directly comparable. Clearly, even as individual lines are increasingly blurred as the detector is withdrawn beyond the Rowland circle, doublets become resolved and the effective spectral resolution improves markedly.

At 96 cm beyond the focusing circle (107 cm behind the crystal) the image plate was employed *in extenso*; Figure 3 displays the image captured after three minutes of exposure. Here the Kα lines from the two symmetric spectra are separated by about 40 cm. The column sums from two indicated regions of interest (ROI) show (a) the extremely high resolving power at Mo Kα (labeled ROI-1) and (b) the entire spectrum including continuum up to the 50 keV end-point energy of the x-ray tube (labeled ROI-2). This latter spectrum, is shown on a logarithmic scale and takes advantage of the large dynamic range of image plate technology. The grayscale contrast on the image has been equalized to reveal the various spectral features, including the absorption edges of barium and iodine that are constituents of the image plate and also noted in the spectrum Figure 3 (b).

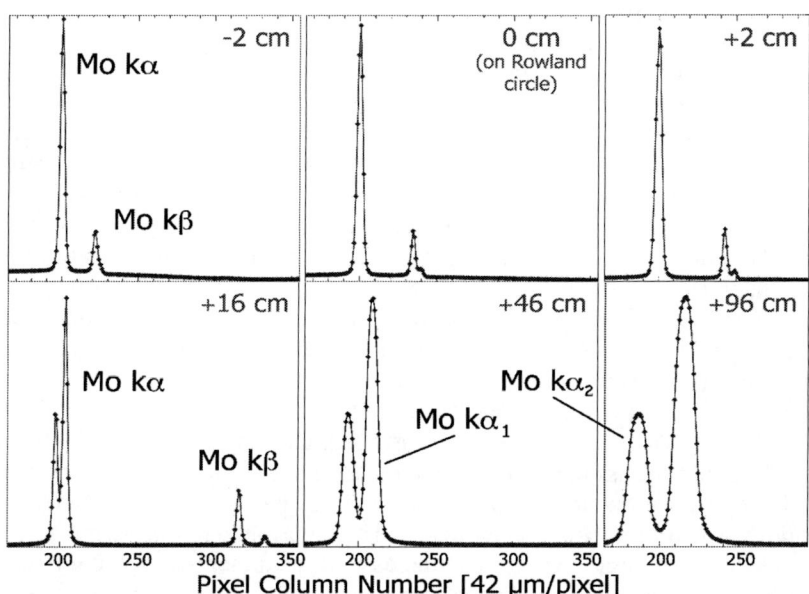

FIGURE 2. Spectral evolution of Mo K-shell emissions imaged at various locations relative to the Rowland circle (distances indicated in cm).

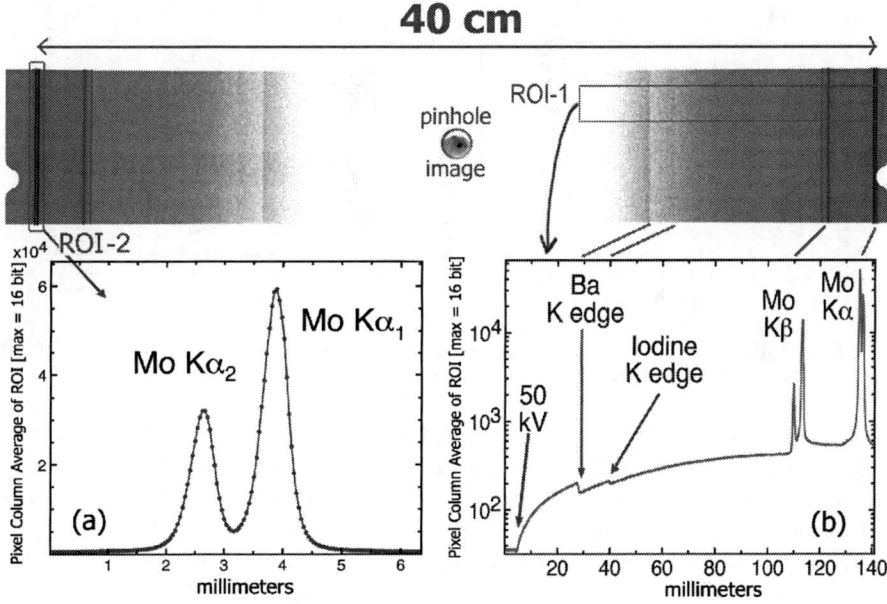

FIGURE 3. Spectrum imaged with a curved-crystal spectrometer and image plate positioned 96 cm beyond the Rowland focusing circle. In this demonstration, the instrumental resolving power is primarily determined by the dispersion rather than optical aberrations, source-size broadening, or detector resolution.

This trend of obtaining improving resolving power off the Rowland circle with high-dispersion crystal geometries is extensible to the higher-energy regimes of interest in many HED studies including the development of Compton radiography using high-Z backlighters produced by ultraintense lasers.[4] Take for example the source-crystal geometry of Figure 1. In general, curved-crystal spectrometers with larger radii of curvature and smaller lattice spacings provide greater dispersion at higher x-ray energies, though often at the expense of sensitivity and bandwidth. In addition, larger bending radii reduce the bent-crystal rocking curve width and therefore the primary source of defocusing in the case of a point-like source. Figure 4(a) illustrates the plate function from 28 keV to 150 keV for the arrangement in Figure 1, both with the detector on the Rowland circle and 20 cm further removed from the crystal. The ordinate axis is the lateral extent along a detector and its zero is the central axis of the spectrometer. Hence, to image both spectra, a detector of 5.6 cm lateral extent is required in the plane of dispersion at the Rowland circle; 13.3 cm is needed at 20 cm. The energy-dependent instrumental resolving power is modeled in Figure 4 (b) under the assumption of a point source, 0.14 mm detector resolution, and negligible defocusing due to rocking-curve width (as motivated above, and as per our observations with a prototype spectrometer). The replacement of the detector to 20 cm beyond the Rowland circle results in an increase in instrumental resolving power of at least 2.3 at all energies with no loss of signal if the area detector is large enough to captured the projected spectral image.

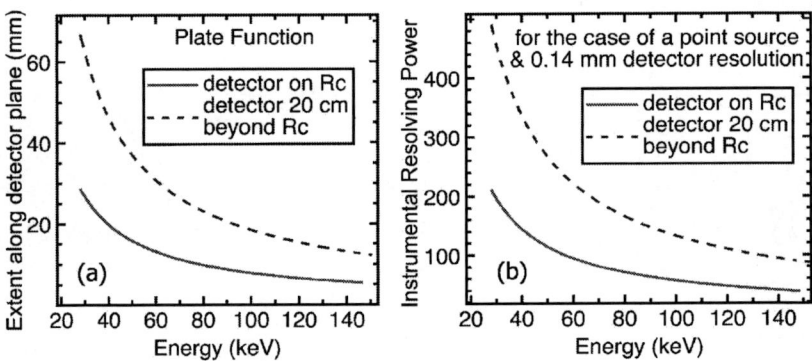

FIGURE 4. (a) Plate functions and (b) resolving power for the spectrometer arrangement of Figure 1, and for detector positioned both at Rowland circle and 20 cm beyond.

Considering now the case of an extended x-ray source with the Cauchois-crystal geometry, one easily sees why conventional wisdom has been to place the detector on the Rowland focusing circle. The spectrum from an extended source focuses near the Rowland circle, but can rapidly diverge from this position. This undesirable effect is greater for larger diffraction angles and smaller radii of curvature, so for clarity of illustration, consider again the quartz crystal used to acquire the data of Figure 2 with a source 0.5 m from the crystal and with lateral extent of 11 cm in the plane of dispersion. The diffraction of 11 keV x rays on just one side of the crystal is illustrated in Figure 5. X rays from source positions "a" to "c" (10° field of view) each diffract from quartz planes in transmission at a Bragg angle of 9.7°, but at unique positions along almost 2 cm of crystal arc, finally converging very nearly, though not exactly, on the Rowland focusing circle. Because of the relative source size and high crystal dispersion, the 11 keV x rays quickly diverge to 5 mm only 25 mm beyond the Rowland circle. In this example, the defocusing would be primarily due to the extended source rather than the crystal rocking curve width that dominates in the case of a point-like source; both mechanisms are present and similar in kind, but can differ in degree depending on the application. Of course, it is not the absolute size of a source that is relevant, but the angular range that it subtends in the plane of dispersion. Finally, it can be noted that it is possible to employ the Cauchois geometry to perform off-Rowland circle spectroscopy of a slice of an extended source by narrowing the crossover slit, thereby reducing the spectrometer field of view.

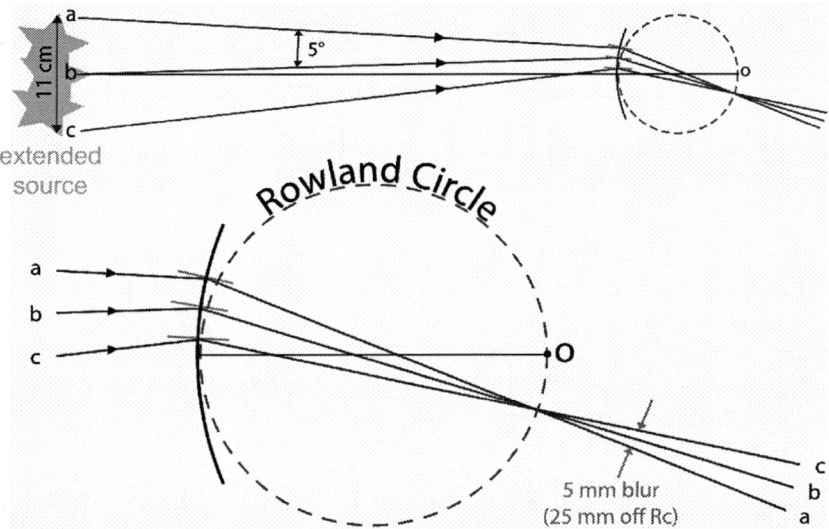

FIGURE 5. Defocusing beyond the Rowland circle caused by an extended x-ray source. In this example, 11 keV x rays are diffracted at 9.7° by a Qz(10-11) crystal that is cylindrically bent to 112 mm radius and positioned 0.5 m from a source with 11 cm lateral extent in the plane of dispersion.

While the ill effects of source-size spectral broadening at smaller diffraction angles and larger crystal radii is not as leveraged as the foregoing example, it can still be a significant detractor from instrumental resolution. We have demonstrated and modeled this trend at higher energies using the prototype spectrometer of Figure 1 in conjunction with an industrial W x-ray source operated at 250 kV and 4 mA and with a source-to-crystal distance of 1.2 m. Here the Kα and Kβ emission lines from tungsten (from 58 keV to 69 keV) were captured with image plates that were positioned 49 cm beyond the Rowland circle or 74 cm beyond the crystal. Figure 6 shows the K-shell spectral lines acquired in this arrangement and with three nominal source-spot sizes: 0.65 mm, 1.8 mm, and 3.5 mm. The expected trend of line broadening with source size is clear.

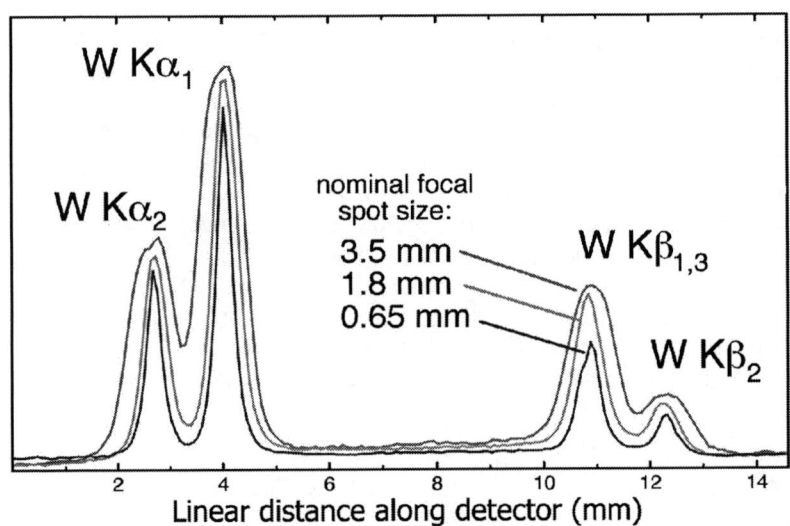

FIGURE 6. W K-shell spectroscopy off the Rowland circle; line widths as a function of nominal spot size. Here a Ge(220) crystal with radius of 25 cm was positioned 1.2 m from an industrial W source at 250 kV. The detector was positioned 49 cm beyond the Rowland focusing circle (74 cm from the crystal).

We have developed an analytical model that addresses quantitatively the qualitative trends that are demonstrated in this proceedings. For the case of symmetric Cauchois diffraction in the small-angle approximation, the model predicts instrumental resolving power taking into account the source, crystal and detector positions and dimensions; the crystal dispersion (lattice spacing and radius of curvature); the detector resolution; and the bent-crystal rocking curve width. These expressions show good agreement with the measured line widths in our testing in a variety of spectrometer-source configurations. This detailed formalism and comparisons with test data are the subject of a forthcoming publication. One insight derived from this work is the ability to predict the optimum detector position relative to the Rowland circle to achieve the highest instrumental resolving power for the types of spectrometer geometries and sources discussed in this paper. An illustration of this for the configuration described in Figure 6 is shown below in Figure 7. Here curves of the resolving power at W Kβ (68 keV) as a function of detector distance beyond the Rowland circle is given for lateral source sizes ranging from 0.1 mm to 6.7 mm. For sufficiently point-like sources, the resolving power only improves beyond the Rowland circle and is limited by other considerations such as detector size and instrumental sensitivity. For sources that subtend larger angles, there is an optimum detector position which, as expected, approaches that of the Rowland circle in the large-source limit.

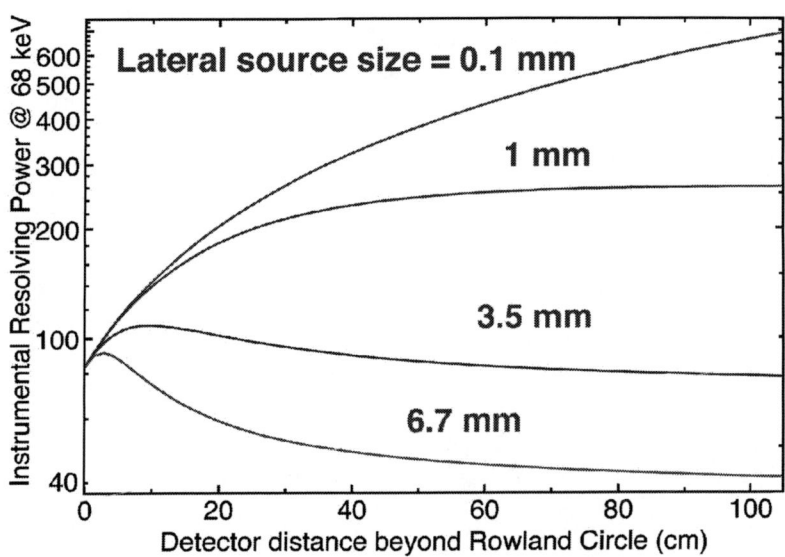

FIGURE 7. Model calculations for the instrumental resolving power as function of detector position, for a variety of source sizes. The source-spectrometer configuration is that used to acquire the data of Figure 6.

SUMMARY

Diagnostics for the HED x-ray sources that are coming on line require designs that are both hardened to the harsh experimental conditions, and in the case of x-ray diagnostics, keep up with the trending interest in higher-energy spectroscopy. We have previously shown the appropriateness of a Cauchois-geometry curved-crystal spectrometer for obtaining high signal in high-noise situations. In this paper we demonstrate the additional benefits that are possible with off-Rowland circle imaging using this type of spectrometer to reclaim the resolving power that is normally lost in high-energy wavelength dispersive spectroscopy.

REFERENCES

1. Y. Cauchois, *J. Phys. Radium* **3**, 320-336 (1932).
2. L. T. Hudson, A. Henins, R. D. Deslattes, J. F. Seely, G. E. Holland, R. Atkin, L. Marlin, D. D. Meyerhofer, and C. Stoeckl, *Rev. of Sci. Instrum* **73**, 2270-2275 (2002).
3. L. T. Hudson, R. Atkin, C. A. Back, A. Henins, G. E. Holland, J. F. Seely, and C. I. Szabó, *Rad. Phys. Chem.* **75**, 1784-1798 (2006) and references therein.
4. R. Tommasini, H-S. Park, P. Patel, B. Maddox, S. Le Pape, S. P. Hatchett, B. A. Remington, M. H. Key, N. Izumi, M. Tabak, J. A. Koch, O. L. Landen, D. Hey, A. MacKinnon, J. Seely, G. Holland, L. Hudson, and C. Szabo, "Development of Compton radiography using high-Z backlighters produced by ultra-intense lasers" in *15ᵗʰ International Conference on Atomic Processes in Plasmas*, edited by J. J. Curry, J. D. Gillaspy, and W. Wiese, AIP Conference Proceedings XXX, New York: American Institute of Physics, 2007, pp. xx - yy. (this volume)

FUSION PLASMAS

Plasma-wall interaction: how atomic processes influence the performance of fusion plasmas

Ralf Schneider

Max-Planck Institute for Plasmaphysics, EURATOM Association, Wendelsteinstr. 1, D-17491 Greifswald, Germany

Abstract. Plasma edge physics is one of the major challenges in fusion plasmas. The need for power and particle exhaust for any reactor inspired a lot of theoretical and experimental work. Understanding this physics requires a multi-scale ansatz bringing together also several physics and numerical models.

The plasma edge of fusion experiments is characterized by atomic and molecular processes. Hydrogenic ions and neutrals hit material walls with energies from several eV up to 1000s of eV. They saturate the wall materials and due to physical or chemical processes neutrals are released from the wall, both atomic and molecular. They determine via interaction with the plasma strongly its properties. These processes can be beneficial for a fusion experiment by using radiation losses to minimize the power load problem of target plates, but also can create severe problems if the dilution of the plasma gets too large or condensation radiation instabilities can be created.

A complete physics model for the plasma-wall interaction processes alone is already rather challenging (and still missing): it requires e.g. inclusion of collision cascades, chemical formation of molecules, diffusion in strongly 3D systems. A full description needs a multi-scale model combining quite different numerical techniques like molecular dynamics, binary collisions, kinetic Monte Carlo and mixed conduction/convection equations in strongly anisotropic systems.

Keywords: Fusion plasmas, plasma-wall interaction, neutrals, impurities
PACS: 52.20.-j, 52.25.Jm, 52.25.Vy, 52.40.Hf, 52.55.Rk

INTRODUCTION

Fusion is one of the options for overcoming the energy problems in this century. Sufficient good energy confinement to achieve ignition is obtained in magnetic fusion by the creation of a magnetic cage of nested flux surfaces using the enormous anisotropy between transport along fieldlines and perpendicular to it. The compression of helium and the helium pumping is important, because one has to avoid too much helium ash which would extinguish the burning plasma in a reactor. A global 0D model for burning D-T plasmas addresses the effect of impurities on ignition conditions [1, 2]. Depending on the assumed concentration of additional impurities, one only gets solutions if the ratio of the global α-particle confinement time and the energy confinement time is less than 10–15. Otherwise, the helium ash just extinguishes the burning plasma.

The scrape-off layer region gets very important in any fusion reactor, because it is directly related to the requirements for power and particle exhaust compatible with the core confinement. One major problem is that the thickness of the layer where the heat flux from the core is transported onto walls is rather small (typically about 1 cm, which is the size of the poloidally projected ion gyro-radius) due to the very strong parallel transport, especially for the required enhanced confinement modes necessary

CP926, *Atomic Processes in Plasmas—15ᵗʰ International Conference on Atomic Processes in Plasmas*
edited by J. D. Gillaspy, J. J. Curry, and W. L. Wiese
© 2007 American Institute of Physics 978-0-7354-0436-6/07/$23.00

for ignition. This results in severe engineering problems due to exceedingly large power loads on the walls, illustrated here for ITER and motivating the need for a detailed understanding of the scrape-off layer physics.

In ITER the amount of power carried by the α-particles P_α has to be exhausted in a controlled way. Doing a simple estimate for the ITER target-plate power load $q_{\perp,tp}$, one gets

$$q_{\perp,tp} = \frac{P_\alpha}{2\pi \cdot R_{X-point} \cdot 2 \cdot \Delta_e}, \tag{1}$$

where Δ_e is the energy decay length at the target-plate and $R_{X-point}$ is the major radius at the X-point. For this estimate one assumes the deposition of the power in a strip of Δ_e on two divertor plates on a ring with length $2\pi R_{X-point}$.

This gives for the planned ITER [3]

$$q_{\perp,tp} = \frac{300\,\text{MW}}{4\pi \cdot 7\,\text{m} \cdot 0.02\,\text{m}} = 175\,\text{MW/m}^2. \tag{2}$$

Introducing further reduction factors due to bulk radiation losses (0.8) and poloidal tilting of target-plates (0.5–0.25) one still get values of about $q_{\perp,tp} = 35\,\text{MW/m}^2$. The reduction factor of (0.5–0.25) by tilting of the target-plates is not as large as expected from flux expansion factors between midplane and target-plates which can easily be 0.1 and lower, because the total surface area does not change as much due to the fact that the intersection angles do not change so strongly. They even increase again if the tilting is very strong, because then the distance to the X-point gets larger in which vicinity the angles are very small.

Realistic values for steady-state operation are below about $5\,\text{MW/m}^2$. This means that additional losses are required to spread the target power load over a wider area. It is important to note here that the final concept has to be compatible then also with the necessity of particle exhaust (helium ash removal) and core confinement.

A divertor is produced by additional coils creating a separatrix intersecting in the X-point. Here, the basic idea is to move the interaction zone away from the core to so-called divertor plates. It is hoped to get a better impurity control due to very good divertor retention and to achieve a reduction of the heat flux to target-plates due to radiation losses.

ELEMENTARY PROCESSES IN THE DIVERTOR

To lower the maximum power load and to broaden the energy deposition profile on the target-plates, one has to use loss channels which are not constricting the energy flow to follow magnetic field-lines. The ideal scenario would be one where these energy losses would be created without additional impurities. For hydrogen three energy exchange channels — charge exchange, atomic radiation and volume recombination — are in principle available (see Fig. 1).

Charge exchange losses. Charge exchange (CX) enery conversions are already used experimentally with great success to convert high energy fluxes of charged particle

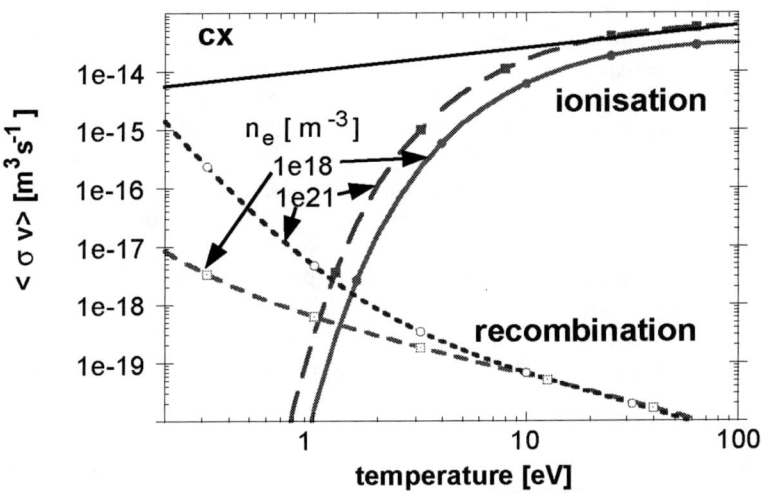

FIGURE 1. Cross sections for charge exchange (CX)), ionization and recombination as a function of electron temperature for hydrogen.

beams into neutrals through neutralizer sections in neutral injector beam lines. In divertors it requires, however, that the plasma be sufficiently thin, so that neutrals can interpenetrate it. For ITER parameters one gets losses of nearly 1 at 660 eV, but only 0.06 at 100 eV and 0.002 at 10 eV. That means, that CX energy losses are only effective at high temperatures. Moreover, CX neutrals from the hot regions pose erosion problems in the main chamber and the transition region from the divertor to the main chamber (baffles, etc.).

Neutral hydrogen radiation. For neutral hydrogen atoms at electron temperatures above about 8 eV, the ratio of an enhanced 'effective' electron ionisation energy loss to the 'true' ionisation energy loss, which is again deposited along field-lines at the target-plate (without volume recombination) is around 2–3.

Only the radiated fraction, which arises during the stepwise ionisation through many excited states of the atom, constitutes a true volumetric loss. Below about 8 eV the radiated power associated with ionisation processes increases significantly. The characteristics of the parallel electron heat conduction ($q_\parallel \sim T^{5/2} \nabla T \sim T^{7/2}$) imply, however, that the volume over which the temperature can be below a certain value at a certain heat flux density is proportional to $T^{7/2}$ along field-lines, and hence dramatically decreases with temperature. Therefore, the low temperature region is very small and cannot contribute to the integral radiation loss.

Volume recombination. For plasma temperatures below about 2 eV, volume recombination processes become important. Recombination of an electron and an ion into a neutral atom needs a second body to account for energy and momentum conservation during the process. This is possible in two ways: in the process of radiative recombination a photon takes care of energy and momentum conservation. The second process is three-body recombination where an additional electron ('spectator electron') is necessary. The recombination rate for this process gives the strong rise below about 2 eV (especially for high densities above 10^{20} m^{-3}). The second process is mainly responsible for the strong effects of volume recombination in the divertor affecting the whole character of the plasma state. Both processes proceed in a ladder-like way through the excited levels of the atoms until the final ground state is reached. Hence hydrogen line radiation is involved with a specific spectral distribution, which is important for diagnosing the onset of strong volume recombination spectroscopically.

The effective electron cooling rate for the recombination rates shows that the absolute direct energy loss for electrons by volume recombination is usually small and can be even a heating term for high densities and low temperatures. That means that the process of volume recombination has the tendency to stabilise the temperature and avoid further cooling. From what was said above it follows that the recombination zone is always located "below" the ionisation zone (i.e. between ionisation zone and plate).

Summarising this discussion of energy losses from hydrogen, it is obvious that for ITER additional radiation losses from impurities are necessary [8].

Divertor geometry effects

The target-plate geometry strongly influences the plasma profiles by controlling the neutral recycling pattern, which in turn has a strong effect on the symmetry and stability of the divertor plasma and finally on the whole edge region. The design of the new divertor configuration called Lyra for ASDEX Upgrade (see Fig. 2) is a good example of this optimisation strategy.

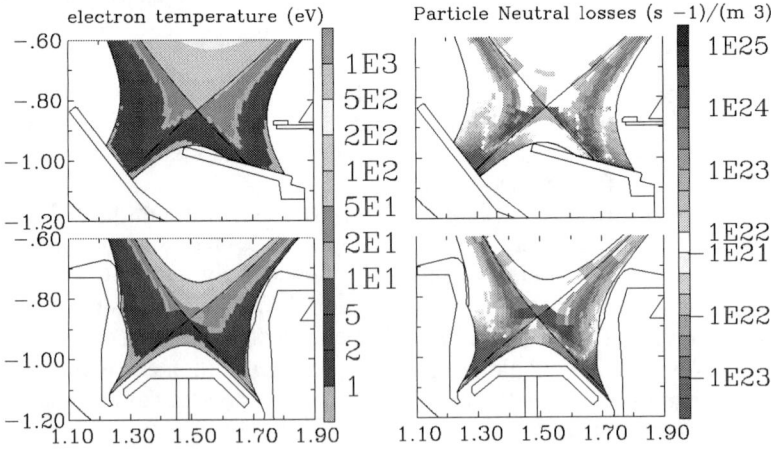

FIGURE 2. Contours of the temperatures and neutral sources as calculated with B2-Eirene. The blue sources are ionisation sources, the red sources are recombination sinks for the plasma.

For Divertor-I, the outer divertor plate reflects the neutrals away from the energy-carrying zone close to the separatrix and thus blocks indirectly the detachment (see

Fig. 2). Therefore, Divertor-I detaches strongly asymmetrically (preferentially on the inner divertor)as can be seen in the temperature and neutral source distributions. Nevertheless, the power load on the outer target-plate is reduced by the large flux expansion close to the X-point. The Lyra has, for the same density conditions, the lowest power loads, because it reflects the neutrals towards the high energy zone at the separatrix and is strongly tilted. Therefore, concerning the operational safety of the machine, it offers the best configuration to start with, allowing for scenarios without external impurities and the largest heating power.

Effect of vibrationally excited molecules

A mechanism, possibly also important in the divertor, is recombination through vibrationally excited molecules. Two possible reaction paths exist. The first one is the creation of negative ions $(e+H_2(v) \rightarrow H_2^-)$, followed by dissociation $(H_2^- \rightarrow H+H^-$ and $H^++H^- \rightarrow H+H^*)$. The second channel is ion conversion $(H^++H_2(v) \rightarrow H+H_2^+)$ and dissociation $(e+H_2^+ \rightarrow H+H^*)$.

The rate coefficients of these two processes lead to a large increase of the total recombination rate coefficient (including also three-body and radiative recombination) by at least one order of magnitude for temperatures below about 5 eV. However, the rate coefficients for ionisation and dissociation increase also, reducing thus the mean free path of the molecules. The recombination source depends on rate coefficient and electron density, but also on the molecular density. Therefore, a large recombination rate coefficient does not necessarily lead to large recombination effects (sources). A detailed discussion of this can be found in [4, 5].

Introducing the different vibrational levels as separate species in the neutral Monte-Carlo-Code Eirene and doing a self-consistent B2-Eirene calculation this recombination channel is only important in a rather thin surface layer close to the plasma boundary, because the neutral molecules do not penetrate as deep as for the case without taking into account the vibrational excitation. The integral contribution to the total recombination rate is also a minor correction.

IMPURITY PHYSICS

After the discussion of the clean plasma, the SOL description must be completed by including the effect of impurities. Fusion plasmas always contain impurities, be it He fusion ash, radiative gases injected (N_2, Ar, Ne, Kr), or wall material sputtered by the plasma (C, W, Fe, Mo, Ni, ...). Therefore, for the understanding of impurity physics in the SOL, one needs a rather complex combination of different aspects. The impurity production process has to be understood, then the effects of impurities in terms of radiation losses have to be included, and finally the impurity transport is necessary.

Impurity production

The radiation losses by impurities are necessary for relaxing the heat exhaust problem by spreading the heat deposition onto a larger surface by radiation. However, radiation losses in the main chamber can be a problem, because a minimum amount of power is necessary in order to sustain sufficient confinement (H-mode threshold) and to sustain ignition one has to avoid too large power losses. Therefore, a scenario is needed where impurity losses (power as well as position) can be (feedback) controlled. Intrinsic impurities (C, W, Be, ...) are produced through erosion (ions, CX neutrals, hot spots, arcs) and melting (disruptions, very large ELMs). The physical sputtering process is a binary collision between surface or bulk atoms and impinging ions (after acceleration in the sheath) or neutrals [6]. Due to the binary collisions, one gets a threshold energy for the impinging particle below which the sputtering yield drops very fast to practically zero. This threshold is therefore higher for heavier bulk particles.

Seeded impurities for controlled impurity losses have to be recycling impurities (noble gases, e.g. Ne, Ar, Kr), because non-recycling impurities cannot be pumped for feedback control but stick to the walls. For the choice of target (and main chamber) materials, one has several candidates representing either low Z (e.g. Be, C) or high Z (e.g. Mo, W) materials, which both have different advantages and disadvantages. Carbon has a very high thermal conductivity, high melting point and low vapour pressure, but is subject to chemical erosion as it readily combines chemically with hydrogen isotopes, leading in particular to tritium wall inventory problems in a reactor environment. Concerning the sputtering by ions (after acceleration in the sheath) and high energy CX neutrals, tungsten has much lower effective sputtering yields than beryllium or carbon. This is due to the advantageous ratio of ionisation-length to ion gyro-radius [7] leading to a high probability of prompt redeposition within the first ion gyration of the tungsten ion after ionisation of the sputtered atom. However, only a very small fraction of W in the main plasma ($2 \cdot 10^{-5}$) would be enough to quench ignition. Molybdenum raises concerns due to its activation under neutron bombardment, leading to the creation of relatively long-lived radioactive isotopes within the wall material. Beryllium has the problem of a low melting point and high toxicity, but excellent vacuum gettering properties.

The radiation losses for the low Z elements are dominated for temperatures below 100 eV by line radiation losses and above several keV by bremsstrahlung. Due to their radiation characteristics, low Z elements will contribute more to SOL and divertor radiation (temperatures below 100 eV), whereas with higher Z more and more radiation will move to the confinement region of closed field lines (Fig. 3). For tungsten, the radiation losses are three orders of magnitude larger than for carbon or beryllium. Therefore, one needs nearly perfect divertor retention for tungsten in contrast to carbon. This gives a very strict limit for main chamber concentration (10^{-5}) for ITER, and up to now it is not clear whether this is achievable.

For carbon one has a different problem to overcome: even at low plasma temperatures, where physical sputtering switches off, sputtering of carbon is still continuing. The reason is chemical sputtering producing methane and other hydrocarbons, which finally break up into carbon ions [9]. This process is a surface effect and has a strong temperature dependence with a maximum yield at about 600 K, characterised by the change of the carbon bonds at a surface from an sp^3 to sp^2 type. This process gives large erosion

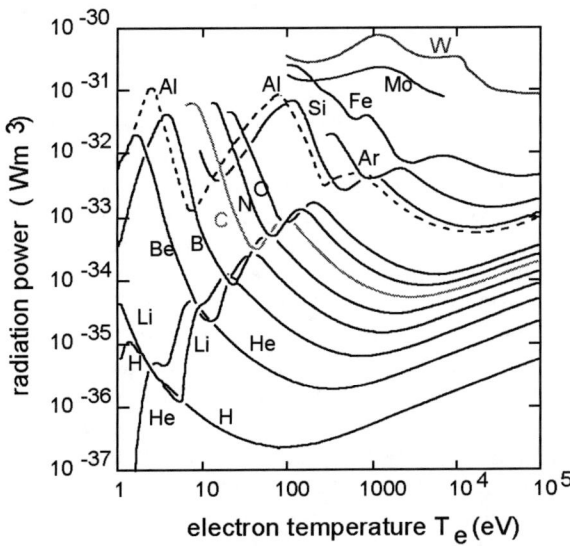

FIGURE 3. Radiation power $\frac{P_{rad}}{Vol \cdot n_e \cdot n_{imp}}$ as a function of electron temperature.

rates for ITER, due to the very large incoming particle fluxes, although there exists a favourable flux dependence of the chemical sputtering yield [10]. Some experimental data also points to a total fluence dependence, but are still unexplained [11]. Possibilities to overcome this problem for carbon are proposed and tested by titanium doping for suppression of the chemical erosion. Other experiments suggest that nitrogen injection may also contribute to reducing chemical erosion [12], likely by a catalytic 'poisoning' of the active carbon sites available.

The hydrocarbons produced by chemical sputtering also tend to form co-deposited layers far away from the plasma like in pump ducts. This results from additional transport through neutrals and/or low temperature plasmas in the periphery. No exact geometrical configuration can be given for these a–C:H films and the material must be characterized in another way. Three of the most important characteristics of an a–C:H are the hydrogen content, sp^2/sp^3 bonding ratio and the density of the sample. Low density and high hydrogen concentration ($\sim 50 - 70$ *at.%*) a–C:H films are generally referred to as polymer or soft a–C:H, whereas, high density, high sp^3 and low hydrogen concentration (up to 30 at.%) hydrocarbon films are usually reffered to as diamondlike carbon (DLC) films or hard a–C:H [15]. These layers pose a severe safety problem for any reactor because they trap a huge amount of tritium. Therefore, understanding of such processes and plasmas is critical for fusion reactor design and will strongly influence the choice of the wall material.

Radiation losses

ITER needs additional radiation losses to overcome the power exhaust problem as discussed before. However, there exists a rigid Z_{eff} limit. Only $\Delta Z_{eff} = 0.2$ is available for seed impurities, otherwise due to too strong bremsstrahlung losses it is not possible to sustain ignition [13]. This results in a fatal fraction f_Z for Be(0.14), C(0.07), Ne(0.025), Ar(0.0054). Using now the planned ITER parameter $n_e \approx 1 \cdot 10^{20}$ m^{-3}, one can estimate the necessary radiation losses on closed fieldlines, which would result in too low power fluxes at the separatrix incompatible with a good confinement. This needs a minimum power flux crossing the separatrix, because H-mode power threshold predictions for ITER are between 70–300 MW.

On open fieldlines one can use different radiators (e.g. neon, carbon and hydrogen, which radiate most strongly at lower and lower temperatures). Using several losses in series (different radiators, like neon, carbon and hydrogen) one gets larger losses, because after each loss the normalised input power gets smaller and this moves the operational point to larger relative losses. Due to the fact that a reduction in the heat flux density increases the spatial extent of the region between two fixed isotherms one gets enhanced total power radiated away by a given impurity at given plasma pressure and impurity concentration. Using this synergy for radiation losses in series e.g. radiation from neutral hydrogen atoms can ultimately become significant (close to the target plate), once the major fraction of the total heating power has been radiated via other upstream loss channels.

Also, for power removal in ITER. one needs enhancements of the radiation loss function beyond coronal equilibrium rates. The main reasons for such non-equilibrium situations are impurity transport and charge-exchange effects. The first (and most important) is the effect of transport, which produces a lag of the actual ionisation degree behind its equilibrium value. This is especially important for the limited residence time of impurities close to the plate, where fast recycling shifts the ionisation balance toward lower charge states. The second effect is CX recombination in the presence of neutral hydrogen ($H^0 + A^{i+} \Rightarrow H^+ + (A^{(i-1)+})^*$), which also shifts the ionisation balance toward lower charge states.

Using multi-fluid SOL transport codes like B2-Eirene to account for these effects, it is demonstrated that these non-equilibrium effects are strong enough that one gets enough radiation losses for sufficient power exhaust for ITER [13].

MARFEs

Another important phenomena is the occurrence of a radiation instability, a so-called MARFE (Multifaceted Asymmetric Radiation From the Edge). This radiation instability build up if one gets higher radiation losses at lower temperature. For the instability, the temperature dependence of the total sum of radiation losses (minus heating power) is important [14]. The radiation function of carbon exhibits (depending on the local heating power), for electron temperatures below about 50 eV, an unstable branch, leading to the formation of a condensation instability at low temperatures. This results in a transition from low radiation level (concentrated in the divertor) to rather large radiation levels (concentrated close to the X-point). The MARFE is then characterised by a low temperature region close to the X-point on closed field lines. Here, the assumption of pressure constancy along field-lines is very well fulfilled, resulting in quite large densities and therefore quite large radiation losses compared to midplane conditions. Due to the relatively low temperatures (below about 50 eV), the counteracting parallel heat conduction, which would try to keep the temperature up at the X-point to midplane values, decreases strongly with temperature and thus facilitates MARFE formation. The location can also be understood, because the cross-field heat source is minimal close to the X-point for a single null case due to the large expansion of flux surfaces in its vicinity. The X-point in a single null is also characterised by a very small effective heat conduction coefficient due to the shallow pitch angle there, creating strong temperature gradients poloidally. In other terms, due to the long connection length in this region the radiation has a tendency to focus here.

Comparing the radiation functions of the different elements it is obvious that one has the possibility to custom-tailor the total radiation function by combining different impurities, like C plus Ne, and by this reduce the tendency for instability. In contrast to the low-Z elements, medium or high-Z do not have this pronounced instability branch in the radiation loss function. Therefore, using e.g. neon as a radiator can reduce the instability formation and allow higher radiation losses before a MARFE is formed. This radiation instability cools the plasma edge which can lead to growth of magnetic islands and finally to a disruption, if the current profile becomes unstable.

PLASMA EDGE MODELING

The basic problem of plasma edge physics is the large range of length (see Fig. 4) and corresponding time scales.

Plasma wall interaction effects introduce microscopic length scales (like the typical interaction distance of about 1 nm between atoms and molecules) and very short time scales (fast momentum transfer processes determining the collision processes occur in 10^{-12} s). These processes are important for material changes in plasma wall facing materials and therefore also for the release of impurities into the plasma (e.g. physical or chemical sputtering). They are studied either by molecular dynamics (MD) or by a simplified binary collision model. The latter strongly reduces the computational effort, though at the price of drastically simplifying the physics. In addition, diffusion in such materials introduces length scales spanning from microns (size of the granules) up to centimeters (size of the tiles). These effects (e.g. diffusion in amorphous materials) are analyzed with Monte Carlo methods (kinetic Monte Carlo with input from molecular dynamics or experiment).

The plasma description again has different levels of complexity. A full kinetic description (including ions, electrons, neutrals and their collisions) is possible for some low temperature plasmas (e.g. electron cyclotron resonance heated methane plasmas) and for qualitative studies of edge plasma effects in fusion edge plasmas. Here, the limitations are given by the fact that the Debye length and the plasma frequency have to be resolved.

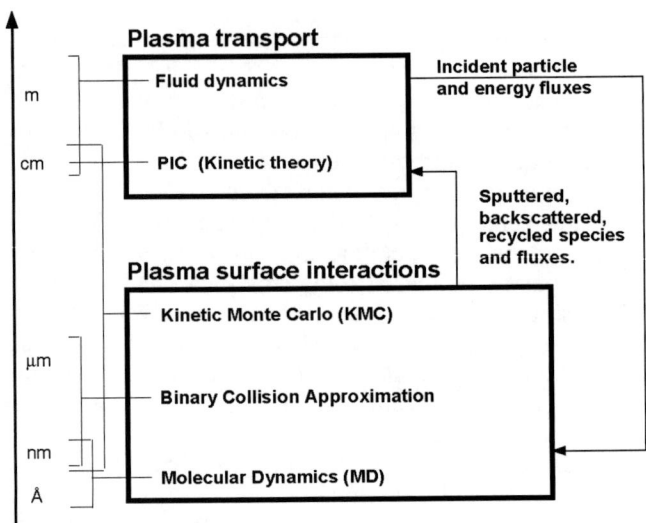

FIGURE 4. The different length scales and methods used for plasma edge modelling.

For the study of the physics of the edge of magnetically confined plasmas (2D tokamaks, tokamaks with ergodic perturbations, 3D stellarators) fluid codes are used for understanding the complex physics in such devices. Depending on the geometrical complexity (2D tokamaks, 3D stellarators) and on the additional effect of ergodicity, different numerical methods (finite volume, finite difference and Monte Carlo methods) are used.

The plasma surface interactions influence the plasma transport through sputtered, back scattered and recycled particles and fluxes. On the other hand, the incident particle and energy fluxes determine the plasma surface interaction.

ACKNOWLEDGMENTS

Ralf Schneider acknowledges funding of the work by the Initiative and Networking Fund of the Helmholtz Association.

REFERENCES

1. D. Reiter and G.H. Wolf and H. Kever, *Nuclear Fusion* **30**, 2141–2155 (1990).
2. R. Behrisch and V. Prozesky, *Nuclear Fusion* **30**, 2166–2169 (1990).
3. ITER physics base editors, ITER physics expert groups, JCT, *Nuclear Fusion* **39**, 2137–2638 (1999).
4. Fantz, U. and Behringer, K. and Gafert, J. and Coster, D.P. and ASDEX Upgrade Team, *Journal of Nuclear Materials* **266–269**, 490–494 (1999).
5. Fantz, U. and Reiter, D. and Heger, B. and Coster, D.P., *Journal of Nuclear Materials* **290–293**, 367–373 (2001).

6. Eckstein, W., "Computer Simulation of Ion–Solid Interactions," in *Springer Series in Materials Science, Vol.10*, Springer, Berlin, 1991.
7. Naujoks, D. and Roth, J. and Krieger, K. and Lieder, G. and Laux, M., *Journal of Nuclear Materials* **210**, 43–50 (1994).
8. Janeschitz, G. and Borrass, K. and Federici, G. and Igitkhanov, Y. and Kukushkin, A. and Pacher, H. D. and Pacher, G. W. and Sugihara, M., *Journal of Nuclear Materials* **220–222**, 73–88 (1995).
9. Küppers, J., *Surface Science Reports* **22**, 249–322 (1995).
10. Roth, J. and Preuss, R. and Bohmeyer, W. and Brezinsek, S. and Cambe, A. and Casarotto, E. and Doerner, E. and Gauthier, E. and Federici, G. and Higashijima, S. and Hogan, J. and Kallenbach, A. and Kirschner, A. and Kubo, H. and Layet, J.M. and Nakano, T. and Philipps, V. and Pospieszczyk, A. and Pugno, R. and Ruggiéri, R. and Schweer, B. and Sergienko, G. and Stamp, M., *Nuclear Fusion* **44**, L21–L25 (2004).
11. Whyte, D.G. and West, W.P. and Doerner, R. and Brooks, N.H. and Isler, R.C. and Jackson, G.L. and Porter, G. and Wade, M.R. and Wong, C.P.C., *Journal of Nuclear Materials* **290–293**, 356–361 (2001).
12. Tabarés, F.L. and Tafalla, D. and Tanarro, I. and Herrero, V.J. and Islyaikin, A. and Maffiotte, C., *Plasma Physics and Controlled Fusion* **44**, L37–L42 (2002).
13. ITER JCT and Home Teams, *Plasma Physics and Controlled Fusion* **37**, 19–35 (1995).
14. Drake, J.F., *Physics of Fluids* **30**, 2429–2433 (1987).
15. Salonen, E., and Nordlund, K. and Keinonen, J. and Wu, C.H., *Phys. Rev. B* **63**, 195415–(1–14) (2001).

Lithium Polarization Spectroscopy: Making Precision Plasma Current Measurements in the DIII-D National Fusion Facility

D.M. Thomas

General Atomics, P.O. Box 85608, San Diego, California 92186-5608, USA

Abstract. Due to several favorable atomic properties (including a simple spectral structure, the existence of a visible resonance line, large excitation cross section, and ease of beam formation), beams of atomic lithium have been used for many years to diagnose various plasma parameters. Using techniques of active (beam-based) spectroscopy, lithium beams can provide localized measurements of plasma density, ion temperature and impurity concentration, plasma fluctuations, and intrinsic magnetic fields. In this paper we present recent results on polarization spectroscopy from the LIBEAM diagnostic, a 30 keV, multi-mA lithium beam system deployed on the DIII-D National Fusion Facility tokamak. In particular, by utilizing the Zeeman splitting and known polarization characteristics of the collisionally excited 670.8 nm Li resonance line we are able to measure accurately the spatio-temporal dependence of the edge current density, a parameter of basic importance to the stability of high performance tokamaks. We discuss the basic atomic beam performance, spectral line-shape filtering, and polarization analysis requirements that were necessary to attain such measurements. Observations made under a variety of plasma conditions have demonstrated the close relationship between the edge current and plasma pressure, as expected from neoclassical theory.

Keywords: plasma diagnostics, lithium beam, polarization spectroscopy, Zeeman effect, current density, tokamak edge confinement
PACS: 52.70.-m, 52.70.Kz, 52.59.Sa, 52.59.Bi, 52.40.Mj,52.20.Hv, 32.60.+I,32.70.Jz, 39.10.+j

1. INTRODUCTION

The edge current density j is an important parameter in toroidal magnetic plasma confinement experiments because it seems to be a key player in the formation and stability of the plasma pedestal. This is a region of enhanced confinement that can be established at the outer edge of the plasma, just inside the transition between open and closed magnetic field lines (Fig. 1). The height of the pedestal pressure appears to be closely related to the performance of the plasma as a whole in terms of overall (core) plasma pressure or effective fusion power. However, the pedestal pressure is limited by various magnetohydrodynamic (MHD) instabilities that can cause the collapse of the good confinement. These MHD modes may be stabilized or destabilized by the existence of neoclassical currents that are self-consistently generated by the pressure gradient in a toroidal geometry. The current and pressure are interrelated in a very complex fashion on very small spatial (10^{-3}–10^{-2} m) and (10^{-3}-10^{-1} s) temporal scales.

CP926, *Atomic Processes in Plasmas—15th International Conference on Atomic Processes in Plasmas*
edited by J. D. Gillaspy, J. J. Curry, and W. L. Wiese
© 2007 American Institute of Physics 978-0-7354-0436-6/07/$23.00

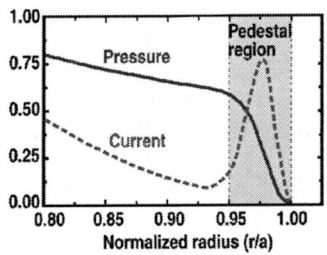

Figure 1. Edge region of a plasma with improved confinement, showing large pressure gradient and associated current peak.

While we can measure the ion and electron components of the plasma pressure with good accuracy using such techniques as Thomson scattering and charge exchange recombination spectroscopy, measurements of j are more difficult. One way to do this is to provide a fine scale profile of the poloidal magnetic field B_{POL} generated by the current. This must be done in the presence of the much larger toroidal magnetic field B_{TOR}. The spatial derivative of B_{POL} can then be used to infer j via Ampere's law. Due to the high power densities in the edges of present-day experiments, measurements using physical magnetic probes are only possible outside of the last closed flux surface and can only yield limited information on the interior structure of the underlying current distribution. More recently, measurements of the plasma internal magnetic fields have been made successfully using the motional Stark effect (MSE) [1] on fast beams of atomic hydrogen injected into the plasma. In the frame of the hydrogen atoms, the equivalent radial electric field due to the beam velocity crossed into the local B field is sufficient to split and polarize the Balmer alpha beam emission, which can then be used to determine the magnetic field components. Unfortunately, this technique is hampered by the existence of large intrinsic radial electric fields, which can result in degenerate or incorrect solutions for the true poloidal magnetic field [2]. In addition, the experiments to date have not had sufficiently fine spatial resolution to accurately analyze the local structure of the current density in the pedestal.

We have overcome these difficulties by using combined polarization and spectroscopic analysis of the resonance emission of an injected beam of lithium atoms. Because of the intrinsic (Tesla-scale) magnetic fields, the Zeeman effect splits and polarizes this resonance radiation in a known fashion. Measurements of the local magnetic field structure can be extracted in various ways from the proper analysis of this radiation. Using this technique we have been successful in making precision measurements of the current density profile in the pedestal region on the DIII-D tokamak. In this paper we will examine the various atomic physics aspects that make lithium such an attractive candidate for these measurements as well as showing some recent results of our investigations on DIII-D.

2. ADVANTAGES OF LITHIUM ATOMIC SYSTEM FOR EDGE PLASMA MEASUREMENTS

In this section we review several specific features of lithium that make it uniquely suited for measurements on modern tokamak edge plasmas.

2.1. Large Excitation Rate

Figure 2 shows a simplified level diagram for the neutral lithium system. For beam energies of a few tens of keV, the resonant 2^2S-2^2P transition is very efficiently pumped by plasma electron and ion collisions [3] (Fig. 3). Since the excitation rate for the lithium resonance line is much higher than for the hydrogen lines (500x in the case of H_α) for electron temperatures typical of the plasma edge ($T_e > \sim 20$ eV), a very small (1 cm-scale) beam can be used for probing the plasma while still providing a reasonable signal level. This can significantly improve the spatial resolution of the measurement. Although the electron loss (ionization and charge exchange) cross sections for lithium are also correspondingly large, the attenuation target thicknesses $\pi = \int n dl$ are on the order of 10^{17}-10^{18} m^{-2}. For typical edge plasma densities of a few 10^{19} m^{-3}, this is more than enough to probe the plasma pedestal region.

2.2. Transition Wavelength

The 670 nm resonance wavelength is in the visible region of the spectrum, simplifying the optical requirements for observing the fluorescence. It is also near the peak of the quantum efficiency curve for common optical detectors. Fortuitously it also resides in a region of the plasma emission spectrum having few competing lines. In addition, the relatively long wavelength results in a correspondingly large Zeeman splitting (below). Finally, because of the exceedingly short time ($\sim 10^{-15}$ s) associated with electron dipole emission for a single atom, for reasonable beam energies the polarization and spectral characteristics are determined by the local plasma conditions at the emission location, with no time-of-flight broadening or averaging.

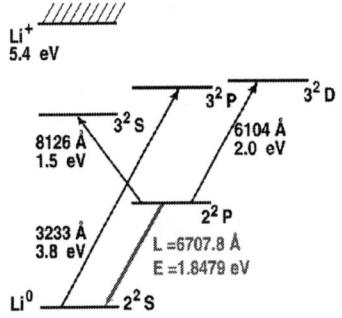

FIGURE 2. Simplified Grotrian diagram for neutral lithium system, showing lowest lying levels.

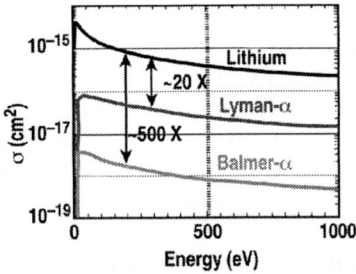

FIGURE 3. Electron excitation cross sections leading to emission for temperatures characteristic of the pedestal region. Black: Li ^{2}P-^{2}S; dark gray: H ^{2}P-^{1}S (Lyman-alpha); light gray: H 3-2 (Balmer-alpha).

2.3. Zeeman Effect

In the magnetic fields typical of most modern tokamaks ($B > \sim 1$ T), the spectral line emission is split and shifted by the Zeeman effect due to spin-orbit interaction with the field, with the shift and polarization of the components being determined by the field strength B and the change in magnetic quantum number Δm [4,5]. At these field levels the lithium ^2P levels are fully mixed (Paschen-Back regime) and the resonance emission forms a Lorenz triplet. The shift in wavelength for the two σ lines is given by $\Delta\lambda_B = \Delta m (\mu/hc)\lambda_0^2 B$ (Fig. 4). For the lithium resonance wavelength the shift is 0.021 nm/T.

For each of the three Zeeman components, the emission has unique polarization characteristics: for emission perpendicular to B, the π line ($\Delta m = 0$) is linearly polarized parallel to the direction of B and the two σ lines ($\Delta m = \pm 1$) are linearly polarized perpendicular to B (Fig. 5). For emission parallel to B, there is no π emission, and the two σ states exhibit circular polarization, with the shorter wavelength $\sigma-$ being left-circularly polarized and the longer wavelength $\sigma+$ being right-circularly polarized. For the general case of emission in an arbitrary direction with respect to **B**, the emission **I** of the [π, $\sigma+$, $\sigma-$] manifold can be expressed in matrix form [6,7] in terms of the Stokes parameters [S0, S1, S2, S3] [8] as a function of two angles α and γ:

$$
I = \left[\, [I\pi] + [I\sigma-] + [I\sigma+] \,\right] = I_0 \left[
\begin{array}{c}
\dfrac{\sin^2\alpha}{2} \\[6pt]
\dfrac{\sin^2\alpha\cos(2\gamma)}{2} \\[6pt]
\dfrac{\sin^2\alpha\sin(2\gamma)}{2} \\[6pt]
0
\end{array}
\right]
+
\left[
\begin{array}{c}
\dfrac{1+\cos^2\alpha}{4} \\[6pt]
-\dfrac{\sin^2\alpha\cos(2\gamma)}{4} \\[6pt]
-\dfrac{\sin^2\alpha\sin(2\gamma)}{4} \\[6pt]
\dfrac{\cos\alpha}{2}
\end{array}
\right]
+
\left[
\begin{array}{c}
\dfrac{1+\cos^2\alpha}{4} \\[6pt]
-\dfrac{\sin^2\alpha\cos(2\gamma)}{4} \\[6pt]
-\dfrac{\sin^2\alpha\sin(2\gamma)}{4} \\[6pt]
-\dfrac{\cos\alpha}{2}
\end{array}
\right]
. \quad (1)
$$

In the above expression α is the angle between the view chord and the field and γ is the projection of the field inclination angle B_{POL}/B_{TOR}.

2.4. Electric Field Insensitivity

Because of the large energy separation of the lithium 2p levels, there is essentially no measurable Stark effect (intrinsic or motional) on the spectra, even for high velocity beams. Thus the polarization and wavelength effects are due strictly to the magnetic field.

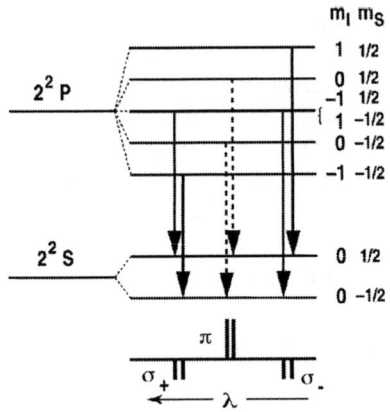

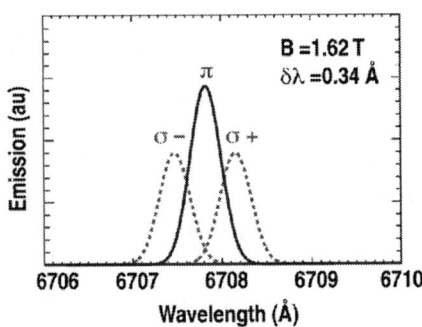

Figure 5. Lithium line emission profile calculated for a position near the outside midplane of DIII-D for a 2.1 T, 1.86 MA plasma. An estimated transverse beam temperature of 1.0 eV was used to calculate the Doppler broadening of the line components. The distance between the two σ peaks for this example (total B = 1.62 T) is 0.068 nm.

Figure 4. Lithium levels and allowed transitions for normal Zeeman (Paschen-Back) regime in lithium. The lower graph shows the polarization associated with each of the transitions.

2.5. Monoenergetic Beam Formation and Neutralization

To take advantage of the polarization behavior, the Doppler broadening of the emission must be minimized compared to the Zeeman splitting. For a transverse beam view this sets stringent limits on beam divergence and monoenergeticity. The existence of effective thermo-emissive lithium ion sources based on aluminosilicates [9,10] permits the creation of high brightness, single isotope beams having a fraction of the energy spread of plasma discharge type sources. In addition, the extremely high efficiency for charge transfer for lithium on sodium vapor allows us to obtain ~90% neutralization efficiency within a very short neutralization distance. This minimizes the space charge emittance growth in the beam that adds to Doppler broadening. Specific techniques are discussed in Sec. 4.

3. METHOD OF INTERPRETATION

There are a number of ways to utilize these atomic properties to infer the local magnetic field components, and hence j, from the fluorescence emission. Prior analysis techniques have included linear polarization analysis of the collisional fluorescence [11,12] enhancement and polarization-dependent pumping using a resonant laser to induce fluorescence [13], line scanning and modulation of the circular polarization to identify parallel field components [14], and precision line profile intensity measurements to take advantage of the known sigma/pi variation with field direction [15,16]. In each case, the goal is to identify ratios of the various terms in Eq. (1).

In the present case, because we wish to achieve the highest possible radial resolution and sensitivity to the poloidal field component, we have chosen a viewing geometry that is: (1) tangent to the magnetic flux surfaces, (2) orthogonal to the beam, and (3) normal to the toroidal magnetic field (Fig. 6). Given this geometry, analysis of one of the sigma states yields the field direction by identifying the ratio of circular to linear polarization:

$$\frac{S_3}{\left(S_1^2 + S_2^2\right)^{1/2}} = \frac{2 \cos\alpha}{1 - \cos^2\alpha} \quad .$$ (2)

This measurement is sufficient to interpret B_{POL}, given the known toroidal field and viewing geometry from a spatial calibration [17].

4. HARDWARE REQUIREMENTS AND THE DIII-D LIBEAM SYSTEM

There are five primary requirements that must be must be satisfied in order to provide a useful measurement based on the above atomic physics: (1) a beam of sufficient intensity to allow the desired time and spatial resolution with minimal Doppler broadening, (2) a suitable array of viewing points along the beam with the requisite angular acceptance, along with an accurate spatial calibration of the viewing geometry, (3) minimization and characterization of any unwanted systematic polarization effects in the optical system, (4) a high efficiency, narrow band optical filter which passes the desired Zeeman component while providing sufficient spectral rejection of the remainder of the line profile to ensure a reasonable level of polarization, and (5) a method of accurately analyzing the resulting time-dependent polarization. The LIBEAM system on DIII-D has been developed to satisfy each of these requirements; the progressive improvements in hardware and performance have been documented in detail in several previous publications [6,7,18-22]. Briefly, the system (Fig. 6) comprises a 30 keV, 10 mA neutral equivalent lithium beam of approximately 1.5 cm diameter and an imaging system which collects 32 channels of finely spaced ($\delta R \sim 0.5$ cm) beam emission from the edge region of DIII-D. The beam is formed using a 5 cm diameter ^{6}Li ion emitter, focused into a 1 cm beam, then neutralized in a small sodium vapor cell. The lighter isotope is used to achieve slightly higher beam velocity and better plasma penetration for a given accelerator voltage. The fluorescent emission from the collisionally excited neutral lithium beam is transmitted through dual photoelastic modulators (PEMs) [23] that modulate the polarization of the input light at multiples of the two drive frequencies (20.1, 23.1 kHz). A linear polarizer immediately after the PEM pair transforms the polarization modulation into amplitude modulation and the fluorescence from each spatial location is transmitted through optical fibers to the detector room. The output of each fiber is filtered using individual doubled etalon narrowband (~0.03 nm) filters that are temperature-tuned to the Doppler-shifted σ- line for each view. The shorter-wavelength σ- line is chosen because the longer wavelength is slightly contaminated by emission from a small amount (~4%) of ^{7}Li in the beam. A 1.0 nm bandpass filter in series with the etalons suppresses the plasma background light. The modulated light signals are then detected using a high quantum-efficiency photomultiplier tube-transimpedance amplifier combination. The 32 individually tuned channel outputs are digitized during each tokamak discharge at up to 300 kHz, along with the two PEM drive frequencies. Details of the digital lock-in analysis used to extract cos α, along with the calibration procedures used to correct for non-perfect filtering of the σ- line by the etalons are given in Ref. 13.

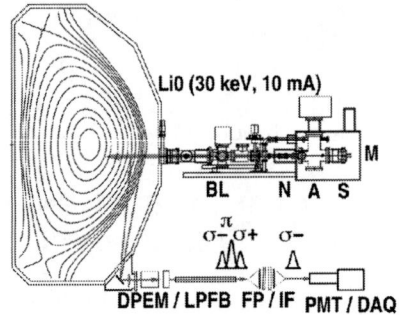

Figure 6. The LIBEAM system on DIII-D, showing the accelerator and schematic of optical system. S=^{6}lithium ion source, A= electrostatic accelerator, N=sodium vapor neutralizer cell, M=magnetic shielding, BL=beamline, DPEM=dual photoelastic modulator, LP=linear polarizer, FB=32 element fiber bundle, FP=dual element, temperature tuned Fabry-Perot etalon (32 channels), IF=1 nm lithium line interference filter, PMT=GaAs photomultiplier tube, DAQ=data acquisition and analysis computer.

5. SOME RECENT RESULTS OF EXPERIMENTS ON DIII-D

In this section we show a few physics results from DIII-D demonstrating the effectiveness of this technique for determining edge current behavior. Figure 7 from Ref. 25 shows the measured poloidal field profile for conditions having low and high edge pedestals. The existence of an edge current is obvious from the large relative change in the field in the region just inside the last closed flux surface. Figure 8 from Ref. 24 shows the inferred current densities using Ampere's law [26] for the two cases. Note the extremely peaked, cm scale current density in the H-mode case.

Calculated values for the expected neoclassical currents [27,28] based on the measured pressure profiles are in good agreement with the measured one. Further measurements [25,29] have confirmed that the location of the current peak is coincident with the peak in the edge pressure gradient, and that the time evolution of

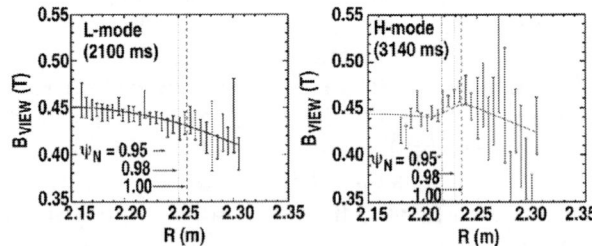

Figure 7. Edge profiles of B_{VIEW} for DIII-D discharge 115114 for times having low and high pedestal pressures. B_{VIEW} is the magnetic field component parallel to the diagnostic sightlines and is essentially the poloidal field because of the arrangement of the viewchords. The location of the last closed flux surface is indicated by the vertical dotted line =1.00, where Ψ_N is the normalized poloidal flux. Adapted from Ref. 25. Solid curves are profiles from an equilibrium reconstruction.

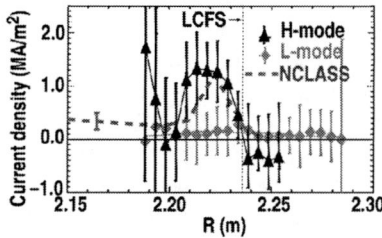

Figure 8. Value of the edge toroidal current density for L-mode (gray) and H-mode (black) on DIII-D shot 115114. Dashed curve is toroidal current density calculated using NCLASS model and measured pressure profiles. LCFS = last closed flux surface. Adapted from Ref. 24.

the current also follows the pressure gradient growth, as expected by theory. Evaluation of the current for different plasma edge densities and temperatures shows a decrease in the edge current density as the plasma collisionality ($\sim n/T^{3/2}$) increases [30], also in accord with theory. Finally, recent efforts to improve the time resolution through conditional averaging techniques have succeeded in showing some features of the magnetic field evolution in between periodic MHD relaxation events in the edge, where plasma confinement is transiently reduced, then restored [31].

6. DISCUSSION AND CONCLUSION

As presently deployed, the LIBEAM diagnostic is a powerful tool for making precise measurements of the local magnetic field structure in the edge of fusion-grade plasmas. This is because of a fortunate coincidence of physical scales, cross sections, and field strengths in existing machines. These measurements have confirmed the existence of large localized currents and are helping to understand the role of these current in pedestal stability and their importance in various toroidal confinement operating regimes. However there are limitations on the time resolution that can be achieved for a given precision. These are primarily set by statistical considerations. In addition, there are several systematic effects — particularly the effect of imperfect spectral filtering and residual polarization — that need to be carefully controlled in order to achieve accurate measurements. We conclude with several improvements that can be made on future experiments that would significantly improve the performance, and hence the utility, of this technique.

First, an improved ion beam current would help on the signal to noise. This can be done through better ion optics design, coupled with a modest increase in beam energy (~30%) that would improve the beam emittance without materially decreasing the effective neutralization efficiency. The higher beam energy would also give better penetration and would decrease the error bars inside of the pedestal region.

Second, beam modulation would help correct for systematic errors in the background light level, which has a small effect on interpreting the proper polarization ratio under low-signal conditions.

Third, an improvement in the quantum efficiency of the detector is desirable. This is straightforward to implement with at least a factor of 3 to 4 improvement possible using silicon detectors. Care needs to be taken to couple the detectors with the appropriate low-noise preamplifiers to maximize the overall improvement.

Next, measurements using complimentary viewing geometry would help. By choosing different views of the beam, the systematic effect of imperfect filtering can be minimized since the new views will yield different values for the ratio of circular to linear polarization. It might be possible to examine strictly linear polarization by arranging views that intersect the beam but are more or less normal (as opposed to tangential) to the flux surfaces. This would essentially eliminate one class of systematic effects. The decrease in radial resolution might be tolerable, depending on the beam thickness and orientation.

Finally, while the existing magnetic field on the outside midplane of DIII-D (~1.5 T) is sufficient to fully split the Zeeman levels, they are marginally resolved by the existing combination of filtering (etalon performance) and beam performance (Doppler broadening). Even a modest (~50%) increase in the magnetic field strength would improve the situation enormously with no other changes to the hardware by decreasing the bleedthrough of unwanted components. This could be achieved on DIII-D by moving from the outside midplane to the top of the machine (because of the ~1/R dependence of B) or by operation on other devices having higher fields.

ACKNOWLEDGMENTS

This work was supported by the US DOE under DE-AC03-89ER51114 and DE-FC02-04ER54698. D.M. Thomas would like to acknowledge the friends and coworkers that have labored on various aspects of this problem for many years: K. McCormick, W.P. West, A.W. Leonard, R.J. Groebner, R.M. Patterson, J. Kulchar, D. Sundstrom, A. Bozek, J.I. Robinson, K.H. Burrell, T.N. Carlstrom, T.H. Osborne, R.T. Snider, D.K. Finkenthal, R. Jayakumar, M.A. Makowski, D.G. Nilson, B.W. Rice, J.J. Peavy, W.P. Cary, D.H. Kellman, D.M. Hoyt, S.W. Delaware, S.G.E. Pronko, T.E. Harris, P.B. Snyder, T.A. Casper, L.L. Lao, P. Gohil, H.W. Mueller, and M.E. Fenstermacher. The enduring support and encouragement of T.S. Taylor and R.D. Stambaugh, is also gratefully acknowledged.

REFERENCES

1. F. M. Levinton, et al., "Magnetic Field Pitch-Angle Measurements in the PBX-M Tokamak Using the Motional Stark Effect," Phys. Rev. Lett. 63, 2060 (1989).
2. B. W. Rice, et al., "Effect of Plasma Radial Electric Field on Motional stark Effect Measurements and Equilibrium Reconstruction," Nucl. Fusion 37, 517 (1997).
3. J. Schweinzer, et al., "Database for Inelastic Collisions of Lithium Atoms With Electrons, Protons and Multiply Charged Ions," At. Data Nucl. Data Tables 72, 239-273 (1999).
4. E. U. Condon and G. H. Shortley, The Theory of Atomic Spectra, Cambridge University Press, 1963, 149 ff.
5. H. A. Bethe and E. E. Salpeter, Quantum Mechanics of One- and Two-Electron Atoms, Plenum, 1977, 205 ff.
6. D. M. Thomas, et al., "Prospects for Edge Current Density Determination Using LIBEAM on DIII-D," Rev. Sci. Instrum. 72, 1023-1027 (2001).
7. D. M. Thomas, "Poloidal Magnetic Field Measurements and Analysis With the DIII-D LIBEAM System," Rev. Sci. Instrum. 74, 1541-1457 (2003).

8. D. S. Kliger, *et al.*, *Polarized Light in Optics and Spectroscopy*, Academic Press, 1990, 75 ff.
9. O. Heinz and R. T. Reaves," Lithium Ion Emitter for Low Energy Beam Experiments," *Rev. Sci. Instrum.* **39**, 964-967 (1968).
10. R. K. Feeney, *et al.*, "Alumino-silicate sources of Positive Ions for Use in Collision Experiments," *Rev. Sci. Instrum.* **47**, 964-967 (1976).
11. K. McCormick, *et al.*, "Temporal Behavior of the Plasma Current Distribution in the ASDEX Tokamak During Lower-Hybrid Current," *Phys. Rev. Lett.* **58**, 491-494 (1987).
12. K. Kadota, *et al.*, "Space-resolved Measurement of Internal Magnetic Field in a Bumpy Torus by Li0-beam Probe Spectroscopy," *Rev. Sci. Instrum.* **56**, 857-859 (1985).
13. W. P. West, *et al.*, "Measurement of the Rotational Transform at the Axis of a Tokamak," *Phys. Rev. Lett.* **58**, 2758-2761 (1987).
14. L. K. Huang, *et al.*, "Safety Factor on the Axis of a Tokamak During Ohmically Heated Sawtoothing Discharges from a Localized Measurement of Circular Polarization of the Li 6708 A Line," *Phys. Fluids* **B2**, 809-814 (1990).
15. A. Korotkov, *et al.*, "Line Ratio Method for Poloidal Magnetic Field Measurements Using Li-multiplet (22S-22P) Emission," in *Advanced Diagnostics for Magnetic and Inertial Fusion*, edited by P.E. Stott, *et al.*, Kluwer Academic/Plenum Publishers, New York, 2002, pp. 209-212.
16. A. A. Korotkov, *et al.*, "Line Ratio Method for Measurement of Magnetic Field Vector Using Li-multiplet (22S–22P) Emission," *Rev. Sci. Instrum.* **75**, 2590-2602 (2004).
17. D. M. Thomas and A. W. Leonard, "Signal Processing Techniques for Lithium Beam Polarimetry on DIII-D," *Rev. Sci. Instrum.* **77**, 10F515-1:4 (2006).
18. D. M. Thomas, *et al.*, "Low-divergence, High Brightness Lithium Ion Source for Plasma Diagnostics," *Rev. Sci. Instrum.* **59**, 1735-1737 (1988).
19. D. M. Thomas, "Development of Lithium Beam Spectroscopy as an Edge Fluctuation Diagnostic for DIII-D," *Rev. Sci. Instrum.* **66**, 806-811 (1995).
20. D. M. Thomas, *et al.*, "Utilization of LIBEAM Polarimetry for Edge Current Determination on DIII-D," in *Advanced Diagnostics for Magnetic and Inertial Fusion*, edited by P.E. Stott, *et al.*, Kluwer Academic/Plenum Publishers, New York, 2002, pp. 319-322.
21. T. N. Carlstrom, *et al.*, "Optical Design for Li Beam Polarimetry Measurements on DIII-D," *Rev. Sci. Instrum.* **74**, 1601-1604 (2003).
22. J. J. Peavy, *et al.*, "Control System for the Lithium Beam Edge Plasma Current Density Diagnostic on the DIII-D Tokamak," Proc. 20th IEEE/NPSS Symp. on Fusion Engineering, IEEE, Piscateway, New Jersey, 2003, pp. 344-346.
23. J. C. Kemp, *Polarized Light and its Interaction with Modulating Devices - A Methodology Review*, Hinds International, Hillsboro, Oregon, 1987.
24. D. M. Thomas, et al., "Measurement of Pressure-Gradient-Driven Currents in Tokamak Edge Plasmas," *Phys. Rev. Lett.* **93**, 0650031-4 (2004).
25. D. M. Thomas, *et al.*, "Measurement of Edge Currents in DIII-D and Their Implication for Pedestal Stability," *Phys. Plasmas* **12**, 056123-1 (2005).
26. D. M. Thomas, *et al.*, "Calculation of Edge Toroidal Current Density Distributions from DIII-D Lithium Beam Measurements Using Ampere's Law," *Rev. Sci. Instrum.* **75**, 4109-4111 (2004).
27. W. A. Houlberg, *et al.*, "Bootstrap Current and Neoclassical Transport in Tokamaks of Arbitrary Collisionality and Aspect Ratio," *Phys. Plasmas* **4**, 3230 (1997).
28. O. Sauter, *et al.*, "Neoclassical Conductivity and Bootstrap Current Formulas for General Axisymmetric Equilibria and Arbitrary Collisionality Regime," *Phys. Plasmas* **6**, 2834 (1999).
29. D. M. Thomas, *et al.*, "Edge Currents and Stability in DIII-D," Proc. 31st EPS Conf., European Physical Society, London, 2004, Paper P2-177.
30. D. M. Thomas, *et al.*, "The Effect of Plasma Collisionality on Pedestal Current Formation in DIII-D," *Plasma Phys. Control. Fusion* **48**, A183-A191 (2006).
31. D. M. Thomas, *et al.*, "Edge Current Growth and Saturation During the Type 1 ELM Cycle," Proc. 33st EPS Conf., European Physical Society, Roma, 2006, ECA Vol. 301, Paper P5-139.

Modeling Nuclear Fusion with an Ultracold Nonneutral Plasma

Daniel H.E. Dubin

Physics Dept., UCSD, 9500 Gilman Drive, La Jolla CA 92093-0319
ddubin@ucsd.edu

In the hot dense interiors of stars and giant planets, nuclear fusion reactions are predicted to occur at rates that are greatly enhanced compared to those at low densities. The enhancement is caused by plasma screening of the repulsive Coulomb potential between nuclei, which increases the probability of the rare close collisions that are responsible for fusion [1]. This screening enhancement is a small effect in the Sun [2], but is predicted to be much larger in dense objects such as white dwarf stars and giant planet interiors where the plasma is strongly correlated (i.e. where the Debye screening length is smaller than a mean interparticle spacing). However, strongly enhanced fusion reaction rates caused by plasma screening have never been definitively observed in the laboratory. This talk discusses a method for observing the enhancement using an analogy between nuclear energy and cyclotron energy in a cold nonneutral plasma in a strong magnetic field. In such a plasma, the cyclotron frequency is higher than other dynamical frequencies, so the kinetic energy of cyclotron motion is an adiabatic invariant. This energy is not shared with other degrees of freedom except through rare close collisions that break this invariant and couple the cyclotron motion to the other degrees of freedom. Thus, the cyclotron energy of an ion, like nuclear energy, can be considered to be an internal degree of freedom that is released only via rare close collisions. Furthermore, it has recently been shown that the rate of release of cyclotron energy is enhanced through plasma screening by precisely the same factor as that for the release of nuclear energy, because both processes rely on close collisions that are enhanced by plasma screening in the same way [3]. Simulations and experiments measuring large plasma screening enhancements for the first time will be discussed, and the possibility of exciting and studying cyclotron burn fronts will also be considered.

Acknowledgments. This work was supported by the National Science Foundation and the Department of Energy under grant numbers PHY-0354979 and PHY-0613740.

1. E. E. Salpeter and H. van Horn, Astrophys. J. **155**, 183 (1969).
2. J. N. Bachall, L. S. Brown, A. Gruzinov, and R. F. Sawyer, A&A **383**, 291 (2002).
3. D. Dubin, Phys. Rev. Lett. **94**, 025002 (2005); M. J. Jensen, T. Hasegawa, J. J. Bollinger, and D.H.E. Dubin, Phys. Rev. Lett. **94**, 025001 (2005).

CP926, *Atomic Processes in Plasmas—15ᵗʰ International Conference on Atomic Processes in Plasmas*
edited by J. D. Gillaspy, J. J. Curry, and W. L. Wiese
© 2007 American Institute of Physics 978-0-7354-0436-6/07/$23.00

PLASMAS IN SMALL TRAPS

Optical Probes of Ultracold Neutral Plasmas

S. Laha, J. Castro, H. Gao, P. Gupta, C. E. Simien, and T. C. Killian

Rice University, Department of Physics and Astronomy, Houston, Texas, 77005

Abstract. We describe the optical diagnostics used to study ultracold neutral plasmas. Imaging and spectroscopy based on both ion absorption and fluorescence provide accurate measurements of ion kinetic energy, plasma size, and the number of ions in the plasma. Absorption measurements yield lower signal-to-noise ratios because they are highly sensitive to laser intensity fluctuations, but the resulting measurement of the number of ions requires no external calibration. Fluorescence measurements of ion number must be calibrated with absorption measurements, but the measurements are less sensitive to technical noise sources. Spatially resolved fluorescence measurements also have the advantage of separating ion kinetic energy due to expansion from thermal kinetic energy.

Keywords: plasma imaging, laser cooling, strong coupling
PACS: 32.80.Pj,52.27.Gr

1. INTRODUCTION

Ultracold neutral plasmas are created by photoionizing laser-cooled atoms just above the ionization threshold [1]. In these systems, electrons have temperatures between 1 and 1000 kelvin, and ions equilibrate at about 1 kelvin. This represent an exotic regime of neutral plasma physics. At such low electron temperatures, processes such as three-body-recombination can be extremely fast [2, 3], and ions behave as a strongly-coupled fluid [4, 5, 6].

These experiments can teach us about plasmas across a much wider energy range. In particular, ultralow temperature allows experiments to access strongly-coupled physics at very low densities compared to high energy-density experiments. Because the densities are low, important time scales of the problem, such as the time between collisions and the inverse of the ion plasma oscillation frequency, are orders of magnitude longer, which greatly simplifies experiments. In addition, optical probes can precisely measure plasma properties, and ultracold plasmas have accurately known and controllable initial density profiles, energies, and ionization states.

In this paper, we will discuss the optical probes used to study ultracold neutral plasmas. In particular, spectroscopy and imaging based on plasma fluorescence is a new diagnostic that offers many powerful capabilities.

CP926, *Atomic Processes in Plasmas—15th International Conference on Atomic Processes in Plasmas*
edited by J. D. Gillaspy, J. J. Curry, and W. L. Wiese
© 2007 American Institute of Physics 978-0-7354-0436-6/07/$23.00

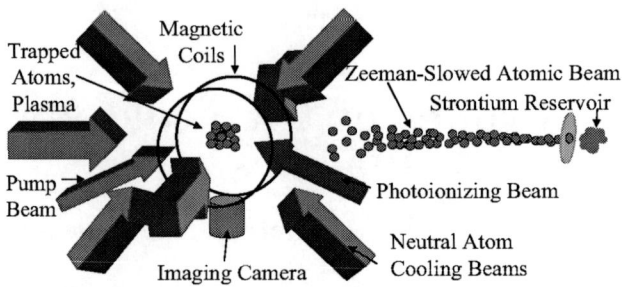

FIGURE 1. Schematic setup of a typical ultracold plasma experiment, exemplified with strontium atoms. Neutral atoms are laser cooled and trapped in a magneto-optical trap operating on the $^1S_0 - {}^1P_1$ transition at 461 nm, as described in [7]. In a second step, $^1P_1$ atoms are ionized by photons from a laser at ~ 412 nm (see Fig. 2A). Finally, the ionic plasma component is imaged using the $^2S_{1/2} - {}^2P_{1/2}$ transition in Sr$^+$ at 422 nm (see Fig. 2B). Probe beams for absorption and fluorescence follow different geometries and are not shown.

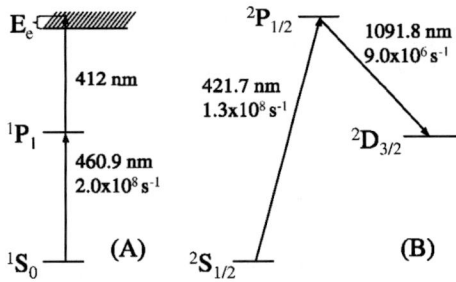

FIGURE 2. Strontium atomic and ionic energy levels with decay rates relevant for ultracold plasma experiments. (A) Creation of an ultracold plasma starts with laser cooling and trapping neutral atoms using the $^1S_0 - {}^1P_1$ transition at 461 nm. $^1P_1$ atoms are then photoionized with photons from a laser at ~ 412 nm. (B) Ions are optically imaged using the $^2S_{1/2} - {}^2P_{1/2}$ transition in Sr$^+$ at 422 nm. $^2P_{1/2}$ ions decay to the $^2D_{3/2}$ state 7% of the time, after which they cease to interact with the probe beam.

2. PLASMA CREATION

The creation of strontium ultracold neutral plasmas starts with laser-cooled neutral strontium. As described in [7], several hundred million strontium atoms are trapped in a magneto-optical trap (MOT) (Fig. 1) based on the $^1S_0 - {}^1P_1$ transition in Sr at 461 nm (Fig. 2A) and cooled to a temperature of a few mK. The atomic density distribution can be adjusted to have a spherical Gaussian form, $n_a(r) = n_a(0)\exp(-r^2/2\sigma^2)$, where the peak density for N_a atoms $(n_a(0) = N_a/(2\pi\sigma^2)^{3/2})$ is typically $10^{16}\,\mathrm{m}^{-3}$ and σ is typically 1 mm. The number of atoms and density of the cloud can be adjusted over several orders of magnitude by tuning the parameters

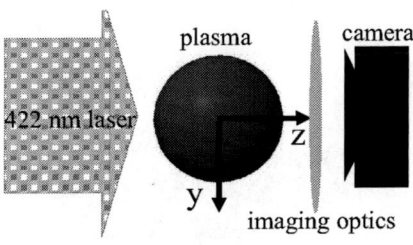

FIGURE 3. Absorption experimental layout. The absorption probe beam propagates along the $\hat{z}$-axis and falls on the camera. Absorption profiles are formed by recording an exposure with and without a plasma present, as described in the text. Spatial resolution can be utilized to isolate particular regions of the plasma, but all measurements integrate along the laser propagation direction.

of the lasers, magnetic fields, and atom-loading flux of the MOT.

Atoms are then photoionized in a two-photon process involving one 461 nm photon to promote atoms to the $^1P_1$ state and a 412 nm photon from a pulsed dye laser (~ 7 ns pulse) to promote electrons into the continuum. The plasma density profile follows the profile of the original laser-cooled atoms.

The ionization process adds energy to the electrons and ions, and the momentum change is negligible [8]. The electrons take away essentially all the excess photon energy as kinetic energy

$$E_e \approx h\nu_{ionize} - E_{IP} \qquad (1)$$

where the ionization potential is E_{IP} and the 412 nm laser frequency is ν_{ionize}. The increase in ion kinetic energy is approximately $E_e m_e/m_i$, which is only on the order of mK even for $E_e = k_B 1000$ K. Within a few tens of nanoseconds after photoionization [9, 10], the electrons collisionally thermalize with themselves at $k_B T_e \approx 2E_e/3$. Establishment of local thermal equilibrium for the ions requires about 1 μs and displays characteristic behavior of strongly-coupled plasmas, such as disorder-induced heating [5, 11] and kinetic energy oscillations [12, 13].

3. ABSORPTION

Absorption by the ions of a near-resonant probe laser has proven to be a powerful diagnostic of ultracold plasmas that has been used to study disorder-induced heating of the ions [5], ion kinetic energy oscillations [12, 13], and plasma expansion [5, 14]. Strontium is a good choice for these experiments because absorption measurements can be performed using the Sr^+ $^2S_{1/2} \rightarrow {}^2P_{1/2}$ transition at 422 nm (Fig. 2B).

To measure the plasma absorption or optical depth, a collimated probe laser beam tuned near resonance with the principal transition in the ions (Fig. 2B) illuminates the plasma. Ions scatter photons out of the laser beam and create a shadow that is recorded by an image–intensified CCD camera (Fig. 3). The

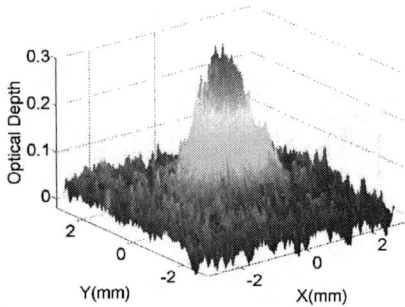

FIGURE 4. Optical depth measurement of a strontium ultracold neutral plasma. Reused with permission from [14]. Copyright 2004, Institute of Physics.

experimentally measured optical depth (OD) is defined in terms of the spatially resolved laser intensity without the plasma (I_0) and with the plasma present (I) as

$$OD(\nu, x, y) \;\; = \;\; \ln\left[I_0(\nu, x, y)/I(\nu, x, y)\right]. \tag{2}$$

Figure 4 shows a typical optical depth measurement. The delay between the formation of the plasma and camera exposure can be varied to study the time evolution of the plasma. Fast time resolution is obtained by gating the camera and the minimum camera exposure time is ~ 10 ns.

The typical absorption is a few percent at most. So the measurement is very sensitive to laser intensity-pattern fluctuations between the images with and without the plasma, which occur because of thermal effects in the nonlinear crystal used to generate the 422 nm light. This is the dominant source of noise and limits the practical regime of usefulness of the technique to plasmas with at least several tens of millions of ions and approximately the first $20\,\mu$s of plasma evolution.

3.1. Absorption Theory

Beer's law allows us to relate the OD theoretically to underlying physical parameters.

$$OD(\nu, x, y) \;\; = \;\; \int dz\, \rho_i(\mathbf{r})\alpha[\nu, T_i(\mathbf{r}), \nu_{exp}^z(\mathbf{r})], \tag{3}$$

where $\rho_i(\mathbf{r})$ is the ion density, and $\alpha[\nu, T_i(\mathbf{r})]$ is the ion absorption cross section at the image beam frequency, ν. The absorption cross section is sensitive to random thermal ion motion and motion arising from plasma expansion because of Doppler broadening. The Doppler shift at $\mathbf{r}$ due to average expansion velocity $\mathbf{u}(\mathbf{r})$ is $\nu_{exp}^z(\mathbf{r}) = \mathbf{u}(\mathbf{r}) \cdot \hat{z}/\lambda$. The local ion temperature also varies with position $T_i = T_i(\mathbf{r})$.

The Voigt profile for the ion absorption cross section reads

$$\alpha(\nu, T_i, \nu_{exp}^z) =$$
$$\int ds \frac{3^*\lambda^2}{2\pi} \frac{\gamma_0}{\gamma_{eff}} \frac{1}{1+4\left(\frac{\nu-s}{\gamma_{eff}/2\pi}\right)^2} \frac{1}{\sqrt{2\pi}\sigma_D(T_i)} \exp\left\{-\frac{[s-(\nu_0+\nu_{exp}^z)]^2}{2\sigma_D(T_i)^2}\right\}, \quad (4)$$

where $\gamma_{eff} = \gamma_0 + \gamma_{ins}$ is the effective Lorentzian linewidth due to the natural linewidth $\gamma_0 = 2\pi \times 20\,\text{MHz}$ of the transition and any instrumental linewidth γ_{ins}. The center frequency of the transition is $\nu_0 = c/\lambda$, and the "three-star" symbol [15] is a numerical factor that accounts for the polarization state of the ions and the imaging light. $\sigma_D(T) = \sqrt{k_B T/m_i}/\lambda$ is defined as is conventional for a Doppler width. We have neglected power broadening of the transition because the probe beam intensity of a few mW/cm^2 is much less than the saturation intensity of the transition. The absorption can be analyzed in different ways that each explore different plasma properties.

3.2. Absorption Imaging

If Eq. 3 is integrated over frequency, the resulting expression is

$$\int d\nu\, OD(\nu, x, y) = \gamma_0 \frac{3^*\lambda^2}{8\pi} \int dz\, \rho_i(\mathbf{r}) \equiv \gamma_0 \frac{3^*\lambda^2}{8\pi} n_{areal}(x, y) \quad (5)$$

which gives a measure of the areal density, $n_{areal}(x, y)$. Experimentally, the integral is performed by summing a series of optical depth measurements taken at equally spaced frequencies that span the absorption resonance. The areal density can be integrated spatially to provide the ion number. Spatial dynamics of the plasma can be studied by recording images at different times.

3.3. Absorption Spectroscopy

The frequency dependence of the ion absorption can be found by integrating the optical depth measurement spatially,

$$S(\nu) = \int dx\, dy\, OD(\nu, x, y). \quad (6)$$

Experimentally, the camera pixels are summed. To analyze this data, Eq. 3 can be used to theoretically relate the frequency dependence of the ion absorption to the density weighted average of the absorption cross section

$$S(\nu) = \int d^3r\, \rho_i(\mathbf{r})\alpha[\nu, T_i(\mathbf{r}), \nu_{exp}^z(\mathbf{r})]. \quad (7)$$

The camera's spatial resolution can be utilized to restrict the volume integral to a part of the plasma, and then the spectrum reflects the properties of the plasma in that region. The variation with position of the Doppler broadening due to thermal motion and plasma expansion complicates Eq. 7, but to a good approximation [14], all Doppler-broadening can be described by introducing an effective ion temperature, which yields

$$
S(\nu) = N_i \frac{3^*\lambda^2}{2\pi} \frac{\gamma_0}{\gamma_{eff}} \int ds \frac{1}{1+4\left(\frac{\nu-s}{\gamma_{eff}/2\pi}\right)^2} \frac{1}{\sqrt{2\pi}\sigma_D(T_{i,eff})} \exp\left[-\frac{(s-\nu_0)^2}{2\sigma_D(T_{i,eff})^2}\right].
$$

(8)

The effective Doppler width, $\sigma_D(T_{i,eff}) \equiv \sqrt{k_B T_{i,eff}/m_i}/\lambda$, which can be characterized by an effective temperature, measures the ion velocity along the laser propagation direction. For simple forms of the plasma expansion and $T_i(\mathbf{r})$, an analytic expression relates the velocity to time-varying plasma properties such as size, electron temperature, and ion temperature [8]. For a self-similar expansion and uniform ion temperature, $\sigma_D(T_{i,eff}) = \sqrt{[\langle(\mathbf{v}\cdot\hat{\mathbf{z}})^2\rangle]_{average}}/\lambda \equiv v_{i,rms}/\lambda$, where $\mathbf{v}$ is the total ion velocity including random thermal motion and expansion, the angled brackets refer to a local integral over the velocity distribution, and the average is over the plasma spatial distribution. In the absence of such simple conditions $\sigma_D(T_{i,eff}) \approx v_{i,rms}/\lambda$ is still a good approximation [14]. Absorption spectra fits also provide an absolute measure of the number of ions.

Fitting experimental absorption spectra to Eq. 8 provides a measure of the ion kinetic energy, but it is not possible to separate contributions from thermal motion and expansion. This restricts study of ion thermal properties using absorption to times less than a few microseconds, after which the expansion dominates the Doppler broadening [8].

4. FLUORESCENCE

Recent experiments have demonstrated the usefulness of fluorescence imaging and spectroscopy. It is intrinsically less sensitive to external noise, such as laser-power fluctuations, and thus is better-suited than absorption probes to study smaller plasmas or plasmas at long delays after formation. As shown below, it has the drawback of not providing an absolute measure of the number of ions in the plasma, so this technique works well in tandem with absorption diagnostics. In [6], monitoring the fluorescence level from a laser-beam focused in a small region of the plasma demonstrated some of the advantages of fluorescence as a probe, but imaging the fluorescence from excitation of the entire plasma provides much more information.

Figure 5 shows a schematic of the fluorescence experiment. A laser beam that is near resonance with the $^2S_{1/2} - {}^2P_{1/2}$ transition in Sr^+ at $\lambda = 422$ nm propagates along $\hat{y}$ and illuminates the plasma. Fluorescence in a perpendicular direction ($\hat{z}$) is recorded by the camera. A typical fluorescence measurement is also shown in Fig. 5. The 422 nm light is typically applied in a $2\,\mu s$ pulse to provide temporal

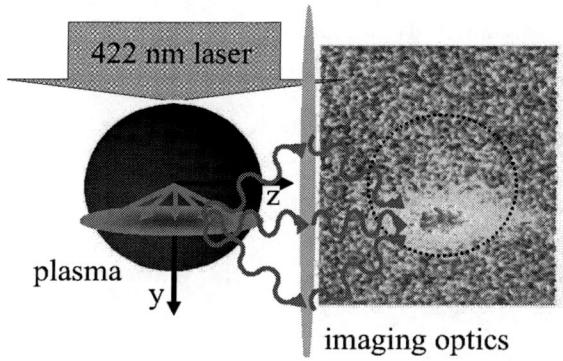

422 nm laser

plasma

imaging optics

FIGURE 5. Recording fluorescence of UNPs. The correlation between position and expansion velocity (red arrows) produces a striped image when the Doppler-shift due to expansion exceeds the Doppler shift associated with thermal ion velocity.

resolution, and the intensity is only a few mW/cm^2, which is low enough to avoid optical pumping to the metastable $^2D_{3/2}$ state.

4.1. Fluorescence Theory

By a procedure similar to the treatment of the absorption measurements, the fluorescence can be related to underlying physical parameters through

$$F(\nu,x,y) \propto \frac{\gamma_0}{2} \int \frac{dz\, n(\mathbf{r})I(\mathbf{r})/I_{sat}}{\sqrt{2\pi}\sigma_D[T_i(\mathbf{r})]} \int ds \frac{\gamma_0/\gamma_{eff}}{1+\left[\frac{2(\nu-s)}{\gamma_{eff}/2\pi}\right]^2} \exp\left\{-\frac{\left[s-(\nu_0+\nu_{exp}^y(\mathbf{r}))\right]^2}{2\sigma_D^2[T_i(\mathbf{r})]}\right\},$$

(9)

where $I(\mathbf{r})$ is the intensity profile of the fluorescence excitation beam. We have neglected power broadening of the transition, so this expression is only valid for $I(\mathbf{r}) \ll I_{sat}$, where the saturation intensity for linearly polarized light, taking Clebsch-Gordon coefficients for the transition into account, is $I_{sat} = 114\,\mathrm{mW/cm}^2$. We have also omitted a multiplicative factor that depends upon collection solid angle, dipole radiation pattern orientation, and detector efficiency. This must be calibrated with absorption measurements in order to derive quantitative information from the amplitude of the signal.

It is important to note the form of the Doppler shift arising from expansion. Due to the directed expansion velocity, the average resonance frequency of the transition for atoms at $\mathbf{r}$ is Doppler-shifted from the value for an ion at rest, ν_0, by $\nu_{exp}^y(\mathbf{r}) = \mathbf{u}(\mathbf{r}) \cdot \hat{y}/\lambda$. For most experimental conditions studied, the plasma expansion is self-similar [3, 9, 16], and $\mathbf{u}(\mathbf{r}) = \gamma(t)\mathbf{r}$, where γ depends upon the plasma parameters. So $\nu_{exp}^y(\mathbf{r}) = \gamma(t)y/\lambda$, which produces a correlation between position in the camera and the Doppler shift due to expansion. This spatial shift allows the Doppler-effect contributions from expansion and thermal motion to be separated.

4.2. Fluorescence Imaging

As with absorption, fluorescence data can be analyzed in different ways, which each measure different plasma properties. Summing a series of camera exposures taken at equally spaced frequencies covering the entire ion resonance is equivalent to integrating Eq. 9 over frequency. This yields

$$\int d\nu \, F(\nu, x, y) \propto \int dz \, I(\mathbf{r}) n(\mathbf{r}). \tag{10}$$

If the plasma is much smaller than the laser beam waist, then the intensity variation is insignificant and an image of the plasma areal density is given by

$$\int d\nu \, F(\nu, x, y) \propto \int dz \, n(\mathbf{r}) = n_{areal}(x, y). \tag{11}$$

If the laser spatial variation cannot be neglected, it distorts the measurement of the plasma width along the $\hat{x}$-axis, but with knowledge of the laser waist size (w), the effect can be accounted for;

$$\int d\nu \, F(\nu, x, y) \quad \propto \quad \int dz \, I_0 \exp\left(-\frac{2x^2 + 2z^2}{w^2}\right) n_0 \exp\left(-\frac{x^2 + y^2 + z^2}{2\sigma^2}\right) \tag{12}$$

$$\propto \quad \exp\left[-x^2\left(\frac{1}{2\sigma^2} + \frac{2}{w^2}\right) - \frac{y^2}{2\sigma^2}\right].$$

Figure 6 shows fluorescence images of an expanding plasma that show the effect of laser intensity variation.

4.3. Fluorescence Spectroscopy

If the laser intensity profile can be approximated as constant, the fluorescence signal can be interpreted as a spatially resolved light-scattering resonance spectrum. In practice, one integrates the signal over some region of the plasma,

$$\int_{reg} dx dy \, F(\nu, x, y) \propto \int_{reg} dx dy \int \frac{dz \, n(\mathbf{r})}{\sqrt{2\pi}\sigma_D[T_i(\mathbf{r})]} \tag{13}$$

$$\times \int ds \frac{\gamma_0/\gamma_{eff}}{1 + \left[\frac{2(\nu-s)}{\gamma_{eff}/2\pi}\right]^2} \exp\left\{-\frac{\left[s - (\nu_0 + \nu_{exp}^y(\mathbf{r}))\right]^2}{2\sigma_D^2[T_i(\mathbf{r})]}\right\}$$

$$\approx N_{i,reg} \frac{1}{\sqrt{2\pi}\sigma_D[T_{i,reg}]} \int ds \frac{\gamma_0/\gamma_{eff}}{1 + \left[\frac{2(\nu-s)}{\gamma_{eff}/2\pi}\right]^2} \exp\left\{-\frac{\left[s - (\nu_0 + \nu_{exp,reg}^y)\right]^2}{2\sigma_D^2[T_{i,reg}]}\right\}.$$

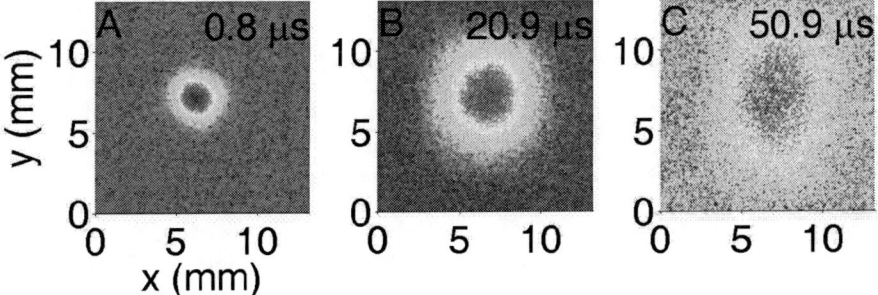

FIGURE 6. Fluorescence images showing expansion of the plasma. The indicated time is time after photoionization. In (A), the plasma is small enough that Eq. 11 is valid. By (C), the finite size of the fluorescence excitation laser elongates the plasma fluorescence even though the expansion is spherically symmetric. This distortion must be accounted for in order to accurately measure the areal density, as in Eq. 12.

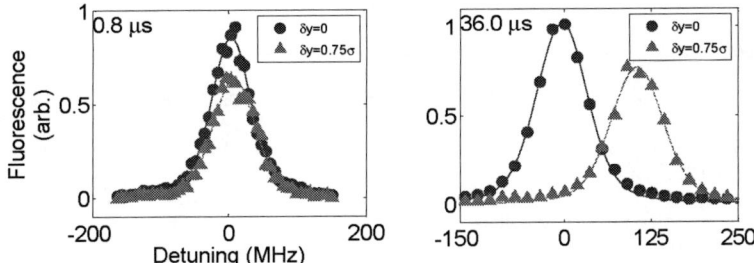

FIGURE 7. Fluorescence spectra of regions with little Doppler shift due to expansion ($\delta y = 0$) and significant Doppler shift ($\delta y = 0.75\sigma$) at two different times after photoionization. δy is the displacement along y of the region being analyzed (Eq. 13).

The z integral necessarily extends over the entire plasma, but $N_{i,reg}$ is the number of ions in the bounded $x - y$ region. $T_{i,reg}$ and $\nu_{exp,reg}^{y}$ are the average ion temperature and expansion-induced Doppler shift for the region. A typical fluorescence signal is so strong that the spatial extent of the region can be a small fraction of the length scale of variation of plasma properties (σ), so the average comes very close to a local value. It is important to note that this allows separation of the spectral influences of expansion and thermal motion because expansion leads to a shift in the spectrum, while thermal motion leads to a broadening. This is shown in Fig. 7.

5. CONCLUSIONS

We have described the optical diagnostics used to study ultracold neutral plasmas. Many of the experiments that have been performed with absorption measurements have been described previously, such as investigation of local thermal equilibration of strongly coupled ions [5, 12, 13], and plasm expansion at short times [5, 14]. Experiments are planned with fluorescence probes that will explore the establishment of global thermal equilibrium for ions and study the expansion dynamics over a longer time scale.

This work was supported by the National Science Foundation (Grant PHY-0355069) and the David and Lucille Packard Foundation.

REFERENCES

1. T. C. Killian, S. Kulin, S. D. Bergeson, L. A. Orozco, C. Orzel, and S. L. Rolston, *Phys. Rev. Lett.* **83**, 4776 (1999).
2. T. C. Killian, M. J. Lim, S. Kulin, R. Dumke, S. D. Bergeson, and S. L. Rolston, *Phys. Rev. Lett.* **86**, 3759 (2001).
3. F. Robicheaux, and J. D. Hanson, *Phys. Rev. Lett.* **88**, 55002 (2002).
4. S. Ichimuru, *Rev. Mod. Phys.* **54**, 1017 (1982).
5. C. E. Simien, Y. C. Chen, P. Gupta, S. Laha, Y. N. Martinez, P. G. Mickelson, S. B. Nagel, and T. C. Killian, *Phys. Rev. Lett.* **92**, 143001 (2004).
6. E. A. Cummings, J. E. Daily, D. S. Durfee, and S. D. Bergeson, *Phys. Rev. Lett* **95**, 235001 (2005).
7. S. B. Nagel, C. E. Simien, S. Laha, P. Gupta, V. S. Ashoka, and T. C. Killian, *Phys. Rev. A* **67**, 011401 (2003).
8. T. C. Killian, T. Pattard, T. Pohl, and J. M. Rost, submitted to *Physics Reports*, http://arxiv.org/abs/physics/0612097.
9. F. Robicheaux, and J. D. Hanson, *Phys. Plasmas* **10**, 2217 (2003).
10. L. Spitzer, Jr., *Physics of Fully Ionized Gases*, Wiley, New York, 1962.
11. M. S. Murillo, *Phys. Rev. Lett.* **87**, 115003 (2001).
12. Y. C. Chen, C. E. Simien, S. Laha, P. Gupta, Y. N. Martinez, P. G. Mickelson, S. B. Nagel, and T. C. Killian, *Phys. Rev. Lett.* **93**, 265003 (2004).
13. S. Laha, Y. C. Chen, P. Gupta, C. E. Simien, Y. N. Martinez, P. G. Mickelson, S. B. Nagel, and T. C. Killian, *Euro. Phys. J. D* **40**, 51 (2006).
14. T. C. Killian, Y. C. Chen, P. Gupta, S. Laha, Y. N. Martinez, P. G. Mickelson, S. B. Nagel, A. D. Saenz, and C. E. Simien, *J. Phys. B: At. Mol. Opt. Phys.* **38**, 351 (2005), (Equations 7, 10, 11, and 17 should be multiplied by γ_0/γ_{eff}.).
15. A. E. Siegman, *Lasers*, University Science Books, Sausolito, California, 1986.
16. T. Pohl, T. Pattard, and J. M. Rost, *Phys. Rev. A* **70**, 033416 (2004).

X-Ray Measurements Using a Microcalorimeter on an Electron Beam Ion Trap

E. Silver[1], G.X. Chen[1,2], K. Kirby[1,2], N.S. Brickhouse[1], J.D. Gillaspy[3], J.N. Tan[3], J.M. Pomeroy[3], and J.M. Laming[4]

[1]Harvard-Smithsonian Center for Astrophysics, [2]ITAMP
60 Garden Street, Cambridge, MA 02138
[3]National Institute of Standards and Technology, Gaithersburg, MD 20899-8421
[4]Center for Space Science, US Naval Research Laboratory, Washington DC 20375

Abstract. The X-ray telescopes and spectrometers flown on Chandra and XMM-Newton are returning exciting new data from a wide variety of cosmic sources such as stellar coronae, supernova remnants, galaxies, clusters of galaxies, active galactic nuclei and X-ray binaries. To achieve the best scientific interpretation of the data from these and future spectroscopic missions and related ground-based observations, theoretical calculations and plasma models must be verified or modified by the results obtained from measurements in the laboratory. Such measurements are the focus of several laboratory astrophysics programs that use an electron beam ion trap (EBIT) to simulate astrophysical plasma conditions. Here we describe our recent spectroscopic measurements of neon-like iron and nickel using a microcalorimeter on the EBIT at the National Institute of Standards (NIST) [1]. We obtain values for the intensity ratios of the well-known lines emitted by these ions and compare the results with new large scale electron-ion scattering calculations. Additional details about our laboratory astrophysics work can be found in some earlier papers [2-4].

REFERENCES

1. G. X. Chen et al., Phys Rev Lett, **97**, 143201 (2006).
2. E. Silver et al., Ap. J., **541**, 495 (2000).
3. J. M. Laming et al., ApJ Lett, **545**, 161 (2001).
4. J. Gillaspy et al., "Visible, EUV, and X-ray Spectroscopy at the NIST EBIT Facility", in *Atomic Processes in Plasmas: 14th APS Topical Conference on Atomic Processes in Plasmas*, edited by J. S. Cohen, S. Mazevet, and D. P. Kilcrease, AIP Conference Proceedings 730, Melville, New York, 2004, pp. 245-254. (2004).

CP926, *Atomic Processes in Plasmas—15th International Conference on Atomic Processes in Plasmas*
edited by J. D. Gillaspy, J. J. Curry, and W. L. Wiese
© 2007 American Institute of Physics 978-0-7354-0436-6/07/$23.00

X-Ray and EUV Spectroscopy of Highly-Charged Ions of Tungsten

Yu. Ralchenko*, J.D. Gillaspy*, J.M. Pomeroy*, J. Reader* and J.N. Tan*

*Atomic Physics Division, National Institute of Standards and Technology, Gaithersburg, Maryland 20899-8422

Abstract.
We report on recent measurements and collisional-radiative simulations of x-ray [1] and EUV spectra from multiply-charged ions of tungsten produced with the Electron Beam Ion Trap (EBIT) at the National Institute of Standards and Technology (NIST). The spectra were recorded in the ranges of 0.3 nm to 1 nm and 4 nm to 20 nm for beam energies varied between 2 and 4.3 keV. A quantum microcalorimeter was used for x-ray measurements while the EUV spectra were recorded with a grazing incidence spectrometer. of 4.08 keV. The uncertainties of our measured wavelengths range from 0.002 to 0.010 nm. Remarkably good agreement between calculated and measured spectra was obtained without adjustable parameters, highlighting the well-controlled experimental conditions and the sophistication of the kinetic simulation of the non-Maxwellian tungsten plasma. This agreement permitted the identification of new spectral lines from W^{39+}, W^{44+}, W^{45+}, W^{46+}, and W^{47+} ions, led to the reinterpretation of a previously known line in the Ni-like ion as an overlap of electric-quadrupole and magnetic-octupole lines, and revealed subtle features in the spectra arising from the dominance of forbidden transitions between excited states. The importance of level population mechanisms specific to the EBIT plasmais discussed as well.

ACKNOWLEDGMENTS

This work is supported in part by the Office of Fusion Energy Sciences of the U.S. Department of Energy.

REFERENCES

1.   Yu. Ralchenko, J.N. Tan, J.D. Gillaspy, J.M. Pomeroy, and E.Silver, Phys. Rev. A**74**, 042514 (2006).

CP926, *Atomic Processes in Plasmas—15th International Conference on Atomic Processes in Plasmas*
edited by J. D. Gillaspy, J. J. Curry, and W. L. Wiese
2007 American Institute of Physics 978-0-7354-0436-6/07/$23.00

A theoretical survey of formation of antihydrogen atoms in a Penning trap

D. Vrinceanu

Theoretical Division, Los Alamos National Laboratory, Los Alamos, NM 87545

Abstract. Numerous antihydrogen atoms are created at CERN, by ATRAP[5] and ATHENA[4] experiments, by bringing together positrons and antiprotons in a magnetic Penning trap. Most of these atoms are created in exotic, highly excited states, such that the magnetic forces on positrons are greater than the Coulomb attraction of antiprotons. This paper presents an overview of the recent progress made toward theoretical understanding of the complicated dynamics which leads to the formation and detection of antihydrogen atoms. There is no formal difference between the plasmas described here and normal, electron-proton, matter plasmas, except the reversed sign of electrical charges. The next generation of experiments need to bring the antihydrogen atoms to the ground state and to cool them to sub-milliKelvin temperature. Only then, high resolution spectroscopy can expose differences between matter and antimatter due to CPT violations. Suggestions are made for possible pathways toward this goal.

Keywords: magnetized plasma, antimatter
PACS: 36.10.-k, 34.80.Lx, 31.15.Qg

INTRODUCTION

Charge, Parity and Time (CPT) symmetry is the cornerstone of the Standard Model and it can be tested only by comparing various properties of particles and corresponding antiparticles. Since the discovery of positrons in 1932, anticipated by Dirac's theoretical work, CPT was tested for many leptons, baryons and mesons. The antihydrogen ($\bar{H}$) atom is the simplest atom made entirely of antimatter and it can provide, in principle, the most accurate CPT test if its 1s-2s transition could be measured with the same precision as the 1s-2s line in hydrogen, currently known with a remarkable 10^{-15} precision [1].

The first $\bar{H}$ atoms were created in accelerator experiments at CERN [2] and Fermilab [3] in $p + \bar{p} \rightarrow \bar{H} + e + p$, high energy ($\sim 6$ GeV) reactions. There is little opportunity to study these $\bar{H}$ atoms due to their relativistic velocities. A different approach has been taken by the ATHENA [4] and ATRAP [5] collaborations which reported in 2002, almost in the same time, the first production of thousands of cold $\bar{H}$ atoms. Both experiments store and recombine cold positrons (e^+) and antiprotons ($\bar{p}$) in similar Penning traps using a combination of strong magnetic (~ 5 T) and quadrupole electrostatic fields. The state analysis by field ionization employed by the ATRAP experiment reveals that antihydrogen is formed (and detected) in highly excited states and at velocities still too large to make them useful for precision experiments. The first laser controlled production of $\bar{H}$ atoms was demonstrated [6] in 2004, in a proof-of-concept experiment, in which laser excited Cs atoms collide with positrons to form positronium atoms which produce $\bar{H}$ atoms in further charge exchange collisions with antiprotons. More recently, the next generation of traps, using a combination of quadrupole magnetic and octupole electric

CP926, *Atomic Processes in Plasmas—15th International Conference on Atomic Processes in Plasmas*
edited by J. D. Gillaspy, J. J. Curry, and W. L. Wiese
© 2007 American Institute of Physics 978-0-7354-0436-6/07/$23.00

fields, has been proposed [7, 8] and advertised as capable of trapping all particles, charged and neutral, in the same place. No H̄ formation has been reported yet in these new kind of traps.

PENNING TRAP

It is not expected that the (eventual) tiny CPT violation to have any effect on the positron plasma dynamics and antihydrogen formation, which are governed by the electromagnetic forces. Therefore, an identical system of electrons and protons will behave in exactly same way as the antimatter counterpart, except the sign of all charges are flipped. However, these plasma systems are less studied at the parameters used in the antihydrogen Penning trap experiments and are quite interesting in their own. Moreover, the antimatter experiments have the advantage of much greater detection efficiency, since any antiparticle annihilates rapidly in contact to matter detectors.

A Penning trap can be built from a stack of cylindrical electrodes aligned coaxially with an uniform magnetic field of $3 \sim 5$ Tesla, which provides transverse confinement. Longitudinal confinement is provided by an electrostatic field generated by applying controlled potentials on these electrodes.

The e^+ loaded in this trap efficiently loose energy by synchrotron radiation (within a time scale of about 0.2 s), to come to thermal equilibrium with the trap walls kept at $4 \sim 50$ K. They move along the field lines like beads on wires. Their cyclotron period ~ 6 ps, is the smallest time scale in the system. The transverse motion of the e^+ is quantized by the Landau levels. For a typical 5.4 T field, the first three Landau levels have energies of 3.63, 10.9 and 18.14 K, which shows that most e^+'s at 4 K are in their ground Landau state. The distance of closest approach between e^+'s is ~ 8 μm, at this temperature, much greater than their corresponding cyclotron radius $\sqrt{\hbar/eB} = 0.011$ μm. It is then very unlikely that a e^+ - e^+ collision can excite higher Landau levels.

Dilute e^+ plasmas, with densities of $10^8 \sim 10^{10}$ cm^{-3}, have been sustained for long periods (minutes to hours) in Penning traps. The typical distance between e^+, or the Wigner-Seitz radius, is ~ 13 μm, and the plasma coupling parameter, which is the ratio between the potential and kinetic energies, is ~ 0.3. Another parameter measuring the correlation in a plasma is the number of e^+ within the Debye sphere. For a Debye radius of ~ 14 μm this parameter is of the order of unity.

Antiprotons are produced in high energy particle accelerator in proton-proton collisions with energy of ~ 6 GeV. In a multi-stage process, their energy is reduced by 10 orders of magnitude, with an efficiency of few percent. Low energy p̄'s are collected on the side wells of the Penning trap and injected in the e^+ plasma, because the part of the trapping potential which confines the e^+'s repels the p̄'s. During its motion through the e^+ cloud, the p̄ is shielded from the exterior electric fields, is slowed down by collisions with e^+'s and eventually captures an e^+. The newly formed neutral H̄ leaves the ion trap and collides with an electrode.

The exact mechanism of e^+ capturing by an p̄ is still debatable. It was with great enthusiasm recognized in an early paper [9] that if the standard three-body recombination law can be scaled down to 4 K and if it is still valid at high magnetic field, then a rate of 7×10^4 s^{-1} can be obtain for H̄ formation. However, later ATHENA experiments [10] measured H̄ formation at various e^+ temperature, up to the room temperatures. Their results shown a much weaker temperature dependence than the trusted $T^{-4.5}$ scaling for the three-body recombination, leading to the conclusion that the radiative recombination, which has a $T^{-0.5}$ dependence, cannot be completely ruled out. At a closer look, it was recognized [11] that the magnetic field has a very strong effect on the recombination dynamics because the Coulomb length scale given by the thermal radius $R_T = e^2/kT \sim$ 4 μm is much greater then the magnetic length scale ~ 0.01 μm. The magnetic field cannot be treated as a perturbation, and as a matter of fact, most of the time, the *Coulomb interaction* has to be regarded as a perturbation of the strong magnetic interaction. Due to the transverse confinement, the magnetic field has a retarding effect on the three-body recombination. On the other hand, it was demonstrated in the ion ring experiments [12] that the magnetic field enhances the radiative recombination at low temperature.

The rate equation for positrons (or electrons) in a field free infinite recombining plasma is

$$\frac{dn_e}{dt} = -\beta n_e^2 n_i$$

where n_e and n_i are electron and, respectively, ion number densities and β is the three-body collisional capture coefficient. Assuming that this is the only process leading to recombination, then essentialy, $\beta \sim$ volume^2/time. Since the Newton's law of motion in a Coulomb field is invariant to the Lie scaling: $\mathbf{r} \to \alpha^2 \mathbf{r}$ and $t \to \alpha^3 t$, then the coefficient β scales as power 9 of the scaling parameter α. The energy scales as α^{-2}, therefore β scales with temperature as $\beta \sim T^{-9/2}$. However, the presence of magnetic field breaks the scaling symmetry (unless one also scales the strength of magnetic field together with the coordinates and time), and there is no reason to assume a power law dependence of the recombination rate coefficient with the temperature!

Thomson [13] arrives at the same basic results by arguing that all electrons which arrive within the sphere with radius R_T and suffer a thermalization collision with a witnessing electron is essentially captured by the ion. The rate of this process, for every ion, is then the flux $n_e v R_T^2$ of electrons incoming in the Thomson sphere, times the number of passive electrons within this sphere $n_e R_T^3$. The capture rate is obtained as $C n_e^2 v R_T^5$, where v is the average velocity of electrons. It is remarkable that after almost a century, this formula is still valid, although the process has been treated with increasingly sophisticated methods, see [14], for example. In elaborated Monte Carlo calculation [15], the coefficient C was obtained as $C \sim 0.7$. Similar calculation [16] including an infinite magnetic field shown that the coefficient C is then reduced by a couple of orders of magnitude. The scaling laws hold again in the limit of infinite magnetic field, which has essentially the effect of reducing the problem to a one-dimensional one. The coefficient C is reduced by a factor of $3^{-9/2}$ to reflect this change, which is in accord to the more complicated calculation of O'Neil [16].

It is possible that the $\bar{p}$ - e^+ recombination in a Penning trap is driven by a combination of three-body, radiative and magnetic field induced processes. For example, when a e^+ approaches a $\bar{p}$, spiraling on a cyclotron trajectory, the cyclotron radius can increase at a short separation, when the Coulomb force increases significantly. With the increase in transverse energy of e^+, its longitudinal energy has to decrease and the e^+ can be "trapped" by the $\bar{p}$, in a quasi-resonance, for quite a long time. Extensive Classical

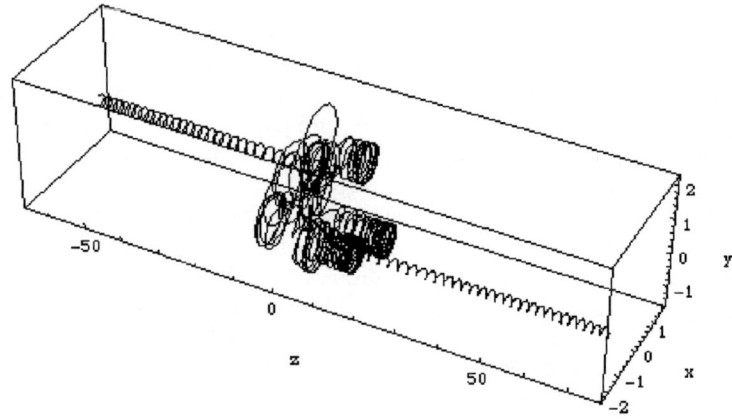

FIGURE 1. A e^+ can be temporarily captured in the field of a $\bar{p}$ when the transverse (cyclotron) energy increases at the expense of the longitudinal energy. This is not possible at B=0.

Monte-Carlo Simulations show that this phenomenon happens many times, especially at low temperature. Figure 1 displays a typical case. Permanent capture, by cyclotron radiation quenching or by stabilizing collision with a second positron, is then more probable as the lifetime of this quasi-resonance increases. In a sense, the magnetic field plays the role of a third body in a collisional capture. Laser induced Landau de-excitation, can stimulate recombination and is possible for two reasons: (a) the distance between adjacent Landau levels, usually $\sim$ 150 GHz, increases in the presence of Coulomb field, and (b) stimulated radiative rates, are on the order of, or less than, the quasi-resonance decay rates.

GIANT DIPOLE STATES

The dynamics of a two charged particle system is fundamentally changed by the presence of a magnetic field. Normally ignored, these effects are enhanced for highly excited systems and at low temperature. A comprehensive review of these phenomena is presented in [17]. The conventional separation of center-of-mass (COM) and relative coordinates is not available in the presence of magnetic field, and the total momentum of the system is **not** conserved. Instead, another quantity, the *pseudomomentum* **K** is conserved, and a continuous flow of energy between the relative and COM coordinates is possible. The notion of the rest frame has to re-analyzed. Exotic atomic states, with permanent dipole moment, are possible when the atom moves across the magnetic field lines.

The Hamiltonian for a system of two charged particles, of charges q_1 and q_2 and masses m_1 and m_2, in an uniform magnetic field B and arbitrary electric field with potential Φ is

$$H = \frac{[\mathbf{p}_1 - q_1\mathbf{A}(\mathbf{r}_1)]^2}{2m_1} + \frac{[\mathbf{p}_2 - q_2\mathbf{A}(\mathbf{r}_2)]^2}{2m_2} + \frac{q_1 q_2}{|\mathbf{r}_1 - \mathbf{r}_2|} + q_1\Phi(\mathbf{r}_1) + q_2\Phi(\mathbf{r}_2)$$

This hamiltonian depends on the choice of a gauge for the vector potential $\mathbf{A}$, but for a gauge transformation $\mathbf{A} \rightarrow \mathbf{A} + \nabla\Lambda$, where Λ is an arbitrary function of position, the Hamiltonian changes by a canonical transformation, so that the equation of motion are invariant. Various gauges can be adopted, but for the moment, it remains unspecified, so that the gauge invariant terms in the Hamiltonian can be easily identified.

Under a change of variables, to relative $\mathbf{r} = \mathbf{r}_1 - \mathbf{r}_2$ and COM $\mathbf{R} = (m_1\mathbf{r}_1 + m_2\mathbf{r}_2)/(m_1 + m_2)$ coordinates, and after a canonical transformation, the Hamiltonian is

$$H = \frac{[\mathbf{P} - Q_0\mathbf{A}(\mathbf{R}) - Q_1\mathbf{B} \times \mathbf{r}]^2}{2M} + \frac{[\mathbf{p} - Q_2\mathbf{A}(\mathbf{r})]^2}{2\mu} + \frac{q_1 q_2}{r} + Q_0\Phi(\mathbf{R}) + Q_1\mathbf{r}\nabla\Phi(\mathbf{R}) \quad (1)$$

where $M = m_1 + m_2$ is the total mass, $\mu = m_1 m_2/M$ is the reduced mass and effective charges are defined as $Q_0 = q_1 + q_2$, $Q_1 = (q_1 m_2 - q_2 m_1)/M$ and $Q_2 = (q_1 m_2^2 + q_2 m_1^2)/M^2$. It is assumed here that the external electric field does not have a large variation over the size of the atom.

The term $Q_1\mathbf{B} \times \mathbf{r}$ in eq. (1) is gauge independent, since the gauge has not been specified yet. If the total charge is 0, for a neutral system, such as $\bar{\mathrm{H}}$ or positronium, and if the electric field is uniform, then the COM position $\mathbf{R}$ is a cyclic coordinate and the associated momentum $\mathbf{P} = \mathbf{K}$ is a constant of motion, called *pseudomomentum*. From the equation of motion for the COM $\dot{\mathbf{R}} = \partial H/\partial\mathbf{P}$, one obtains

$$M\dot{\mathbf{R}} = \mathbf{K} + Q_1\mathbf{r} \times \mathbf{B}$$

This equation shows that the pseudomomentum $\mathbf{K}$ is identical to the mechanical momentum of the COM along the magnetic field, but it has an extra component which depend on the relative separation, in the transverse direction. It is easy to prove that $\mathbf{K} = \mathbf{k}_1 + \mathbf{k}_2$, where the particle pseudomomenta $\mathbf{k}_{1,2}$ are related to the positions of the guiding centers of the free charged particles. The separation between the guiding centers is given by the transverse component $K_\perp$ of the pseudo-momentum and is conserved during the motion. By defining this distance as $\rho_0 = (\mathbf{K} \times \mathbf{B})/Q_1 B^2$, then the total energy separates as

$$E = \frac{K_\parallel^2}{2M} + \frac{\mu v^2}{2} + \frac{Q_1^2 B^2}{2M}(\mathbf{r}_\perp - \rho_0)^2 + \frac{q_1 q_2}{r} - Q_1\mathbf{r} \cdot \mathbf{F} = \frac{K_\parallel^2}{2M} + \frac{\mu v^2}{2} + V_{\mathrm{eff}}(\mathbf{r})$$

where $\mathbf{F}$ is the strength of the electric field and $K_\parallel$ is the projection of $\mathbf{K}$ along the magnetic field direction. This is equivalent with the motion in relative coordinates of a particle with mass μ in a magnetic field $\mathbf{B}$ and an effective potential given by

$$V_{\mathrm{eff}}(\mathbf{r}) = \frac{Q_1^2 B^2}{2M}(\mathbf{r}_\perp - \rho_0)^2 + \frac{q_1 q_2}{r} - Q_1\mathbf{r} \cdot \mathbf{F}$$

The COM - relative coordinates separation succeeded only partially here since the effective potential, and the relative motion, depends on the motion of the atom as a whole through the parameter ρ_0.

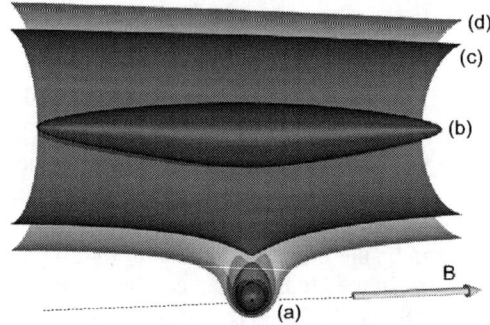

FIGURE 2. Surfaces of equal energy for $V_{eff}(\mathbf{r})$ showing the transition from spherical symmetry, at short distances, to cylindrical symmetry at large separations. Electric field $\mathbf{F}$ is zero.

The effective potential $V_{eff}(\mathbf{r})$ can be interpreted as a Coulomb attraction together with a two-dimensional harmonic oscillator transversally displaced by an amount ρ_0 proportional to the transverse COM motion and along a direction perpendicular to it. Three surfaces of equal energy for the effective potential are shown in fig. 2. Interesting topologies arise due to the mixing between spherical Coulomb symmetry (a) and cylindrical harmonic oscillator symmetry (d). For large enough $K_\perp > K_{crit} = 3e[MB/16\pi\varepsilon_0]^{1/3}$ a potential well can form, separated from the origin (surface (b) in fig. 2 has negative energy). The states localized within such well have a permanent dipole moment and are called *giant dipole* states. Surface (c) has (barely) positive energy, and extends along the magnetic field direction to infinity. This is the surface along which the magnetized Rydberg atom can ionize.

Energy E and pseudomomentum $\mathbf{K}$ are the only globally conserved quantities for the two charged particle system. When there is no transverse COM motion ($K_\perp = 0$), then the projection of the angular momentum on the magnetic field direction is also conserved. The cyclotron motion (or the magnetic moment) is an adiabatic invariant providing the perturbation, which in this case is the Coulomb force, does not vary much during a gyro-period. This leads to the approximate conservations of the magnetron and axial adiabatic invariants, making the system completely separable and therefore fully integrable. This approximation is called the Guiding Center Drift approximation [18].

MOLECULAR DYNAMICS SIMULATIONS

The recombination process of e^+'s with $\bar{p}$'s in a Penning trap is not only complicated by the strong magnetic field, but also by the finite size effects of the e^+ cloud, the significant deviation from thermal equilibrium between $\bar{p}$ and e^+ and the complicated electric field configurations. In these circumstances, there is little hope that simple, steady state, infinite plasma recombination models would accurately describe the $\bar{H}$ experiments.

We performed Molecular Dynamics Simulations [19] and [20], for $\bar{H}$ formation in a strongly magnetized e^+ plasma with the following goals in mind. We wanted to reproduce as closely as possible the typical experimental conditions. The full classical dynamics has to be faithfully followed, free of the guiding center approximation, which is valid for infinitely strong magnetic field but fails at close distances, where the Coulomb force dominates. The dynamics of the particles has to be followed for long enough times ($\sim \mu$s) in order to exhibit the initial collisional capture and the subsequent relaxations. A large number of recombination events allows a detailed investigation of forming mechanisms and analysis of formed states.

A population of $4000 \sim 8000$ e$^+$ and $1000 \sim 2000$ p̄ with densities similar to those in experiments were followed for long simulations times $0.5 \sim 1$ ms. A special symplectic integrator was utilized in order to accurately capture both the fast cyclotron motion and the slow capture of e$^+$ by p̄ . The simulation was run on a parallel computer using MPI. Due to the magnetic field confinement the plasma expands only along the magnetic field. Hundreds of capturing events were detected and carefully analyzed. Surprisingly, most of the H̄ atoms formed in a giant dipole state. Unfortunately, the integration time was not long enough to exhibit the relevant relaxation stage so that the recombination rate obtained in simulations cannot be reliably compared with the experimental rates. A small number of positive H̄$^+$ ions, were also detected in our simulation, proving the stabilizing effect of the magnetic field on these short lived ions.

ACKNOWLEDGEMENTS

This work has been supported by the U.S. Department of Energy through a grant to the Los Alamos National Laboratory.

REFERENCES

1. G. Gabrielse, *Advan. Atom. Mol. Opt. Phys.* **50**, 155 (2005);
2. G. Baur *et al.*, *Phys. Lett. B* **368**, 251 (1996);
3. G. Blanford *et al.*, *Phys. Rev. Lett.* **80**, 3037 (1998);
4. M. Amoretti *et al.*, *Nature (London)* **419**, 456 (2002);
5. G. Gabrielse *et al.*, *Phys. Rev. Lett.* **89**, 213401 (2002); **89**, 233401 (2002);
6. C. H. Storry *et al.*, *Phys. Rev. Lett.* **93**, 213401, 263401 (2004);
7. G. Andersen *et al.*, *Phys. Rev. Lett.* **98**, 023402 (2007);
8. G. Gabrielse *et al.*, *Phys. Rev. Lett* **98**, 1133002 (2007);
9. G. Gabrielse, S. L. Rolston and L. Haarsma, *Phys. Lett. A* **129**, 38 (1988);
10. M. Amoretti *et al.*, *Phys. Lett. B* **583**, 59 (2004);
11. D. Vrinceanu *et al.*, *Phys. Rev. Lett.* **92**, 133402 (2004)
12. M. Hoerndl *et al.*, *Phys. Rev. A* **74**, 052712 (2006);
13. J. J. Thomson, *Phil. Mag. S.* **23**, 449 (1912);
14. E. Hinnov and J. G. Hirschberg, *Phys. Rev.* **125**, 795 (1962);
15. P. Mansbach and J. Keck, *Phys. Rev.* **181**, 275 (1969);
16. M. E. Glinsky and T. M. O'Neil, *Phys. Fluids B* **3**, 1279 (1991);
17. P. Schmelcher and L. S. Cederbaum, *Atoms and Molecules in intense fields* **86**, 27 (1997);
18. S. G. Kuzmin, T. M. O'Neil and M. E. Glinsky, *Phys. Plasmas* **11**, 2382 (2004);
19. S. X. Hu *et al.*, *Phys. Rev. Lett* **95**, 163402 (2005);
20. D. Vrinceanu *et al.*, *Phys. Rev. A* **72**, 042503 (2005).

ASTROPHYSICAL PLASMAS

Plasma Astrophysics – Cosmology and the Growth of Cosmic Structure

Richard Mushotzky

National Aeronautics and Space Administration, Goddard Space Flight Center, Greenbelt, Maryland, 20771

Abstract. I will present some of the ways that x-ray spectroscopy can be utilized to determine cosmological parameters focusing on 5 methods : the gas fraction in clusters, the use of the Sunyaev-Zeldovich effect, the detection of resonance scattering in clusters , the use of resonance absorption and emission in background sources and the growth of structure. All of these techniques except the S-Z effect rely heavily on high resolution x-ray spectroscopy and require the next generation of x-ray spectroscopic missions such as Constellation-X. The promise of these techniques is great and they have the potential for precision cosmology with errors similar to those of other precision techniques such as type Ia supernova. If time permits I will also talk about how we can learn about how active galaxies strongly influence the growth of cosmic structure and how broad band high resolution x-ray spectra are necessary to measure the effects of AGN and how much energy they input into the universe and the role of new atomic physics calculations in interpreting these results. A related discussion can be found in a previously published manuscript [1].

REFERENCES

1. R. Mushotzky, "Clusters of galaxies: a cosmological probe," in Frontiers of X-ray Astronomy, edited by A. C. Fabian, K. A. Pounds, and R. D. Blandford, Cambridge University Press, Cambridge, (2004), pp. 149-164.

CP926, *Atomic Processes in Plasmas—15th International Conference on Atomic Processes in Plasmas*
edited by J. D. Gillaspy, J. J. Curry, and W. L. Wiese
2007 American Institute of Physics 978-0-7354-0436-6/07/$23.00

Probing the Cassiopeia A Supernova Explosion

Una Hwang* and J. Martin Laming†

*NASA Goddard Space Flight Center, Code 662, Greenbelt MD 20771, USA; and Johns Hopkins
University, 3400 Charles Street, Baltimore MD 21218, USA
†Naval Research Laboratory, Code 7674L, Washington DC 20375, USA

Abstract. X-ray observations of the Cassiopeia A supernova remnant reveal explosive nucleosynthesis products such as Si and Fe, and thus provide a unique window into the core-collapse explosion that formed the remnant 330 years ago. We review current progress using X-ray spectra extracted on arcsecond angular scales from a 10^6 s Chandra observation of Cas A, in conjunction with models that follow the remnant's hydrodynamical evolution and treat the relevant plasma microphysics. We address questions related to the explosion such as the degree of explosion asymmetry, the nature of the jets, the nature of the circumstellar environment, and extent of radial mixing of the Fe ejecta.

Keywords: Astronomical Observations; Supernovae, Supernova Remnants
PACS: 95.85.Nv, 97.60.Bw, 98.38.Mz

INTRODUCTION

Supernovae and their progenitors produce most of the heavy elements in the Universe, but core-collapse supernovae remain a perplexing, unsolved problem in modern astrophysics. Though successful explosions occur in nature, they have eluded decades of effort on theoretical models. The general picture is that after a massive star exhausts its nuclear fuel upon formation of an Fe core, the core collapses to nuclear density. The ensuing "bounce" initiates a shock wave that works its way outward, depositing energy into the outlying stellar layers and enabling further nucleosynthesis. Direct observables of the collapse are gravitational waves, which have never been detected from any source, and neutrinos, of which some 20 electron antineutrinos were detected from SN 1987A [e.g. 1]. Though an experimental triumph, this detection did not have real diagnostic value. Fortunately, many details of the core-collapse mechanism are also imprinted in the ejecta synthesized in the innermost layers of the star during the explosion. These clues may be read many years later in ejecta observed in the supernova remnant.

Theoretical calculations of core-collapse supernovae have improved dramatically in recent years, with multi-dimensional simulations showing how the development of instabilities can help to revive the stalled supernova shock front toward explosion [e.g., 2, 3, 4, 5, 6]. At the same time, observations have also improved greatly. The most important advance here is the exquisite $0.5''$ angular resolution provided in X-rays by the mirrors on the Chandra Observatory. The angular resolution is crucial to identify individual compact ejecta fragments, and is fully exploited by the 2004 10^6 s observation of the Cassiopeia A supernova remnant (SNR). Never before attained in X-rays, Chandra's sharp X-ray focus will not be rivalled during our research lifetimes by any of the X-ray missions now being planned.

The ejecta in SNRs become accessible to observation in X-rays after they pass through

CP926, *Atomic Processes in Plasmas—15ᵗʰ International Conference on Atomic Processes in Plasmas*
edited by J. D. Gillaspy, J. J. Curry, and W. L. Wiese

an inner (reverse) shock that starts at the interface (contact discontinuity) between the ejecta and the interstellar material swept up by the primary outer blast wave. The shock energetics make X-rays the characteristic emission of plasmas heated by the passage of these shocks. In young remnants, the reverse-shocked ejecta often dominate the X-ray emission by virtue of their high densities and enriched composition, making the supernova ejecta accessible for study hundreds or thousands of years after the explosion. Because of its low density, the remnant gas is typically evolving towards ionization equilibrium so that the ionization structure must be characterized to interpret the X-ray spectrum (most simply by an ionization age $n_e t$, the product of electron density and time since shock heating).

The Galactic supernova remnant Cassiopeia A is an ideal candidate for study of ejecta. It has a dynamical age of 330 yr [7], making it young enough that the ejecta dominate the X-ray emission, and old enough that a substantial fraction of the ejecta have passed through the reverse shock. It has been extremely well observed at all wavelengths, and in particular, the recent Chandra 1 Ms observation is unequalled in soft and medium energy X-rays for any SNR. Chandra provides moderate resolution X-ray spectra at the $0.5''$ scale of the Chandra mirrors for nearly the entire $5'$ extent of the supernova remnant. Of equal importance, the X-ray observations reveal the inner ejecta layers in Cas A, in particular the Si and Fe that were synthesized during the explosion.

HYDRODYNAMIC STATE AND MODELS

Cas A has long been known to have had a massive progenitor on the basis of optical emission dominated by O, since O is synthesized hydrostatically by massive stars during their lifetimes. During their evolution, massive stars also undergo mass loss that generates a circumstellar wind environment (often idealized as a $\rho \propto r^{-2}$ density profile), before exploding as core-collapse supernovae (Type Ib/Ic with loss of H and/or He and variants of II). They generally leave a compact stellar remnant behind, either neutron star or black hole, while exhibiting a wide range of properties that reflects the diversity of their progenitors [e.g., 8].

Cas A's progenitor probably had a main sequence mass of 15-25 $M_\odot$ (at the lower end of Wolf Rayet masses; Massey et al. [9] see arguments given by Laming & Hwang [10] and Chevalier & Oishi [11]; optical evidence in Fesen et al. [12]. The progenitor mass was reduced to about 3-4 $M_\odot$ at the time of explosion by extensive mass loss, mainly in the red supergiant (RSG) phase, possibly with the aid of a binary companion. A binary progenitor becomes compelling for Cas A when observational constraints are weighed in the context of three-dimensional explosion calculations and stellar models for both single and binary stars [13]. There is also speculation that Cas A actually exploded in a circumstellar bubble, which we discuss further below, but for the most part, the remnant's blast wave has been propagating through the remains of this slow, dense RSG wind. About $2 M_\odot$ of Cas A's ejecta have now been shock-heated to emit X-ray (and optical) emission. This hydrodynamical model [derived by Laming & Hwang [10] based on 14] also accounts for the forward shock velocity and radius [15], the separation between reverse and forward shocks [16], the emission measure and mass of the shocked circumstellar plasma [17], and the temperatures and ionization ages of the ejecta knots.

The apparently high degree of mass-loss incurred by the Cas A progenitor helps to explain the unusual clumpiness of the X-ray emitting ejecta in Cas A, which are organized into compact knots and filaments (see Figure 1). Clumps formed during the explosion are better able to survive in progenitors that have shed more of their mass prior to explosion because they have weaker supernova reverse shocks, which suppresses the Rayleigh-Taylor instabilities that would otherwise shred the clumps completely [18, 19].

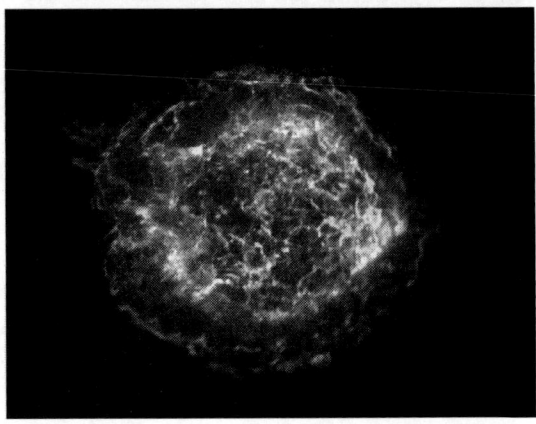

FIGURE 1. Three-color 1 Ms Chandra ACIS image of Cas A with red=Si He α, blue=Fe K, green=4-6 keV continuum from the blast wave. The clumpiness of the ejecta is evident.

The small size of Cas A's ejecta knots and properties related to their stability following shock passage are used to advantage by Laming & Hwang [10] and Hwang & Laming [20]. The shock crossing time is 15-30 years for a 500-1000 km s^{-1} reverse shock passing through knots that are $1''$ (5×10^{16} cm) across, and is similar to the observed lifetimes of optical knots in Cas A Thorstensen, Fesen, & van den Bergh [7]. The short shock crossing time and the constraints on the density contrast allow a knot spectrum to be fitted with a single ionization age $n_e t$, which can be plotted along with the fitted temperature T_e and compared with the results of models. This method allows Lagrangian mass coordinates (i.e. location) within the ejecta to be assigned to particular features, as well as inference of the time of reverse shock passage.

Modeling is essential in deriving element abundances from X-ray spectra because of the large density gradients found in supernova remnants, particularly those like Cas A that are expanding into their presupernova stellar winds. To this end we have built a simulation code BLASPHEMER[1] [21, 10] which follows a Lagrangian plasma element through either the forward or reverse shock, integrating equations for the ionization balance, and electron and ion temperatures within a framework of analytic hydrodynamics based on self similar solutions for the SNR evolution. Ions and electrons undergo temperature jumps at the shock in proportion to their masses, followed by Coulomb equilibration. Faster collisionless equilibration appears to be more effective at lower velocity, or more properly, lower Alfvén Mach number shocks [22, 23, 24, 25, 26, 27, 28, 29].

---

[1] BLASt Propagation in Highly EMitting EnviRonment

EXPLOSION ASYMMETRY

Quantitative inferences about the degree of asymmetry in Cas A are made by Laming & Hwang [10] from constraints on the shape of the ejecta density profiles derived using early X-ray Chandra observations. The ejecta are taken to be distributed with uniform density in the core, and a power-law envelope $\propto r^{-n}$. The outer ejecta density slope is found to be shallower in the directions close to the "jet" axis in Cas A (but not actually in the jet). This implies up to a factor of 2 more kinetic energy and 10% less ejecta mass imparted in this direction by the explosion if the forward shock speed and radius are to be the same everywhere.

Asymmetries in the SN ejecta are not only observed in core-collapse supernovae that follow substantial mass loss [8], but are also predicted by theoretical scenarios. The degree of asymmetry inferred for Cas A is at the low end of those predicted by theoretical models involving jet-induced explosions [30] and rotating progenitors [31, 32]. It is also similar to that which initiates model A1 of Nagataki et al. [33], which gives $1.8 \times 10^{-4} M_\odot$ ejected mass of ^{44}Ti and $0.06 M_\odot$ of ^{56}Ni, similar to what Cas A is likely to have produced [34, 35]. The asymmetry estimates in Laming & Hwang [10] assume Cas A to be expanding into a circumstellar wind with density $\propto 1/r^2$. A phase of expansion into a circumstellar bubble would likely reduce the degree of explosion asymmetry inferred from the ejecta knot temperatures.

Since Cas A exhibits prominent "jet"-like structures in the ejecta, both in optical and X-ray emission (Figure 2), Laming et al. [36] use the Chandra Ms data to compare the spectral properties of the jet knots with those predicted by models for evolution of a spherical remnant into circumstellar cavities and for evolution of a simplified jet. They conclude that the so-called "jet" regions are truly due to jets originating from the center of the explosion. Limits on the isotropic energy and the jet-opening angle give a total kinetic energy in the stronger northeast jet of about 10^{50} ergs, which is lower than the inferred energy for "typical" GRBs, though it is more in line with the weaker long-duration GRBs.

The total explosion energy of $\sim 2 \times 10^{51}$ ergs estimated by Laming & Hwang [10] for Cas A is also low compared to hypernovae that have been associated with GRBs. The nearby GRB 060218 associated with SN 2006aj, however, has an inferred energy and ejecta mass very similar to Cas A [37]. Moreover, the prompt radiated energy release of GRB 060218 is even lower than the energy that we inferred for the northeast jet in Cas A, at $10^{49} - 10^{50}$ ergs [38]. Other low-energy GRBs include GRB 980425 associated with SN 1998bw and GRB 031203 associated with SN 2003lw [39]; these possibly constitute a separate population of long duration GRBs.

To explore the evolution of ejecta knots in a small bubble surrounded by circumstellar wind, Chandra spectra were sampled for radial series of knots in various locations of Cas A (some examples are shown in Figure 3), and models fitted to determine the temperature and ionization age of the X-ray emitting gas. Typically, two spectral components are needed to describe the spectra (with a separate component required for the Fe emission). For this purpose, however, we focus on the emission from ejecta other than Fe. Models were computed for ejecta with uniform density core and outer density profile power-law slopes $n = 9.7$ [from 40] evolving into bubble radii of 0, 0.2, and 0.3 pc within an otherwise $1/r^2$ circumstellar wind, based in part on analytic results

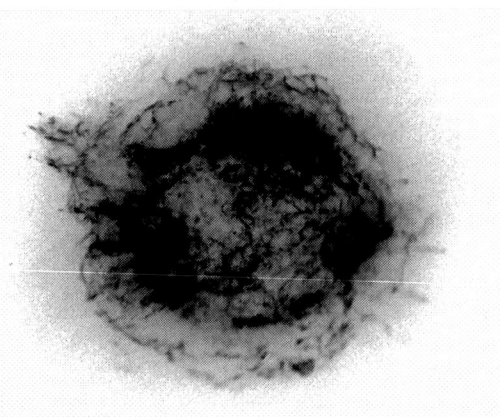

FIGURE 2. Broadband Chandra Ms X-ray image of Cas A. The contrast has been adjusted to bring out the low surface brightness features, in particular the jet like filaments in the northeast (upper left) and southwest (lower right).

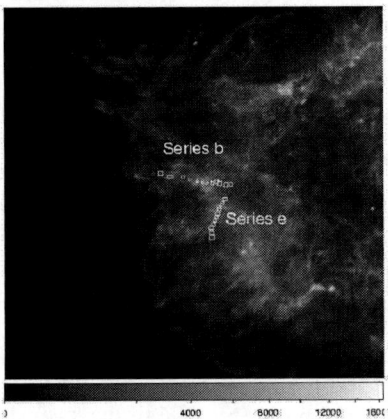

FIGURE 3. Locations of knot series on the east limb of Cas A.

by Chevalier & Liang [41]. However these models have no swept up mass shell or bubble boundary, but transition directly to a $1/r^2$ wind density profile. Hence the implied explosion dates, which match the currently observed blast wave velocity and radius, are 1660, 1670, and 1680 AD, respectively. Recalling that the earliest possible explosion date is 1671 ± 1 [7], a pure $1/r^2$ circumstellar medium would appear to be ruled out. More recent studies [42], accounting for possible deceleration of some of the optical knots, move the explosion date even later, albeit with larger uncertainties. The locus of temperature and ionization age evolution for each of the models is plotted in Figure 4 along with the values determined from spectral fits to the data for the series b. The size of the box for each data point indicates the 90% confidence error for the measured values.

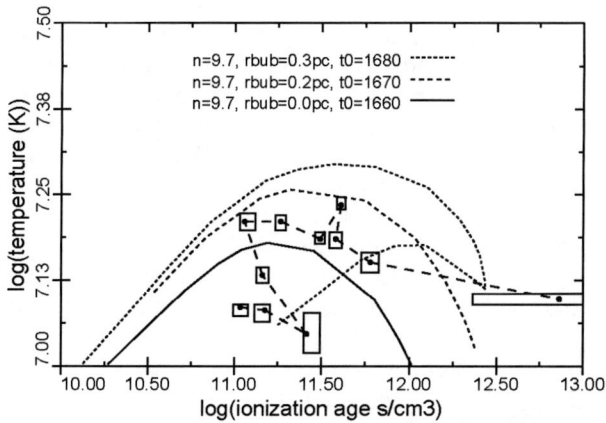

FIGURE 4. Locus of temperature against ionization age for series b. Models with ejecta envelope power law of 9.7 for various bubble radii, compared with data points determined from single ionization age fits to knot spectra. Projection effects are likely for the first four regions starting at log $n_e t = 11$.

In evaluating the figure, projection effects must be kept in mind. In general, such complications are expected to be worse in the interior regions of the remnant, for which the lines of sight through the remnant are longer. The first four regions of series b on the east limb are in the interior, and are expected to have the lowest ionization age. However, in Figure 4, they do not fit well with the trend of increasing ionization age seen for the other knots, and are likely to suffer problems from projection. The remaining knots in the series are in fairly good agreement with a small bubble of 0.2 pc or slightly smaller. Compared to the models without a bubble (as were presented in Laming & Hwang [10] and Hwang & Laming [20], the inclusion of a small bubble tends to require a narrow range of steeper power laws ($n = 10, 11$ compared to $n = 6, 7$ and $n = 9, 10$), reducing the inferred degree of asymmetry in the remnant. The bubble models also allow for X-ray emitting knots with very high ionization age, at least some of which are indeed observed. Moreover, the 0.2 pc model also gives the explosion date of 1671 AD that is determined from the dynamics of the optically emitting ejecta.

FE EJECTA

The explosion that produced Cas A left its imprint on the composition of the supernova ejecta. Of the well-studied core-collapse remnants, Cas A most clearly reveals the inner ejecta layers of Si and Fe that were synthesized during the explosion. These elements have large-scale differences in their distribution, with the Fe rich products of complete Si burning being *exterior* to the lower Z elements in the southeast [see Figure 1 43, 44]. This indicates that some convective overturn has occurred between the ejecta layers. X-ray Doppler measurements of Si and Fe with *XMM-Newton*[2] [17] show that the Si and Fe emission in the northwest are kinematically distinct, even though they overlap on the sky, with the Fe having a higher redshift and therefore being located behind (i.e., exterior to) the Si in that direction. Similar overturns are observed between the deeper S-, and overlying N-rich, optically emitting ejecta in the jet regions [45], but the overturns do not occur throughout the remnant since O- and N-rich optical knots are found external to the X-ray emitting Fe outside the jet [42].

In explosively synthesizing Fe, Cas A underwent α-rich freezeout, wherein Si burns at the highest temperatures and low densities. The α-rich freezeout ashes are essentially pure ^{56}Ni (decaying to ^{56}Fe) along with residual He and trace amounts of certain elements such as radioactive ^{44}Ti, which decays via ^{44}Sc to ^{44}Ca [e.g., 46, 47, 48]. Cas A is the only source for which both decay products of ^{44}Ti have been clearly detected in gamma-rays and hard X-rays, giving an inferred ^{44}Ti mass of $1.8 \times 10^{-4} M_\odot$ [50, 51, 34, 35, 52]. More recently, Renaud et al. [49] revise this to $1.6^{+0.6}_{-0.3} \times 10^{-4} M_\odot$, from an analysis of *INTEGRAL/IBIS/ISGRI* data. Early *Chandra* observations reveal a faint knot in Cas A that is evidently made of nearly pure Fe, and is thus a possible product of α-rich freezeout(see Figs 3 and 4 in Hwang & Laming [20]. The position of this Fe in the ejecta (its mass coordinate) can be inferred from its ionization age and is similar to those for the other Fe knots. The Chandra Ms spectrum of the knot confirms its pure Fe composition (see Figure 5).

The distribution of the Fe ejecta also gives clues to the mixing processes that occur during supernovae. An initial quantitative measure of the mixing is given by the analysis of the Fe knot spectra in early Chandra data [20]. The mass coordinates inferred (albeit with large uncertainties for the time being) are consistent with mixing of the Fe (and therefore ^{44}Ti) from the α-rich freezeout zone out to the O/Si interface, assuming a 25 $M_\odot$ progenitor. This extent of mixing is consistent with the formation of X-type SiC (silicon carbide) grains which have elevated ^{44}Ca abundances from ^{44}Ti formed during α-rich freezeout [e.g., 53].

The Fe knots studied by Hwang & Laming [20] are notably absent above a maximum ionization age of about 10^{12} cm^{-3} s (see Figure 6). In our models, Fe knots that were shocked sufficiently early on at high densities (and consequently have high ionization ages) undergo radiative instability so that they will not be detectable in X-rays. The nondetection of Fe knots in X-rays with high ionization ages at large radii therefore does not necessarily indicate that they do not exist. Indeed, the transport of Ti (i.e., α-rich freezeout ash) to far outlying ejecta layers is implied by the existence of presolar graphite grains with embedded TiC and metal subgrains. To form these grains, the Ti must be transported all the way to the graphite layers without undergoing substantial mixing with other nucleosynthesis layers, a feat that Lodders [54] suggests might be accomplished by jets. There is little Fe detected in Cas A that is associated with the "jets", but some of the α-rich freezeout products were presumably transported to the outer graphite layers in the jet if these SN type graphite grains were formed. There they would then have already undergone radiative instability to temperatures below the X-ray regime, and might still be detectable through the innershell transition lines of the

---

[2] at lower angular resolution than with *Chandra*

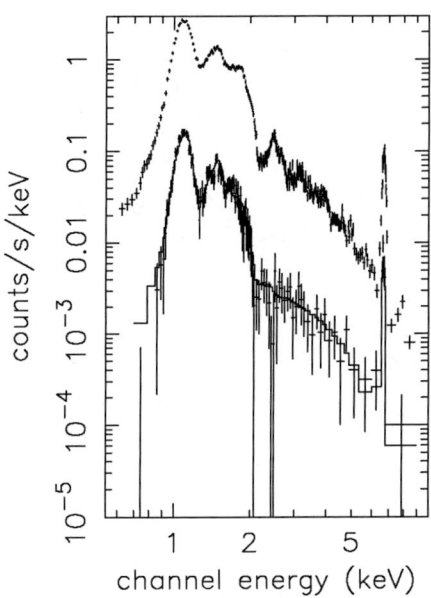

FIGURE 5. Chandra spectra of a nearly pure Fe knot from 1 Ms observations (top; scaled by a factor of 10) and earlier observations (bottom). Except for a small contribution from Si and S, the spectrum can be completely explained by Fe lines and continuum.

daughter products of ^{44}Ti, namely ^{44}Sc and ^{44}Ca.

Aside from the compositional evidence given by Lodders [54], initial estimates of the ^{44}Ti mass in Cas A were sufficiently high that they led by themselves to suggestions that an asymmetric explosion involving jets might be required to sufficiently enhance the efficiency of α-rich freezeout [33], though that requirement has been somewhat mitigated.

CONCLUSIONS

We believe that the 1 million second *Chandra* observation of the Cassiopeia A SNR, together with various supporting observations in other wavebands, represent a significant new body of data with which to explore the physics of core-collapse explosions. While the unusally high degree of presupernova mass loss makes Cas A unlikely to have been produced by a "normal" core-collapse event, it gives significant advantages in our analysis. First, the reverse shock has penetrated much further into the inner ejecta than would otherwise be the case, and we can hope to see in these ejecta traces of processes at work during the explosion itself. Second, the mass loss allows clumps of inner ejecta survive to be observed in the remnant today, rather than being mixed with other ejecta by reverse shocks during the explosion. This greatly simplifies the analysis of the X-ray spectra. Eventually, we hope to address the three-dimensional distribution of the ejecta, particularly Fe, and compare that distribution to the mass and momentum of the neutron

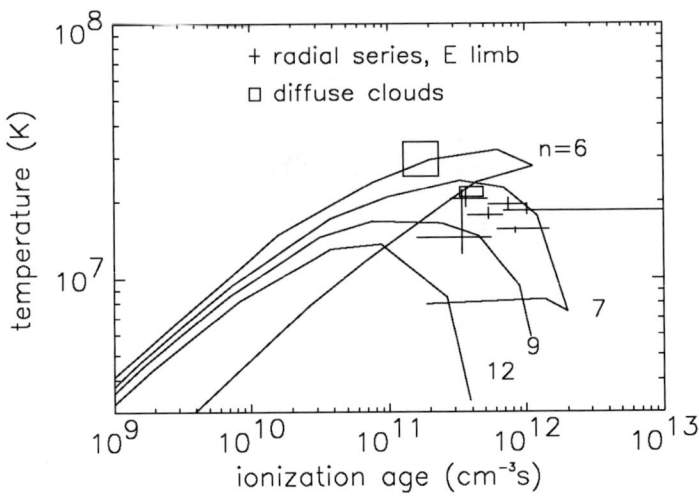

FIGURE 6. Locus of temperature and ionization age for Fe-rich knots from Hwang & Laming [20]. The lower branch for each model with outer envelope power law *n* corresponds to reverse shock propagation through this envelope, the upper branch to reverse shock propagation through the uniform density ejecta core. The place where these meet at highest $n_e t$ is the ejecta core-envelope boundary. For $n > 7$, this region is radiatively unstable. Note the absence of knots with high ionization ages above $\sim 10^{12}$ cm^{-3}s, consistent with the thermal instability of knots shocked at early times in the models.

star, but we expect a complete characterization of the ejecta emission to take several years of spectrum extraction, fitting and modeling.

This work was supported by the *Chandra* GO Program, the NASA LTSA Program, and by basic research funds of the Office of Naval Research.

REFERENCES

1. Bahcall, J. N. 1989, Neutrino Astrophysics, (Cambridge: Cambridge University Press)
2. Burrows, A., Livne, E., Dessart, L., Ott, D. C., & Murphy, J. 2006, ApJ, 640, 878
3. Scheck, L., Plewa, T., Janka, H.-Th., Kifonidis, K., & Müller, E. 2006, A&A, 457, 963
4. Kifonidis, K., Plewa, T., Scheck, L., Janka, H.-T., & Müller, E. 2006, A&A, 453, 661
5. Blondin, J. M., & Mezzacappa, A. 2006, ApJ, 642, 401
6. Blondin, J. M., Mezzacappa, A., & DeMarino, C. 2003, ApJ, 584, 971
7. Thorstensen, J. R., Fesen, R. A., & van den Bergh, S. 2001, AJ, 122, 297
8. Filippenko, A. V. 2001, Proc. of 11th Astrophysics Conference, "Young Supernova Remnants", eds. S. S. Holt & U. Hwang (Melville NY: AIP), page 40
9. Massey, P., DeGioia-Eastwood, K., & Waterhouse, E. 2001, AJ, 121, 1050
10. Laming, J. M., & Hwang, U. 2003, ApJ, 597, 347
11. Chevalier, R. A., & Oishi, J. 2003, ApJ, 593, L23
12. Fesen, R. A., Becker, R. H., & Goodrich, R. W. 1988, ApJ, 329, L89
13. Young, P. A., et al. 2006, ApJ, 640, 891
14. Truelove, J. K., & McKee, C. F. 1999, ApJS, 120, 299, erratum ApJS, 128, 403

15. DeLaney, T. A., & Rudnick, L. 2003, ApJ, 589, 818
16. Gotthelf, E. V., Koralesky, B., Rudnick, L., Jones, T. W., Hwang, U., & Petre, R. 2001, ApJ 552, L39
17. Willingale, R., Bleeker, J. A. M., van der Heyden, K. J., Kaastra, J. S., Vink, J. 2002, A&A, 381, 1039
18. Kifonidis, K., Plewa, T., Janka, H.-T., & Müller, E. 2000, ApJL, 531, L123
19. Kifonidis, K., Plewa, T., Janka, H.-T., & Müller, E. 2004, A&A, 408, 624
20. Hwang, U., & Laming, J. M. 2003, ApJ, 597, 362
21. Laming, J. M., & Grun, J. 2002, Phys. Rev. Lett., 89, 125002
22. Laming, J. M., Raymond, J. C., McLaughlin, B. M., & Blair, W. P. 1996, ApJ, 472, 267
23. Laming, J. M. 1998, ApJ, 499, 309
24. Laming, J. M. 2001a, ApJ, 546, 1149
25. Laming, J. M. 2001b, ApJ, 563, 828
26. Ghavamian, P., Raymond, J. C., Smith, R. C., & Hartigan, P. 2001, ApJ, 547, 995
27. Ghavamian, P., Winkler, P. F., Raymond, J. C., & Long, K. S. 2002, ApJ, 572, 888
28. Rakowski, C. E., Ghavamian, P., & Hughes, J.P. 2003, ApJ, 590, 846
29. Ghavamian, P., Laming, J. M., & Rakowski, C. E. 2006, ApJ Letters, in press, astro-ph/0611306
30. Khokhlov, A. M., Höflich, P. A., Oran, E. W., Wheeler, J. C., Wang, L., & Chtchelkanova, A. Yu 1999, ApJ, 524, 107L
31. Fryer, C. L., & Heger, A. 2000, ApJ, 541, 1033, Core-Collapse Simulations of Rotating Stars
32. Fryer, C. L., & Warren, M. S. 2004, ApJ, 601, 391, The Collapse of Rotating Massive Stars in Three Dimensions
33. Nagataki, S., Hashimoto, M., Sato, K., Yamada, S., & Mochizuki, Y. 1998, ApJ, 492, L45
34. Vink, J., Laming, J. M., Kaastra, J. S., Bleeker, J. A. M., Bloemen, H., & Oberlack, U. 2001, ApJ, 560, L79
35. Vink, J., & Laming, J. M. 2003, ApJ, 584, 758
36. Laming, J. M., Hwang, U., Radics, B., Lekli, G., & Takács, E. 2006, ApJ, 644, 260
37. Mazzali, P. A., et al. 2006, Nature, 442, 1018
38. Campana, S., et al. 2006, Nature, 442, 1008, The association of GRB 060218 with a supernova and the evolution of the shock wave
39. Cobb, B. E., Bailyn, C. D., van Dokkum, P. G., & Natarajan, P. 2006, ApJL, 645, L113, SN 2006aj and the Nature of Low-Luminosity Gamma-Ray Bursts
40. Matzner, C. D., & McKee, C. F. 1999, ApJ, 510, 379
41. Chevalier, R. A., & Liang, E. P. 1989,ApJ, 344, 332
42. Fesen, R. A., Hammell, M. C., Morse, J., Chevalier, R. A., Borkowski, K. J., Dopita, M. A., Gerardy, C. L., Lawrence, S. S., Raymond, J. C., & van den Bergh, S. 2006, ApJ, 636, 859
43. Hughes, J. P., Rakowski, C. E., Burrows, D. N., & Slane, P. O. 2000, ApJL, 528, L109
44. Hwang, U., Holt, S. S., & Petre, R. 2000 ApJ, 537, L119
45. Fesen, R. A. 2001, ApJS, 133, 161
46. Arnett, D. 1996, Supernovae and Nucleosynthesis, (Princeton: Princeton University Press)
47. Thielemann, F.-K., Nomoto, K., & Hashimoto, M. 1996, ApJ, 460, 408
48. The, L.-S., Clayton, D. D., Jin, L., & Meyer, B. S. 1998, ApJ, 504, 500
49. Renaud, M. et al. 2006, ApJ, 647, L41
50. Iyudin, A. F., et al. 1994, A&A, 284, L1
51. Iyudin, A. F., Diehl, R., Lichti, G. G., et al. 1997, ESA SP-382, 37
52. Rothschild, R. E., & Lingenfelter, R. E. 2003, ApJ, 537, 904
53. Clayton, D. D., Meyer, B. S., The, L.-S., & El-Eid, M. F. 2002, ApJ, 578, L83
54. Lodders, K. 2006, ApJL, 647, L37

The Role of Atomic Physics in Understanding Physical Processes in High Energy Astrophysics

Nancy S. Brickhouse

Harvard-Smithsonian Center for Astrophysics, 60 Garden St. MS 15, Cambridge, MA 02138

Abstract. X-ray grating spectra from *Chandra* and *XMM-Newton* have provided new insights into many of the physical processes present in astrophysical sources. For example, (i) shocks produced by magnetic accretion onto stellar surfaces cool as the material flows down, with density and temperature diagnostics providing tests of the accretion models; (ii) many active galactic nuclei (AGN) produce winds or outflows, detectable through X-ray absorption; (iii) active cool stars have coronal pressures several orders of magnitude larger than found on the Sun.

The diagnostics used to determine temperatures, densities, elemental abundances, ionization states, and opacities require extremely accurate atomic data. At the same time, we must have a fairly complete database in order to ensure that the diagnostics are not blended or otherwise compromised. The best spectra are from bright objects with long exposures (days), but the information contained allows us to infer the location(s) of the emitting and absorbing plasmas and understand the physical properties. We will give examples to illustrate the role of atomic physics in our analyses of such spectra and the quality of data required.

Keywords: galactic and intergalactic astronomy; solar corona; stellar systems; X-ray spectra
PACS: 32.30.Rj; 96.60.P-; 98.

INTRODUCTION

With the launches of *Chandra* and *XMM-Newton* nearly eight years ago, X-ray grating spectroscopy has brought new capabilities for studying the physical processes in astrophysical plasmas. With resolving powers $\lambda/\Delta\lambda \sim 500$ to 1000 [1, 2, 3], X-ray grating observations now provide line flux diagnostics of the physical conditions of the source plasma. Cosmically abundant elements from C to Ni produce absorption and emission lines in the 0.3 to 10 keV (1 to 40 Å) region. Emission lines from H-like and He-like ions are routinely observed with the gratings, as are lines from the Fe L-shell ions from Fe XVII to XXIV. In addition to these transitions, inner-shell lines are often observed in sources with a strong continuum superimposed with absorption features. For example, absorption attributed to oxygen free-bound transitions before we had obtained high resolution spectra from *Chandra* and *XMM-Newton*, is now understood to be inner-shell line absorption from Fe M-shell ions.

For the strongest lines from these ions the demand for accurate models is high: currently systematic uncertainties in the calculation of H-like and He-like atomic collision strengths and charge balances constitute the largest uncertainties in the interpretation of their line ratios in collisionally ionized astrophysical plasmas. For weaker diagnostic lines, blending must also be taken into account. Continuum levels may be incorrectly determined when weak lines are unaccounted for in the models. Even when weak lines are present in the models, their wavelengths may be of sufficient uncertainty to prevent

CP926, *Atomic Processes in Plasmas—15th International Conference on Atomic Processes in Plasmas*
edited by J. D. Gillaspy, J. J. Curry, and W. L. Wiese
2007 American Institute of Physics 978-0-7354-0436-6/07/$23.00

reliable assessment of blending. Of course, some of the most interesting diagnostic lines require the calculation of multiple processes, including direct excitation, cascades from directly excited upper levels, and cascades driven by recombination. The presence of inner-shell lines in both emission and absorption spectra broadens the charge state range available from the X-rays to include ions usually associated with ultraviolet transitions. While this is in principle an advantage, inner-shell atomic data have only recently been calculated in response to astrophysics needs [e.g. 4, 5]. Laboratory benchmarks, including wavelength measurements, are sorely needed.

New results from X-ray astrophysics illustrate the need for accurate spectral models. Unlike laboratory plasmas, astrophysical plasmas can only be understood by the photons they produce. X-ray photons are particularly precious: the *Chandra* grating spectrum of the AGN NGC 3783 (discussed below) has a 900 ksec exposure time (10 days exposure, or 2 weeks of satellite time on one source). The investment made to obtain such spectra must be supplemented by strong theoretical and experimental atomic physics efforts as well as dedicated spectral modeling infrastructure.

STRUCTURE OF THE ACCRETION COLUMN IN A CATACLYSMIC VARIABLE

A cataclysmic variable is a binary system consisting of a white dwarf which is accreting material from a late-type secondary star. In some systems the white dwarf has a strong magnetic field which can either directly channel the flow of material from the secondary or can channel the flow from the inner region of an accretion disk. EX Hydrae (EX Hya), a system believed to be of the latter type, is viewed nearly edge-on to the disk (Figure 1) and thus produces eclipses. By examining phase-dependent spectra, we can establish the geometry of the system.

The *Chandra* High Energy Transmission Grating (HETG) X-ray spectrum of EX Hya shows a wealth of emission lines superimposed on a strong bremsstrahlung continuum. The continuum emission has an electron temperature (T_e) of $\sim 200 \times 10^6$ K and is produced as the accreting material shocks above the surface of the white dwarf. From the shock to the surface, the gas cools and produces the observed line emission. Models of the shock and resulting cooling shock column predict the T_e and electron density (N_e) structure of the plasma as a function of height above the white dwarf surface [6]. The observed spectrum provides quantitative tests of those models.

N_e-sensitive line ratios of Fe XVII $\lambda 17.10/\lambda 17.05$ ([gnd $^1S_0 - 2p^5(^2P)3s\ ^3P_2$]/[gnd $^1S_0 - 2p^5(^2P)3s\ ^3P_1$]) and Fe XXII $\lambda 11.92/\lambda 11.77$ ([$2p\ ^2P_{3/2} - 3d\ ^2D_{5/2}$]/[gnd $2P_{1/2} - 3d\ ^2D_{3/2}$]) indicate that N_e is $\sim 10^{14}$ cm^{-3} at $5 - 10 \times 10^6$ K [7, 8]. The two diagnostics are currently consistent with each other, given all the uncertainties, and with general model predictions. Detailed tests of the models, however, require N_e measurements for a range of different ions, as the models predict a rapid increase in N_e near the surface of the white dwarf. The most popular N_e diagnostic line ratios from He-like ions (described in the section on stellar coronae) are potentially contaminated by photoexcitation from the black body spectrum of the white dwarf, and thus are not currently useful.

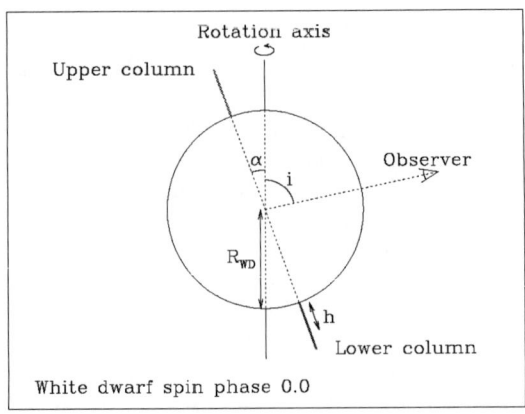

FIGURE 1. Schematic of EX Hya, showing the rotation axis, upper and lower accretion columns and the observer's view. The white dwarf has a spin period of 67 min, while the binary orbital period is 98 min). Courtesy of R. Hoogerwerf.

A new technique using line-based light curves to test the structure models has recently been demonstrated. Using spin-phased light curves of Mg XI lines (formed at $\sim$ 6×10^6 K), we have been able to locate the source of the emission of Mg XI relative to the surface of the white dwarf [9]. Most of the time only one of the shock columns is visible, as the white dwarf occults the other column; however, Mg XI is formed at a height such that both columns are visible for brief periods of time near zero phase. The Mg XI light curve thus shows two spike corresponding to these two phases. The time width of the spikes and the gap between them are determined by the height and latitude of the emission, given the known orientation of the system. Both density and height measurements provide tests of the structure model.

Chandra is observing EX Hya for 500 ksec in May 2007, an order of magnitude increase in the exposure time. The new data will allow us to map the T_e and N_e structure of the cooling shock column with numerous T_e, N_e, and line-based light curve diagnostics. Given T_e as a function of height above the white dwarf, contamination of the He-like N_e diagnostics by photoexcitation from the white dwarf's black body spectrum can be removed. In order for meaningful tests of the structure model to be carried out, the theoretical atomic data must be accurate enough that we can determine densities to within a factor of two or less. Laboratory benchmarks are needed to ensure this level of accuracy. In addition to testing the simple accretion structure model, additional physical effects may be observed, such as thermal instabilities, resonance scattering from the shock column, or turbulence.

LOCATION OF THE IONIZED ABSORBER IN ACTIVE GALACTIC NUCLEI (AGN)

In AGN the central continuum source associated with the massive black hole photoionizes the surrounding medium. A highly ionized outflow with velocity of order 1000 km/s is observed in about half of all Seyfert 1 type AGN and is believed to be ubiquitous in quasars [e.g. 10], though the wind is not always observable such as along sight lines obscured by the accretion disk. The role of this highly ionized outflow (also known as the "warm absorber") in enriching the surrounding galaxy or driving the galaxy's evolution is of great interest. A fundamental question is to determine the origin of this wind. Suggestions in the literature for its distance from the central source range over six orders of magnitude.

The 900 ksec *Chandra* HETG spectrum of the Seyfert 1 AGN NGC 3783 shows a powerlaw continuum spectrum with more than 100 absorption features from a range of charge states. The spectrum is well fit by two components with different ionization parameters[1] and column densities but the same velocity (determined by Doppler shifts of ~ 750 km/s) and pressure [11]. The higher ionization parameter component produces absorption from H- and He-like Si, Mg, Ne, and Fe L-shell ions and from O VIII, among others, while the lower ionization parameter component produces absorption by O VII and Fe M-shell ions from Fe XII and lower charge states. Similar models appear appropriate for other warm absorber spectra as well [12]. (A small number of forbidden lines are in emission in the spectrum, and may not be associated with the same material.) The high quality of this spectrum has stimulated a great deal of work not only on AGN wind modeling but also on the production of new atomic data, including calculations for inner-shell lines [e.g. 4, 13, 14], measurements of low T_e dielectronic recombination [15], and revision of the ionization balance models at low ionization parameter [16].

Given the relative success of these models, one can begin to address the key question of where the warm absorber is located. The ideal diagnostic is to measure the time-dependent response of the absorber to changes in the ionizing continuum [17]. Our group has recently used the *XMM-Newton* observation of NGC 4051 to set tight constraints on the warm absorber's location [18], taking advantage of *XMM-Newton*'s capability to measure simultaneously the high resolution grating spectrum as well as the high throughput, lower resolution CCD spectrum. First, we have fit the total grating spectrum, confirming that this source has two ionization components with similar pressure and velocity. We then assume that the column density and location of the wind do not change in response to changes in the ionizing flux. We finally divide the CCD spectrum into ~ 20 spectra, corresponding to different time bins, and fit each one, ensuring sufficient signal-to-noise in each of the spectra to obtain a good fit to the CCD data. Since in photoionization equilibrium, with fixed N_e and R, the ionization parameter is proportional to L_X, agreement with the prediction gives an upper limit to the equilibrium timescale and thus a lower limit on N_e. In most cases the change in the ionization parameter tracks well with the change expected based on the variation in the continuum

---

[1] The ionization parameter $\eta \equiv L_X/(R^2 N_e)$, where R is the distance to the central source.

flux. For a few cases, the high ionization parameter does not quite reach its expected equilibrium ionization value. In those cases, we have an upper limit on N_e. In both cases the limits on N_e break the degeneracy with R. It is interesting to note that even though the individual absorption lines cannot be resolved with the CCDs, the centroids of broad features such as the Fe L-shell or Fe M-shell unresolved transition arrays (UTA) are sensitive to the charge state and thus measureable at the low CCD resolution.

We have found that the absorber is located quite close to the black hole (0.5 light-day for the high ionization parameter and less than a few light-days for the lower ionization parameter). This puts the location close to the inner edge of the accretion disk. This wind is not energetic enough to influence the galaxy evolution unless further acceleration occurs. Whether higher velocities observed in optical spectra are related remains speculative.

CORONAL STRUCTURE IN COOL STARS

The X-ray emission from cool stars like the Sun primarily originates in the magnetically confined coronal plasma. Typical solar active regions have $N_e \sim 10^{10}$ cm^{-3} and $T_e \sim$ 2-3 $\times 10^6$ K with flares showing N_e up to 10^{13} cm^{-3} at $T_e > 10 \times 10^6$ K. By stellar standards the Sun is not considered an active star, since L_X can exceed solar values by several orders of magnitude, even under apparently quiescent conditions. One such active system is Capella, a binary composed of two evolved G stars. Capella is the brightest cool star in X-rays and is thus the primary in-flight grating calibration target for the X-ray satellites.

Capella is never observed to flare, and shows only small variability on long time scales, and thus the spectra serve as useful benchmarks of atomic data. (Stellar spectra are not a replacement for laboratory measurements under controlled conditions, but can point to inconsistencies in diagnostics that might need further investigation.) The broadband *Chandra* Low Energy Transmission Grating allows simultaneous measurement of the Fe XVIII EUV and X-ray lines at 100 and 14 Å, respectively. Even for the strongest resonance lines, there are discrepancies in the EUV/X-ray line ratios, suggesting atomic data problems [19].

As an another example, we consider the He-like Ne IX system. In He-like systems the ratio of the forbidden ($^1S_0 - 2s\ ^3S_1$) to intercombination ($^1S_0 - 2p\ ^3P_1$; $^1S_0 - 2p\ ^3P_2$) lines, known as the R-ratio, is sensitive to N_e. Using the Capella observations, we have shown that significant blending occurs in the Ne IX spectral region, especially important for the weaker intercombination line [23]. Most of the blends are from Fe XIX and other Fe ions. Astrophysical models for the spectra of these ions rely on public atomic databases such as ATOMDB[2] that incorporate the best wavelengths available for a complete set of lines [20]. With laboratory measurements [21] accurate to 5 to 10 mÅ and theoretical calculations [22] accurate to a few mÅ, the blending can be assessed and the full potential of the line ratios can be exploited (Figure 2).

Astrophysics benchmarks are also useful to identify potential problems with line

---

[2] http://cxc.harvard.edu/atomdb

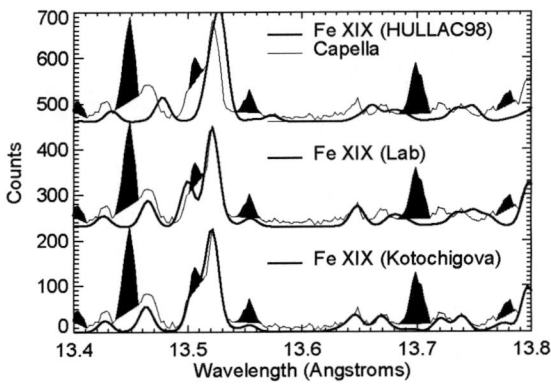

FIGURE 2. *Chandra* spectrum of Capella in the Ne IX spectral region [19], compared with models from ATOMDB [20]. *Top:* models with original calculated wavelengths (labeled HULLAC98). *Middle:* models with laboratory measurements [21]. *Bottom:* models with state-of-the-art theoretical calculations [22]. Solid shaded regions show the Ne IX lines. Hash-shaded regions show lines other than Fe XIX. The alignment of the new Fe XIX model wavelengths with the observation is remarkable.

fluxes. The He-like G-ratio, the ratio of the forbidden and intercombination lines described above to the resonance line ($^1S_0 - 2p\ ^1P_1$), is in principle a useful T_e diagnostic; however, the G-ratio models used in astrophysics are not consistent with other T_e diagnostics for many stars [23, 24]. Fully relativistic R-matrix calculations for Ne IX increase the G-ratio by 20 to 25% [25] over the older models, increasing the T_e derived from this ion for Capella by a factor of 2 and bringing it into better agreement with the charge state distribution (Figure 3). The new calculations are also in better agreement with laboratory experiments [26]. Additional work is still needed for the other important He-like ions.

Another system of great interest in astrophysics is that of Fe XVII, a system which poses considerable theoretical challenges [27, 28, 30, 29, 31]. Here we consider only one line ratio, 3C (gnd $^1S_0 - 2p^5(^2P)3d\ ^1P_1$) / 3D (gnd $^1S_0 - 2p^5(^2P)3d\ ^3D_1$), with lines at 15.01 and 15.26 Å, respectively. This ion is formed over a broad T_e range. In Capella, with peak emission at 6×10^6 K, the line ratio is in good agreement with laboratory measurements [32, 33]; however, the solar 3C/3D ratios from active regions formed near 2.5×10^6 K fall in a range well below the predicted value, suggesting that perhaps solar active regions are losing 3C photons to resonance scattering out of the field of view [34]. We find instead that the solar 3C/3D ratio can be brought into agreement with the optically thin plasma assumption by deblending the 3D line from an inner-shell Fe XVI line, using a previously unidentified neighboring Fe XVI line [35]. The solar ratios are thus explained as blending at the lower temperatures characteristic of solar active regions. Without experimental measurements [36] accurate deblending would not be possible.

Resonance scattering has also in the past been invoked to explain "fuzzy" EUV Fe XV images from the *TRACE* satellite. Instead we now know that Fe XV peaks at the

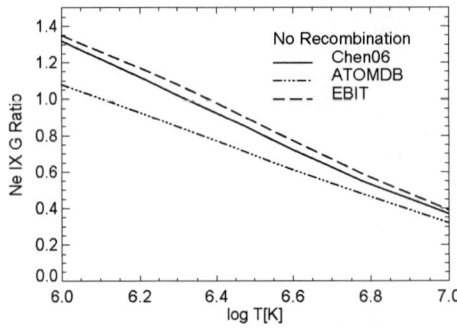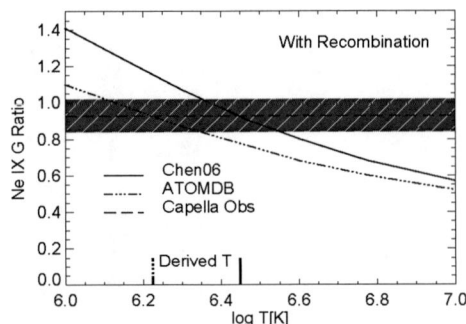

FIGURE 3. *Left:* The Ne IX G-ratio ($[i+f]/r$) shown from older atomic data (*dash-dotted*) in ATOMDB [20] and from new (*solid*) theory [25]. These models do not include recombination cascades for comparison with experimental measurements on an electron beam ion trap (EBIT) [26]. *Right:* Same as *Left* except that these models include estimates of recombination cascades as well as direct collisional excitation. The observed Capella ratio [23] is represented across the T_e range by the *dashed* line with errors in the shaded area. Note that the derived T_e increases by 0.2 dex with the new theory. While this temperature is still below the peak of the Capella emission measure distribution at 6×10^6 K (log T_e = 6.8), different abundances in the coronae of the two stars of the binary might account for the remaining difference [23].

T_e where we expect the most coronal loops and the images appear fuzzy because they are simply not spatially resolved. This case shows that even for imaging instruments, accurate and complete spectral models are needed [35].

Reasonable samples (a few dozen) of active stellar corona X-ray spectra are now obtained at high resolution. Such spectra help determine the general properties of coronal structure useful to understand stellar dynamos and heating processes.

CONCLUSIONS

We have discussed several spectral diagnostics used to determine the properties of magnetic structures such as accretion shocks and stellar coronae and to determine the physical sizes and locations of emission regions of various astrophysical systems. The best spectra are used to benchmark the models, identify potential problems with the atomic data, and motivate improvements. Although many (though not all) of the diagnostics have been proposed and utilized in solar and laboratory plasmas, the demands on the models from astrophysics are perhaps greatest. Without experimental knobs to control parameter space, we require both completeness and accuracy of the atomic data underlying spectral models in order to maximize the potential of the high resolution astrophysics data.

ACKNOWLEDGMENTS

The author acknowledges useful discussion and collaboration with Randall Smith, Ronnie Hoogerwerf, Chris Mauche, Yair Krongold, Fabrizio Nicastro, Martin Elvis, Smita Mathur, Joan Schmelz, Guo-Xin Chen, and Svetlana Kotochigova. This work was supported in part by the NASA grant NAS8-39073 to the Smithsonian Astrophysical Observatory for the Chandra X-ray Observatory Center and by the Smithsonian Institution's Atherton Seidell Grant Program.

REFERENCES

1. Canizares, C. R. et al. 2000, ApJ, 539, L41
2. Brinkman, A. C. et al. 2000, ApJ, 530, L111
3. Brinkman, A. C. et al. 2001, A&A, 365, L324
4. Behar, E., Sako, M., & Kahn, S. M. 2001, ApJ, 563, 497
5. Badnell, N. J. & Seaton, M. J. 2003, J. Phys. B, 36, 4367
6. Aizu, K. 1973, Prog. Theor. Phys., 49, 1184
7. Mauche, C. W., Liedahl, D. A., & Fournier, K. B. 2001, ApJ, 560, 992
8. Mauche, C. W., Liedahl, D. A., & Fournier, K. B. 2003, ApJ, 588, L101
9. Hoogerwerf, R., Brickhouse, N. S., & Mauche, C. W. 2006, 643, L45
10. Elvis, M. 2000, ApJ, 545, 63
11. Krongold, Y. et al. 2003, ApJ, 597, 832
12. Krongold, Y. et al. 2005, ApJ, 620, 165
13. Pradhan, A. K., Chen, G. X., Delahaye, F., Nahar, S. N., & Oelgoetz, J. 2003, MNRAS, 341, 1268
14. Gu, M. F., Holczer, T., Behar, E., & Kahn, S. M. 2006, ApJ, 641, 1227
15. Schmidt et al. 2006, ApJ, 641, L157
16. Netzer, H. 2004, ApJ, 604, 551
17. Nicastro, F., Fiore, F., Perola, G. C., & Elvis, M. 1999, ApJ, 512, 136
18. Krongold, Y. et al. 2007, ApJ, in press
19. Desai, P. et al. 2005, ApJ, 625, L59
20. Smith, R. K., Brickhouse, N. S., Liedahl, D. A., & Raymond, J. C. 2001, ApJ, 556, L91
21. Brown, G. V., Beiersdorfer, P., Liedahl, D. A., Widmann, K., Kahn, S. M., & Clothiaux, E. J. 2002, ApJS, 140, 589
22. Kotochigova, S. A., Kirby, K. P., Brickhouse, N. S., Mohr, P. J., & Tupitsyn, I. I. 2005, in X-Ray Diagnostics of Astrophysical Plasmas, AIP Conf Proc, 774, 161
23. Ness, J.-U., Brickhouse, N. S., Drake, J. J., & Huenemoerder, D. P. 2003, ApJ, 598, 1277
24. Testa, P., Drake, J. J., Peres, G., & DeLuca, E. E. 2004, ApJ, 609, L79
25. Chen, G. X., Smith, R. K., Kirby, K., Brickhouse, N. S., & Wargelin 2006, Phys. Rev. A, 74, 042709
26. Wargelin, B. J. 1993, PhD Thesis, LLNL, unpublished
27. Chen, G. X., & Pradhan, A. K. 2002, Phys. Rev. Letters, 89, 3202
28. Doron, R. & Behar, E. 2002, ApJ, 574, 518
29. Loch, S. D., Pindzola, M. S., Ballance, C. P., & Griffin, D. C. 2006, J. Phys. B, 39, 85
30. Gu, M. F. 2003, ApJ, 582, 1241
31. Chen, G. X. 2006, preprint
32. Brown, G. V., Beiersdorfer, P., Kahn, S. M., Liedahl, D. A., & Widmann, K. 1998, ApJ, 502, 1015
33. Laming, J. M. et al. 2001, ApJ, 545, L161
34. Schmelz, J. T., Saba, J. L. R., Chauvin, J. C., & Strong, K. T. 1997, ApJ, 477, 509
35. Brickhouse, N. S., & Schmelz, J. T. 2006, ApJ, 636, L53
36. Brown, G. V., Beiersdorfer, P., Chen, H., Chen, M. H., & Reed, K. J. 2001, ApJ, 557, L75

Radiative Shocks And Plasma Jets As Laboratory Astrophysics Experiments

M. Koenig[1], B. Loupias[1], T. Vinci[1], N. Ozaki[1], A. Benuzzi-Mounaix[1], M. Rabec le Goahec[1], E. Falize[2], S. Bouquet[2], C. Michaut[3], G. Herpe[3], P. Baroso[3], W. Nazarov[4], Y. Aglitskiy[5], A. YA. Faenov[6], T. Pikuz[6], C. Courtois[2], N.C. Woolsey[7], C.D. Gregory[7], J. Howe[7], A. Schiavi[8], S. Atzeni[8].

[1] Laboratoire pour l'Utilisation des Lasers Intenses, UMR7605, CNRS – CEA - Université Paris VI - Ecole Polytechnique,, 91128 Palaiseau Cedex, FRANCE
[2] CEA/DIF/ □BP 12 □91680 Bruyères-le-Châtel, France.
[3] Laboratoire de l'Univers et de ses Théories, UMR8102, Observatoire de Paris, 92195 Meudon, France.
[4] University of St Andrews, School of Chemistry, Purdie Building, North Haugh, St Andrews, UK
[5] Science Applications International Corporation, McLean, Virginia 22102, USA
[6] Joint Institute for High temperatures of RAS, Izhorskaya 13/19, Moscow, 125412, Russia
[7] Department of Physics, University of York, Heslington, York, YO10 5DD, United Kingdom
[8] Dipartimento di Energetica, Universita' di Roma "La Sapienza" and CNISM, Italy

Abstract. Dedicated laboratory astrophysics experiments have been developed at LULI in the last few years. First, a high velocity (70 km/s) radiative shock has been generated in a xenon filled gas cell. We observed a clear radiative precursor, measure the shock temperature time evolution in the xenon. Results show the importance of 2D radiative losses. Second, we developed specific targets designs in order to generate high Mach number plasma jets. The two schemes tested are presented and discussed.

Keywords: Laboratory astrophysics-radiative shocks- Plasma jets.
PACS: 52.35.Tc, 52.50.Jm

INTRODUCTION

During the last 30 years, high power laser facilities have been developed worldwide to pursue ICF (Inertial Confinement Fusion) research. These experimental facilities enable the formation of matter in states of extreme conditions of from example temperature, density, plasma flow. One of the challenges is to properly diagnose these HED (High Energy Density) regimes allowing models and simulations to be tested. A significant advantage of laboratory experiments, is the freedom to change initial conditions and determine how these changes influence the time evolution of data. Laboratory astrophysics[1] is a rapidly developing field of research that started about 15 years ago, the use of laser facilities to simulate specific aspects of astrophysical plasmas was discussed about 40 years ago.

CP926, *Atomic Processes in Plasmas—15th International Conference on Atomic Processes in Plasmas*
edited by J. D. Gillaspy, J. J. Curry, and W. L. Wiese
© 2007 American Institute of Physics 978-0-7354-0436-6/07/$23.00

At the LULI laboratory, over the last few years, we have actively developed laboratory astrophysics experiments to study radiative shocks (RS) and plasma jets. In this paper, some of the results obtained on the new LULI2000 facility in the area of RS will be presented. These are compared to 2D radiative simulations highlighting the importance of radial radiative losses.

Plasma jets are often observed for Young Stellar Objects (YSO), particular features are the collapse to form collimated plasmas, and the bipolar outflows (jets) which end as emission lobes (bow shocks). The objective of our experiments was to try new target designs in order to generate plasma jets and to characterize their time evolution. In this paper we discuss these designs and show some of the results obtained.

RADIATIVE SHOCK

Radiative shocks (RS) combine both hydrodynamics and radiation physics in a non-trivial way. The major effects occurring during the evolution of various astrophysical objects are driven by these two processes and, although each of them has been widely studied, their coupling through the RS is still a source of numerous questions[2, 3]. It is a difficult regime to explore in laboratory systems, as experiments require facilities capable of heating moderate-density material to relatively high temperatures. The experiment must be large enough extent to allow the radiation transport to affect the hydrodynamic properties of the system. From the experimental point of view, recent laboratory studies were performed using high power and/or energy lasers[4-6]. However, these RS experiments were limited by the lack of simultaneous measurements on the same shot. This prevented the characterization of the radiative shock in a consistent way. The main goal of experiments reported here is achieving RS conditions while performing simultaneous measurements of fundamental parameters, and then varying the initial conditions.

In order to define hydrodynamic parameters, such as shock velocity, and achieve RS conditions, one can compare the radiative flux and pressure to the corresponding thermal quantities, as developed in the literature[2, 3, 7, 8]. This implies comparing radiative flux in a full black body approximation $F_r = \sigma T^4$ (σ is the Stefan-Boltzman constant, T the downstream temperature) to the thermal flux $F_{th} = \rho_0 c_v T u_s$ (ρ_0 is the initial density, c_v the heat capacity at constant volume, u_s the shock velocity). At the same time, one can evaluate the radiative pressure $P_r = 4\sigma T^4/(3c)$ (c is the speed of light) with respect to the thermal pressure P_{th}. The various regimes that one can encounter in the RS situation has been well described by Drake[8].

It is shown in a previous publication[7] that RS conditions can be more easily achieved by propagating a shock in a low density medium having a high atomic number. This is the reason we use xenon as our RS medium. For xenon we are required to define the velocity threshold at which the experiment enters the radiative regime. This velocity threshold is about 30 km/s for a 1 mg/cm^3 gas density. The shock is launched by the laser irradiation of a solid target, which acts as a piston. The laser, piston, and xenon gas cell are optimised to reach the highest radiative regime according to our laser characteristics.

In this article, an intermediate regime of the RS was studied. Here the radiation flux is greater than thermal flux, whilst the radiative pressure remains well below material pressure. We performed a series of experiments in which both the shock and precursor velocities and temperatures, the electron density in the precursor and the radial expansion were measured. Measurement of these quantities show significant signatures of the importance of radiation in the behavior and the properties displayed by the radiative shock. This is especially true of the downstream, here radiation losses can reduce the temperature. The result is the downstream material is compressed between 50 and 100 times the density in the upstream. The key parameters (velocities, temperature, and electron density) provide a consistent set of measurements in order to make progress in the understanding of the physical phenomena taking place in the RS. These measurements strongly constrain the radiative hydrodynamic model (opacity, radiative transfer), and test theory. To generate a radiative shock, we performed a series of experiment on the laser facilities at the *Laboratoire d'Utilisation des Lasers Intenses* (LULI) laboratory. These were performed on the new LULI2000 laser facility, two beams providing $E_{2\omega} \approx 1$ kJ with a square pulse ranging from 1 to 5 ns. We used Phase Zone Plates in order to eliminate large scale spatial intensity modulations and obtain a flat-top intensity profile in the focal spot[9] (≈ 300 μm), corresponding to a maximum laser intensity $I_L \leq 10^{14}$ W/cm^2.

The two laser beams were focused on an ablator-pusher foil to generate a shock into a gas cell. The pusher design was optimized using 1D radiative hydrodynamic simulations (MULTI[10]) according to the laser characteristics. The pusher was made of three layers (20 μm CH-3 μm Ti- 30 μm CH) . The gas cell was filled with Xenon gas at two initial pressures 0.1 and 0.2 atmosphere. As mentioned in previous experiments[11], a transverse probe beam was used in order to determine both the precursor density and the shock velocity in Xenon. In this new experiment, shock velocities in Xenon were ranging from 50 to 80 km/s depending on the laser intensity.

Among the methods existing for the determination of the temperature[12, 13], we adopted an absolute photon counting technique. The thermal emission was observed using a narrow range (using a 10 nm wide blue filter centred at 450 nm). In order to determine the shock temperature from the CCD counts, we performed an absolute spectral and energy calibration of the optical system using a white lamp (OL 455 from Optronic Lab. Inc.). Then the streak camera itself was energy calibrated with a low energy (< 200 μJ) pulsed laser at 532 nm. We also measured its spectral response in the visible domain (from 400 to 800 nm) and compared it to the known one for a S20 tube. This enables us to link the total energy emitted by the shock to the number of CCD counts recorded. For the CH-Ti-CH pusher target, the last CH layer can be considered as a witness plate; from this we were able to measure the shock velocity from VISAR[14] and self-emission at the same time.

We observe in Fig. 1 the first emission with a low intensity followed by a much brighter emission. This is due first to the shock self-emission in CH then emission from the shock front in xenon. The measured intensity in the CH layer is associated to the reflectivity given from the VISAR data. This gives a value for the shocked plastic layer temperature. According to our self-emission diagnostic calibration, thetemperature is T≈ 6.3 eV (the emissivity is taken into account through the factor 1-R given by the VISAR diagnostic, where R is the CH shock front reflectivity). We have previously verified the validity of SESAME table for plastic[15]. The measurement of the shock velocity in CH coupled with the SESAME EOS allow us to determine all the thermodynamic parameters (ρ, T,...) for this layer (T ≈ 7 eV). This is then compared to the deduced temperature from the self-emission. Therefore, we have an *in situ* calibration for the temperature, with a 10% error, thanks to the last CH layer acting as a witness plate.

In addition, there is *in situ* cross checking of the shock temperature measurements in xenon. In Figure 2, we observe that the temperature reaches a maximum (16 eV) 2ns after the breakout in xenon and then decays slowly to 14 eV at 8ns. The maximum temperature is well reproduced by both 1D (---) and 2D (— —) simulations performed with the DUED code[16, 17];

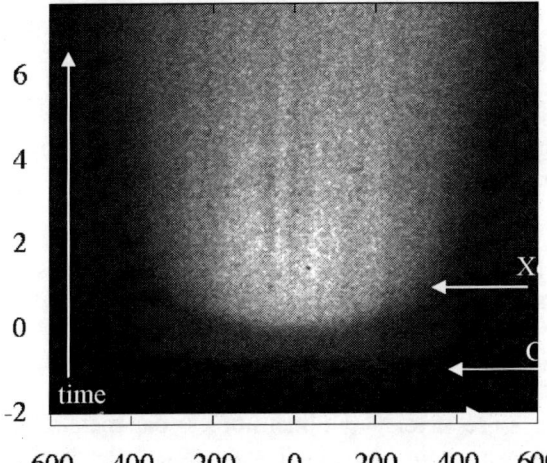

FIGURE 1. Shock temperature time evolution,(——) experimental data ,(---) 1D & (— —) 2D simulations.

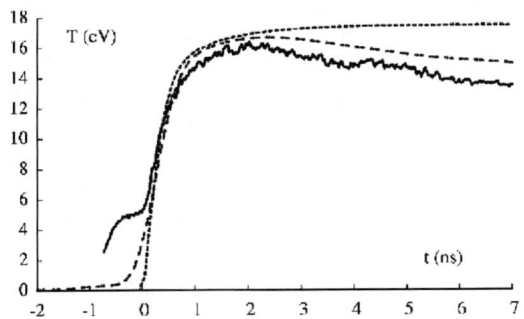

FIGURE 2. Shock temperature time evolution,(——) experimental data ,(---) 1D & (— —) 2D simulations.

However in the 1D case the temperature is almost constant after 1.8ns whereas experimental results show that the temperature decreases. This trend is well reproduced by the 2D simulations. In order to check what is responsible for these energy losses (radiation or thermal conduction), we performed dedicated 1D and 2D pure hydrodynamic simulations (without radiation) which the same temperature trend. This demonstrates clearly that the temperature decrease is due to radial radiative losses, i.e. the radiative diffusion coefficient plays and important role in the time evolution of the shock characteristics.

PLASMA JETS

Astrophysical jets[18] are one of the most fascinating astronomical objects observed. The variety of jets is enormous. For example the spatial scale extends from 10^{17}cm for young stellar object jets to 10^{24}cm for jets from quasars. Broadly, astrophysical jets are associated in astronomical systems exhibiting accretion disks. That's why jets are observed associated with young stellar object, supernovae, pulsars, active galactic nuclei and so forth. Many phenomena such as radiation, magnetic fields, and the interaction with the interstellar medium affect the plasma jets flow. To try to understand these phenomena, astrophysical jets have been the subject of elaborate studies in theory, simulation and observations. Nevertheless, they still raise unanswered questions such as the high collimation, the effect of radiation loses, knot formation and the interaction with the ambient medium leading to bow shocks. Jet experiments offer an additional approach to improve our understanding of the physical processes that occur during the jet propagation. In laboratory experiments the ability to create and to diagnose the plasma jet flow allow the freedom to perform measurements and to investigate several physical models and validate jet simulations and theory. According to scaling laws one can compare astrophysical objects and laboratory experiments[19] through the use of relevant dimensionless parameters. For the jet we use: the jet-to-ambient density ratio, the Mach number and the cooling parameter to describe the jets global properties. In addition, for the ideal fluid description of the plasma, the localization parameter (ion-ion mean free path/jet width) must be small, whilst the Reynolds number and Peclet number must be large.

The objective of our experiments is to address some important issues related to jets and formation of Herbig Haro (HH) objetcs. Jets with HH objects are thought to be non-relativistic collimated bipolar outflows emerging from accretion disks during the star formation process. Detailed observations of evolutionary change show the motion and velocity of the knots as they propagate along the plasma jets. The typical velocity of jets are few hundreds km/s corresponding to Mach numbers ranging from 10 to 50. In our experiment, we aimed to produce high velocity collimated plasma jet flows using high energy, intense lasers to study jet formation and propagation in vacuum. Dedicated diagnostics to obtain, in a single shot, the main jet parameters (velocity, radius time evolution, internal Mach number, …) were implemented.

Experimental set up

The experiment was performed on the LULI2000 (Laboratoire pour l'Utilisation des Lasers Intenses) laser facility. We used the two kJ beams, converted to the second harmonic focused on a 500 μm focal spot diameter with a 1.5 ns pulse duration giving a maximum laser intensity of $I_L \sim 10^{14}$ W/cm2.

In order to achieve high velocity plasma flows, we used two different target schemes: a foam filled cone and a V foil target (fig. 3). The first target is composed of a three layer pusher and a cone filled with foam (figure 3(a)). We used two different foam densities, 50 mg/cc, 100 mg/cc and sometimes a high-Z material doped foam to study radiative effects on the jet. For some of the targets, we added a 100μm length washer at the cone end to increase the plasma jet collimation. The second target scheme (V foil) is composed by two aluminum foils at a 45° angle. The rear side plasma flow collides on the symmetry axis generating a hot high velocity jet. One of the key points about those designs is to create the jet at the rear side of the target this is to allow the propagation of the jet in to an ambient medium without any interference with laser beam as some previous front side experiments[20].

Our main goal, was to measure the jet characteristics and explore the best target design to form a collimated plasma jet and match the scaling. Transverse to the jet propagation direction we used shadowgraph and interferometer (VISAR) of a probe beam at 532nm (fig. 4). The transverse VISAR allowed us to determine the jet tip evolution at the critical density $N_C = 4.10^{21} cm^{-3}$. This gives data on the jet velocity V_{jet} and the electronic density gradient evolution with a VISAR sensitivity of 5.4×10^{19} cm^{-3}.ns^{-1} per fringe. We used two Gate Optical Imagers (GOI) with a 120ps integration to record the 2D jet shape evolution.

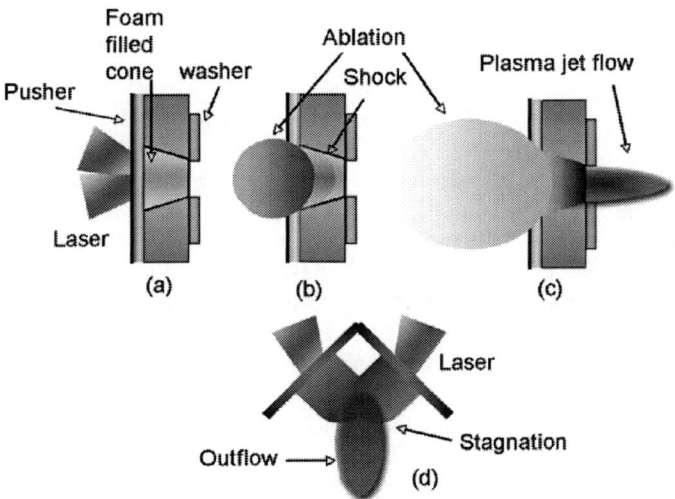

FIGURE 3. The two target schemes and the plasma jet generation. (a) The laser is focused onto the pusher (0.16μm Al/20μmCH/3μm Ti). (b) It drives a shock into the foam. (c) The shock propagates trough the cone launches a high velocity plasma jet flow. (d) V-foil target.

The rear side diagnostic was Self-emission Optical Pyrometry (SOP), the same diagnostic used for the radiative shock described above. With this we measured the jet radial evolution (radius R_{jet} and velocity V_R) and the temperature of the jet (T_{jet}). In order to infer the core density of the jet, we performed a transverse 2D X-ray monochromatic shadowgraph of the jet at 5.4 keV (He-alpha V line) using a spherical bent crystal[21].

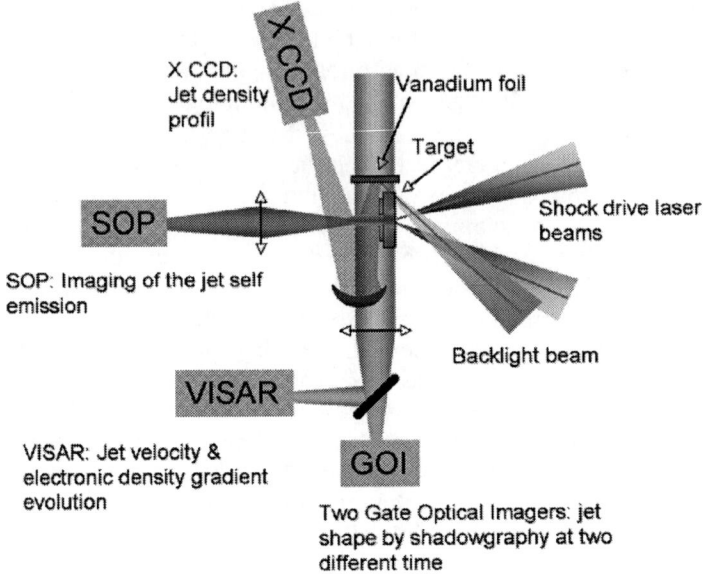

FIGURE 4. Schematic view of the experimental setup.

Results and discussion

For the foam cone target, the main measured jet parameters are given in Table 1. We were able to generate high velocity jets (100-145 km.s^{-1}) with an internal jet Mach number relevant to those of the astrophysical jets. We notice that with the washer the jet velocity is increased while the radial expansion is slightly reduced. This point clearly shows that plasma flow guiding in the washer is important and results in enhanced collimation of the jet. This collimation was observed both in the shadowgraphy and self-emission diagnostics (figure 5). These diagnostics show that the radial evolution of the plasma flow is slower than its propagation velocity. With the washer (table 1) we obtain a faster but cooler jet.

TABLE 1. Experimental jet parameters.

Foam density (mg/cc) (* with washer)	T_{Jet} (eV)	V_R (km.s^{-1})	V_{Jet} (km.s^{-1})	L_{Jet}/R_{Jet}	M
50	3.2	52	122	2.4	9.2
50*	3	45	145	3.2	11
100	2	33	100	2.3	9.4
100*	1.9	32	110	2.7	10.7

From our measurements, we find that during the jet propagation the jet evolves an envelope structure. At the early time we find the jet radius as inferred by shadowgraphy and self-emission are similar, yet at later time these differs. This is suggestive of a polytropic expansion of the plasma jet. The emitting jet is surrounded by a cooler and less dense plasma envelop. In order to confirm this hypothesis, we performed dedicated shots using the X-ray shadowgraphy diagnostic.

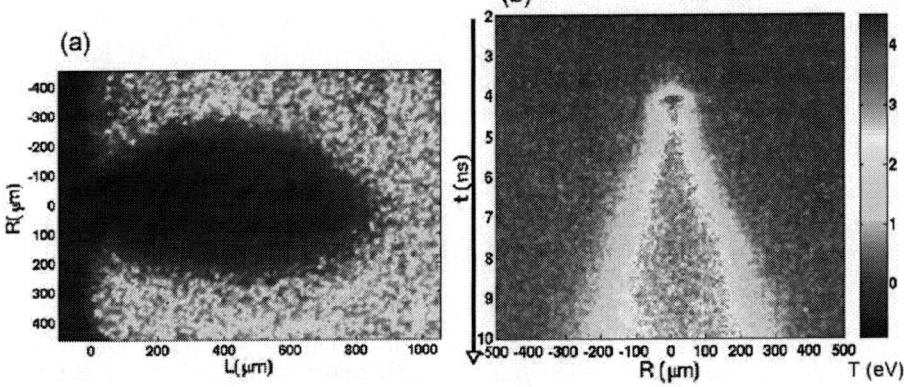

FIGURE 5. (a) Transverse shadowgraph of 50mg/cc foam density cone target with washer taken 9.5 ns after the main laser. (b) Rear side self-emission of the jet. The time origin is the main laser time.

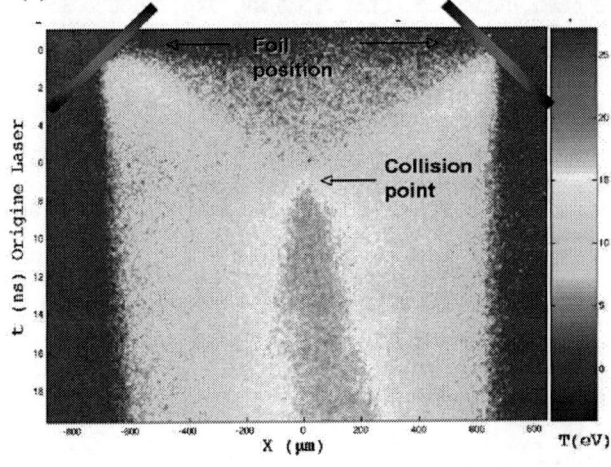

FIGURE 6. Rear side SOP for 5µm aluminum V foil target at 45°.

Radiography indicates that the radius of the dense region in the jet is smaller than that extracted from self-emission. Moreover, we were able to infer from this diagnostic the jet core density from the X-ray absorption. ($N_e \sim 2.5 \times 10^{22} cm^{-3}$) whereas the jet envelope (fig. 5a) has a density around $N_c/10$ (roughly $10^{20} cm^{-3}$). The present laboratory jet parameters are representative of astrophysical YSO jets. Measured Mach number ($M \sim 10$), the Peclet number ($Pe \sim 2.8 \times 10^5$) and the Reynolds number ($Re \sim 1 \times 10^8$) but differ for the density contrast ($\eta \gg 1$) and cooling parameter ($\chi \gg 1$). In future experiments, new target parameters will be investigated to obtain all the parameters relevant to astrophysical jets.

In figure 6, we present the rear side self-emission of the V foil target. We can observe the plasma expanding from the foils and converging in the collision point. We were able to access to a temperature of about 20eV. We measured a velocity, in the transverse diagnostic, of 100km/s. As a consequent the useful dimensionless parameters for the astrophysical scaling are: Mach number ~ 5, $Re \sim 1e5$, $Pe \sim 30$. In comparison with the jet from the foam cone target we were able, using the plasma collision from the foils, to reach higher temperature with a high flow velocity. Nevertheless from geometric consideration the plasma outflow from a V foil target is not axially symmetric. In one direction the jet radius seems to stay collimated, with a jet size $\sim 450\mu m$ (figure 6). Whereas, in the other direction, it is the plasma expansion from the original foils which determine the jet dimension.

CONCLUSIONS

In this paper, we describe some laboratory astrophysics experiments performed at the LULI laboratory. In the RS regime, we observed strong radiative precursor of a shock propagating in a xenon gas cell. Evidence of 2D radiative losses have been pointed. More recent jet like experiments have been pursued exploring two targets design. Both of them led to high velocity plasma flow (~150 km/s). Most of the relevant jet parameters have been measured. The upcoming experiments will be done with an ambient gas medium.

REFERENCES

[1]B. A. Remington, et al., Reviews of Modern Physics 78, 755 (2006).
[2]Y. B. Zeldovich and Y. P. Raizer, Physics of shock waves and high temperature hydrodynamic phenomena (Academic Press, New York, 1967).
[3]D. Mihalas and B. W. Mihalas, Foundations of radiation hydrodynamics (Dover Publications, N.Y., 2000).
[4]J. C. Bozier, et al., Phys. Rev. Lett. 57, 1304 (1986).
[5]J. C. Bozier, et al., Astroph. J. Supp. 127, 253 (2000).
[6]P. A. Keiter, et al., Phys. Rev. Lett. 89, 165003 (2002).
[7]S. Bouquet, et al., Astroph. J. Supp. 127, 245 (2000).
[8]R. P. Drake, Astrophysics and Space Science 298, 49 (2005).
[9]M. Koenig, et al., Phys. Rev. E 50, R3314 (1994).
[10]R. Ramis, et al., Comp. Phys. Comm. 49, 475 (1988).
[11]S. Bouquet, et al., Phys. Rev. Lett. 92, 225001 (2004).
[12]G. W. Collins, et al., Phys. Rev. Lett. 87, 165504 (2001).
[13]T. Hall, et al., Phys. Rev. E 55, R6356 (1997).

[14]P. M. Celliers, et al., Applied Phys. Lett. 73, 1320 (1998).
[15]M. Koenig, et al., Physics of Plasmas 10, 3026 (2003).
[16]S. Atzeni, Comput. Phys. Comm. 43, 107 (1986).
[17]S. Atzeni, et al., Comput. Phys. Commun. 169, 153 (2005).
[18]B. Reipurth and J. Bally, Annual Review of Astronomy and Astrophysics 39, 403 (2001).
[19]D. D. Ryutov, et al., Physics of Plasmas 8, 1804 (2001).
[20]D. R. Farley, et al., Phys. Rev. Lett. 83, 1982 (1999).
[21]T. Pikuz, et al., Laser and Particle Beams 22, 289 (2004).

Line spectra and profiles for ultracool substellar objects

Christine M. S. Johnas and Peter H. Hauschildt

Hamburger Sternwarte, Gojenbergweg 112, 21029 Hamburg, Germany

Abstract. The pressures in the line forming regions of cool stellar and substellar objects increase dramatically with lower effective temperatures. This causes strong pressure broadening of the few remaining atomic lines, with damping wings more than $0.5\,\mu m$ wide, dominating the emitted spectrum. Therefore, there is an essential need for reasonably accurate line profiles for these lines under high-pressure conditions. We show the results of model atmosphere calculations using detailed line profiles for a number of alkali resonance lines and discuss the need for additional and improved line profile for stellar and planetary atmosphere simulations.

Keywords: line profiles, stellar atmospheres, modeling, radiation transport
PACS: 97.10.Ex,95.30.Jx,95.30.Ky,32.70.Jz

THEORETICAL BACKGROUND

In the atmospheres of brown dwarfs, pressure broadening is prevalent. With a semi-classical approach, it is possible to compute the profile function of an alkali perturbed by H_2 or He. The alkali absorption line profiles are represented by two terms, one describing the line core with the impact approximation [1], [2] and the other describing the line wing with the one–perturber approximation [2]. For a detailed derivation see Allard and Kielkopf [3] and Allard et al. [4]. More detailed calculations of the perturbed alkali absorption doublets of Na I, Li I, K I, and Rb I have been supplied [5], [6]. Using the multi-purpose stellar atmosphere code PHOENIX, atmosphere models have been constructed for a range of effective temperatures and surface gravities typical for low-mass stars and brown dwarfs, tracing the behavior of the alkali absorption profiles under the low temperatures and high gas pressures prevalent in these atmospheres.

MODELS AND SYNTHETIC SPECTRA

When comparing typical synthetic spectra calculated with PHOENIX (AMES-Cond models, [7]) and varying approximations of the pressure broadened profiles of the alkalis, differences are noticeable.

The synthetic spectra calculated with the impact approximated van der Waals broadened profiles [7] underestimate observed line strengths out to $\sim 9000\,\text{Å}$ (for the K I lines), but produce too strong absorption at larger distances from the line core. The synthetic spectra calculated with the accurately calculated alkali profiles provide more opacity in the intermediate region, but then drop off rapidly as seen in the observations. The wings of the two most abundant alkalies, Na I and K I, form the pseudo continuum

CP926, *Atomic Processes in Plasmas—15th International Conference on Atomic Processes in Plasmas*
edited by J. D. Gillaspy, J. J. Curry, and W. L. Wiese

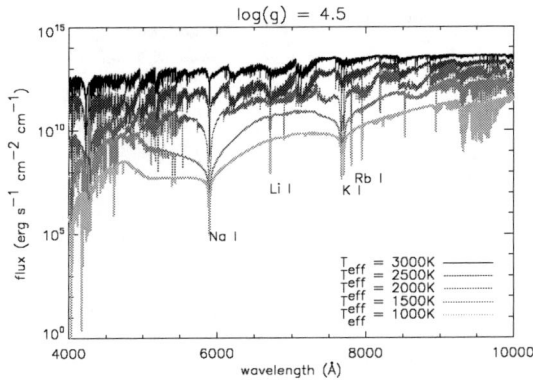

FIGURE 1. Effective temperature dependence of the synthetic spectra for T_{eff} from 1000 K to 3000 K.

of Li I and Rb I and thus also affect line formation analyses of these species.

EFFECTIVE TEMPERATURE DEPENDENCE OF THE LINE WINGS

Figure 1 shows the effective temperature dependence of the synthetic spectra for $T_{\text{eff}} = 1000$ K to 3000 K. Towards lower effective temperatures, the alkali absorption lines and especially the far wings of the most abundant alkalies increasingly dominate the spectrum. This behavior is contrary to what might be expected from the profiles introduced in [5], [6] and [8], in which the HWHM and the contributions of the far wing profile increase towards higher gas temperatures. However, analyzing the pressure relations in the different model atmospheres, one finds a decreasing gas pressure in the line forming regions towards higher effective temperatures. Consequently, the observed line broadening decreases with higher effective temperatures. This effect is augmented by increasing background opacity from molecular band absorption, mostly due to TiO, VO, CaH and FeH, which masks much of the alkali line wings at $T_{\text{eff}} = 3000$ K. Furthermore, when studying the relative abundances of the three most abundant species (H_2, He, and H I) in the synthetic spectra, as depicted in Fig. 2, a change of the relative abundances is visible. While He remains at its cosmic abundance in all parts of the atmosphere, almost all hydrogen is distributed between H_2 and H I according to dissociation equilibrium. At low effective temperatures, H_2 thus is the most abundant species, followed by He. Only in the innermost part of the atmosphere, H I forms in appreciable amounts. However, towards higher effective temperatures, the abundance of H I increases drastically. At $T_{\text{eff}} = 3000$ K, H I represents the most abundant species followed by H_2 from $\tau_{\text{std}} = 1.0$ on inwards. Keeping this in mind and taking a look at the flux contribution function $\mathscr{C}_F$, which approximately measures the extend of the line forming region, [9], [10], studies show, that the wings of spectral lines form farther inside the atmosphere than the cores of the line. For a more detailed study of the formation of the line wing

121

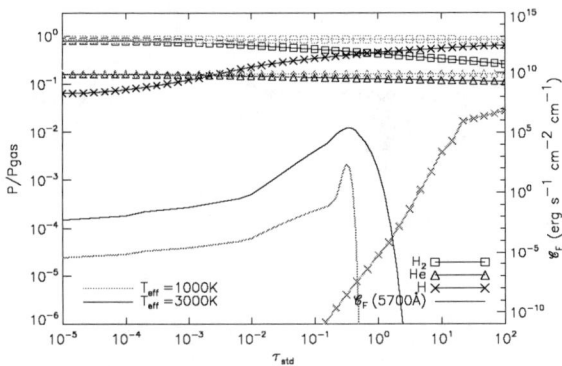

FIGURE 2. Relative abundance of H_2, He and H I in the atmosphere at $T_{eff} = 1000\,K$ and $3000\,K$ with $\log(g) = 4.5$ (left ordinate). Besides the flux contribution function at a wavelength of $5700\,\text{Å}$ is depicted (right ordinate).

of $Na\,I\,D_2$, we compare the run of $\mathscr{C}_F$ for $T_{eff} = 1000\,K$ and $3000\,K$ at a wavelength of $5700\,\text{Å}$ in Fig. 2. Its maximum marks the location of the formation of the line in the atmosphere. For $T_{eff} = 1000\,K$ the maximum of $\mathscr{C}_F$ is not located close to regions of significant H I partial pressure. This relation changes towards higher effective temperatures. At $T_{eff} = 3000\,K$, the locations in the atmosphere where H I becomes dominant and the line wing forms are very close. Hence, at higher effective temperatures, it will be more important to consider also collisional broadening due to H I in order to obtain a more accurate line profile.

DATA NEEDS FOR COOL STELLAR ASTROPHYSICS

We have shown that individual line profiles are very important when generating accurate models of cool stars and substellar objects. To synthesize the spectra properly, atomic data for each line are required. As shown in Figure 2, it is very important that all important perturbers (H I, H_2, and He) are included and that the temperature range covered is large (ideally, from 100 K to 4000 K) in order to have a smooth transition in the deep atmosphere. However, since the observations are noisy, the accuracy required for the line profiles is fortunately not very high.

ACKNOWLEDGMENTS

This work was supposed in part by the DFG via Graduiertenkolleg 1351. Some of the calculations presented here were performed at the Höchstleistungs Rechenzentrum Nord (HLRN), and at the National Energy Research Supercomputer Center (NERSC), supported by the U.S. DOE, and at the computer clusters of the Hamburger Sternwarte, supported by the DFG and the State of Hamburg. We thank all these institutions for a generous allocation of computer time.

REFERENCES

1. M. Baranger, *Phys. Rev.* **111**, 494–504 (1958).
2. A. Royer, *Phys. Rev. A* **3**, 2044–2049 (1971).
3. N. F. Allard, and J. Kielkopf, *Rev. Mod. Phys.* **54**, 1103–1182 (1982).
4. N. F. Allard, R. A., J. F. Kielkopf, and N. Feautrier, *Phys. Rev. A* **60**, 1021–1033 (1999).
5. N. F. Allard, F. Allard, and J. F. Kielkopf, *A&A* **440**, 1195–1201 (2005).
6. N. F. Allard, and F. Spiegelmann, *A&A, accepted* (2005).
7. F. Allard, P. H. Hauschildt, D. R. Alexander, A. Tamanai, and A. Schweitzer, *ApJ* **556**, 357–372 (2001).
8. N. F. Allard, F. Allard, C. M. S. Johnas, and J. F. Kielkopf, *A&A, submitted* (2006).
9. P. Magain, *A&A* **163**, 135–139 (1986).
10. B. Fuhrmeister, C. Short, and P. H. Hauschildt, *A&A, accepted* (2006).

Some Ionization and Recombination Data Needs for Cosmic Atomic Plasmas

Daniel Wolf Savin

Columbia Astrophysics Laboratory, MC 5247, 550 West 120th Street, New York, NY 10027, USA

Abstract.
Cosmic atomic plasmas can be divided into two broad classes: electron ionized and photoionized. Electron-ionized plasmas are formed in objects such as the sun and other stars, supernova remnants, galaxies, and the intercluster medium in clusters of galaxies. Photoionized plasmas are formed in objects such as planetary nebulae, H II regions, X-ray binaries, and active galactic nuclei. Understanding the spectral and thermal properties of these objects requires an accurate knowledge of the ionization level of the gas. This in turn depends on a reliable understanding of the underlying ionization and recombination processes which determine the ionization balance. Here we review some of the various atomic collision processes which determine the charge state distribution in a cosmic atomic plasma and briefly discuss some of the recent theoretical and experimental advances in generating the needed atomic data. We close by describing the relevant atomic data needs for the near future.

Keywords: elementary processes in cosmic atomic plasmas, electron impact ionization, electron-ion recombination, collisionally-ionized plasmas, photoionized plasmas, ionization balance calculations
PACS: 34.70.+e, 34.80.Dp, 34.80.Kw, 34.80.Lx, 52.20.-j, 52.20.Hv, 52.25.Jm, 52.72.+v, 95.30.Dr, 95.30.Ky, 95.55.-n, 98.38.Bn, 98.58.Bz

I. INTRODUCTION

Astrophysics drives much of the study into atomic processes in plamas. This is because spectral observations can be used to infer properties of the cosmos. In fact, in many cases spectroscopy is the only way to do this; and even in cases where imaging or timing observations are possible, plasma properties such as abundances, temperatures, and densities are only accessible through spectroscopy. To do all this this we must improve our knowledge of atomic physics in plasmas to the point where discrepancies between observations and models tell us something about the astrophysics of the observed sources and cannot be attributed to errors in the atomic data used in the models.

In this review we focus on the atomic data needed to calculate the ionization structure of cosmic plasmas. Specifically we will highlight some of the various ionization and recombination data needed for the different types of cosmic objects encountered. The rest of this paper is organized as follows: In Sec. II we discuss the two classes of atomic plasmas encountered in astrophysics: electron-ionized (sometimes called collisionally-ionized) and photoionized. This section also discusses the atomic data needed to calculate the ionization balance for each class of plasma. Section III reviews the current status of electron impact ionization (EII) data used in astrophysics. Recent advances in the dielectronic recombination (DR) data used by astrophysicists are reviewed in Sec. IV. Section V presents recent changes in fractional ionic abundance calculations

CP926, *Atomic Processes in Plasmas—15ᵗʰ International Conference on Atomic Processes in Plasmas*
edited by J. D. Gillaspy, J. J. Curry, and W. L. Wiese
© 2007 American Institute of Physics 978-0-7354-0436-6/07/$23.00

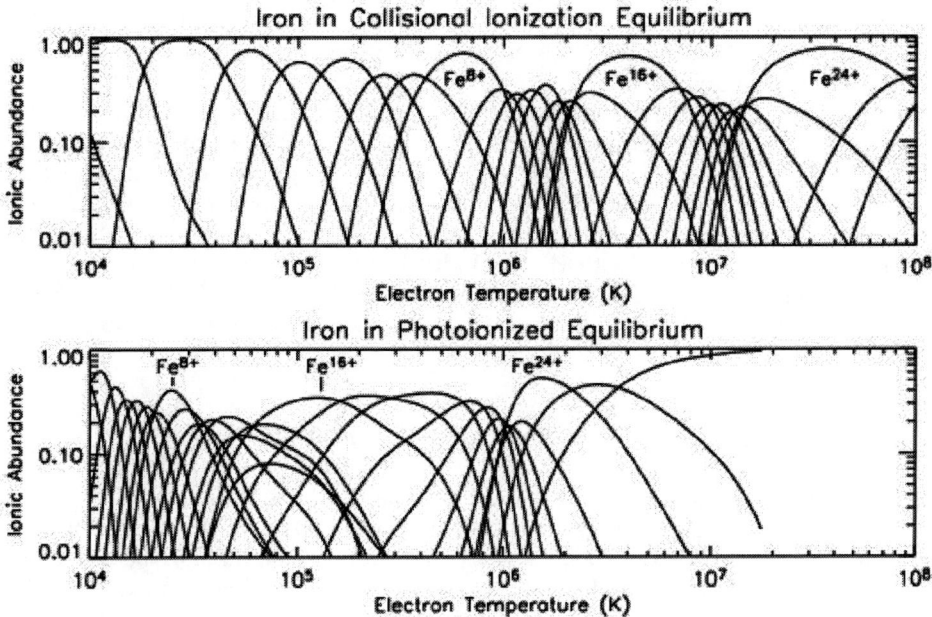

FIGURE 1. Calculated fractional ionic abundances for Fe. The top plot shows the results for collisional ionization equilibrium (CIE) [1]. The bottom plot gives the results for photoionization equilibrium (PIE) of gas with cosmic abundances illuminated by a 10 keV bremmstrahlung ionizing spectrum [4].

due to improvements in the DR and radiative recombination (RR) data available. Results are presented for collisional ionization equilibrium (CIE) and photoionization equilibrium (PIE). Lastly, Sec. VI lists remaining ionization and recombination data needed for modeling cosmic atomic plasmas.

II. TYPES OF COSMIC ATOMIC PLASMAS

Cosmic atomic plasmas can be divided into two broad classes: electron-ionized and photoionized. Electron-ionized plasmas are formed in objects such as the sun and other stars, supernova remnants, galaxies, and the intercluster medium in clusters of galaxies. Photoionized plasmas are formed in objects such as planetary nebulae, H II regions, X-ray binaries, and active galactic nuclei (AGNs).

In electron-ionized gas, the ionization structure at low densities is a function solely of the electron temperature. In CIE, a given ion forms at an electron temperature roughly half that of the ionization potential for that ion (cf., Fig 1). As a result the dominant electron-ion recombination mechanism for most ions is high temperature DR [1]. For temperatures of $\sim 10^4$ K, where significant abundances of H^{0+} and He^{0+} can exist, charge transfer can be an important ionization and recombination process involving systems up to four-times ionized [2]. In some cases EII leading to multiple electron loss can be important. This is particularly true for EII of inner shells leading to the Auger emission of multiple electrons [3].

For photoionized gas, the ionization structure is determined by a number of factors including the shape of the ionizing spectrum, the metallicity of the gas, additional heating and cooling mechanisms, etc. [4, 5]. In PIE, a given ion forms at an electron temperature roughly one-twentieth that of the ionization potential for that ion (cf., Fig 1). As a result, the dominant electron-ion recombination for most ions is low temperature DR [4, 5]. Charge transfer with H^{0+} and He^{0+} can be an important ionization and recombination mechanism for systems up to four-times ionized [6]. Photoionization of innershell electrons and and subsequent Auger emission can also be important [7, 8, 9, 10].

III. ELECTRON IMPACT IONIZATION (EII)

EII can occur through either direct ionization or indirect processes such as excitation-autoionization (EA). Direct ionization is a non-resonant process in which an incident electron transfers energy to a bound electron of the ion, promoting it to the continuum. EA occurs when an incident electron collisionally excites an ion to a state that then decays by autoionization rather than radiative decay and can enhance ionization cross sections by a factor of five or more over direct ionization for ions with certain electron configurations, such as those with one or two valence electrons.

There has been no significant updating of the EII database for astrophysics since 1992. Astrophysicists use the recommended data of Arnaud and Rothenflug [2] for H, He, C, N, O, Ne, Na, Mg, Al, Si, S, Ar, Ca, and Ni, and of Arnaud and Raymond [11] for Fe. The other important set of recommended EII data are those from The Queen's University of Belfast group [12, 13] for elements from H up to and including Ni. The recommended EII data from each group are based on basically the same theoretical and experimental results. Yet surprisingly the recommended rate coefficients are in poor agreement. Differences of up to a factor of 2-3 as has been noted by Kato et al. [14] and Savin [15].

There are a number of challenges to generating reliable EII data. On the experimental side, EII measurement are typically carried out using ion beams of unknown metastable fractions. The resulting lack of unambiguous benchmark measurements has complicated comparison with theory. On the theory side, the infinite number of final states has made the problem computationally challenging. These two difficulties are part of the reason why there has been no significant improvement to the EII database in the last $\sim$ 15 years.

Recently, there have been several promising advances in laboratory studies of EII. Loch et al. [16] measured EII and DR of C^{2+} and used the DR measurements in combination with DR theory to infer the ion beam metastable fraction. Here we refer to the charge of the initial system before ionization or recombination. Fogle et al. [17] have used a gas attenuation cell to directly measure the metastable fraction of their ion beams. With this approach they have carried out EII measurements for C^{2+}, N^{3+}, and O^{4+}. These laboratory advances have also been accompanied by theoretical advances which are discussed in more detail in Refs. [16, 17] and references therein.

IV. DIELECTRONIC RECOMBINATION (DR)

DR is a two-step recombination process that begins when a free electron approaches an ion, collisionally excites a bound electron of the ion and is simultaneously captured into a Rydberg level. The intermediate state, formed by simultaneous excitation and capture, may autoionize. The DR process is complete when the intermediate state emits a photon which reduces the total energy of the recombined ion to below its ionization limit.

Conservation of energy requires that for DR to go forward $E_k = \Delta E - E_b$. Here E_k is the kinetic energy of the incident electron, ΔE the excitation energy of the initially bound electron in the presence of the captured electron, and E_b the binding energy released when the incident electron is captured onto the excited ion. Because ΔE and E_b are quantized, DR is a resonant process. High temperature DR is important in CIE and typically occurs for $E_k \sim \Delta E$. Low temperature DR is important in PIE and typically occurs for $E_k \ll \Delta E$.

Reliable theoretical DR rate coefficients have been a challenge for many years [18], but recently there have been significant advances in DR theory. Modern calculations now exist for K-shell, L-shell, and Na-like ions for all elements from H through Zn [19, 20, 21]. K-shell ions have been extensively benchmarked using EBITs and storage rings and agreement between experiment and theory is typically $\sim 20\%$ [18].

For L-shell ions, there have also been major improvements in DR theory. For high temperature DR theory and experiment agree to within $\sim 35\%$ for the few systems studied. However, the situation is rather mixed for low temperature DR. Difficulties in atomic structure calculations limit the accuracy of DR resonance energies for center-of-mass collision energies $\lesssim 1 - 3$ eV (see Bryans et al. [1] for a more detailed discussion). Uncertainties in theoretical resonance energies can lead to factors of 2 or more errors in the calculated rate coefficients. Laboratory measurements are often the only reliable way to generate accurate low temperature DR rate coefficients (e.g., [23]). The bottom line is that for L-shell DR in general, little experimental work exists for B-, C-, N-, O-, F-, and Ne-like ions and that theoretical DR rate coefficients for temperatures $\lesssim 3 \times 10^4$ K must be used with caution whether they are for collisionally-ionized or photoionized plasmas.

For ions with an open M-shell, DR theory is even more challenging than for L-shell ions. For example, for Fe^{13+} and Fe^{14+} modern theory does a poor job of predicting DR resonance energies and strengths for collision energies $\lesssim 20$ eV [24, 25]. This is especially an issue for low temperature DR. For these cases, difference of up to a factor of two are found between experimentally-derived and theoretical DR rate coefficients [24]. A significant amount of experimental and theoretical work remains to generate reliable DR data for M-shell ions of all cosmically abundance elements.

Reliable DR data for Fe M-shell ions are particularly important for analyzing spectra from AGN warm absorbers. Recent *Chandra* and *XMM-Newton* satellite observations of these sources have detected absorption between 15-17 Å due to Fe M-shell ions [26, 27]. The blend of numerous absorption lines, due mainly to $2p - 3d$ photoexcitation, form an unresolved transition array (UTA). The shape, central wavelength, and equivalent width of the UTA can be used to diagnose the properties of these AGN warm absorbers [28].

AGN models, however, which fit absorption features from second and third row elements are unable to reproduce correctly the observed UTAs. The models appear to predict too high an ionization level for iron. Netzer et al. [29] attribute this to an underestimate of the low temperature DR rate coefficients for Fe M-shell ions. Subsequent modeling studies support this hypothesis [30, 31].

These errors in the DR data used should not be surprising. The data used are nearly 20 years old, calculated when computational power was a small fraction of what it is now. Additionally the M-shell DR data had been calculated with a focus on high temperature plasmas [11, 32]. Moreover, underestimated low temperature DR rate coefficients should have been expected based on experimental work predating the observations [33, 34].

Communication between the astrophysics and atomic physics communities is sometimes poor. But issues like this have strengthened some of the ties between the two fields. As a result, in the last couple of years there has been a concerted experimental and theoretical effort to generate reliable DR data for Fe M-shell ions [24, 25, 35, 36].

V. IONIZATION EQUILIBRIUM CALCULATIONS

Plasmas encountered in astrophysics are often optically-thin, low-density, dust-free, and in steady-state. Under these conditions, ionization and recombination balance one another and the fractional ionic abundances can readily be calculated. But the accuracy of these ionization equilibrium calculations is determined by the reliability of the ionzation and recombination used in the models. Hence the recent publication of state-of-the-art DR data [19, 20, 21] and RR data [37, 38] for K-shell, L-shell, and Na-like ions of all elements from H through to Zn represents a significant advance.

Bryans et al. [1] used these modern DR and RR data to generate new CIE results for H through Zn. The solid curve in Fig. 2 shows their fractional ionic abundances for Fe using the data of Badnell and colleagues [19, 38]. Similar results are found using the data of Gu [20, 21, 37]. The dashed curve in Fig. 2 show the results for Fe from Mazzotta et al. [39] which represented the state-of-the-art for CIE calculations until their work was superceded by that of Bryans et al.

Significant differences are found between the results of Bryans et al. [1] and those of Mazzotta et al. [39]. Peak fractional abundances for the various elements differ by up to 60%. At 0.1 fractional ionic abundances, differences between the two datasets of up to a factor of 5 are found. At 0.01, these differences can grow to a factor of 11. The temperature of peak formation shifts for some ions by up to 20%. Lastly, ions with particularly large differences include Mg, Al, Ca, Fe, Co, and Ni.

The new DR and RR data of Badnell and colleagues have also been incorporated into CLOUDY, a commonly used code for modeling photionized gas [4]. The solid curve in Fig. 3 shows the calculated PIE fractional abundances for Fe in a gas of cosmic abundances illuminated by a 10 keV bremmstrahlung ionizing spectrum. The dashed curve shows the same results but using the old recommended DR and RR data of Mazzotta et al. [39]. As can be readily seen, differences for some charge states can approach a factor of 50.

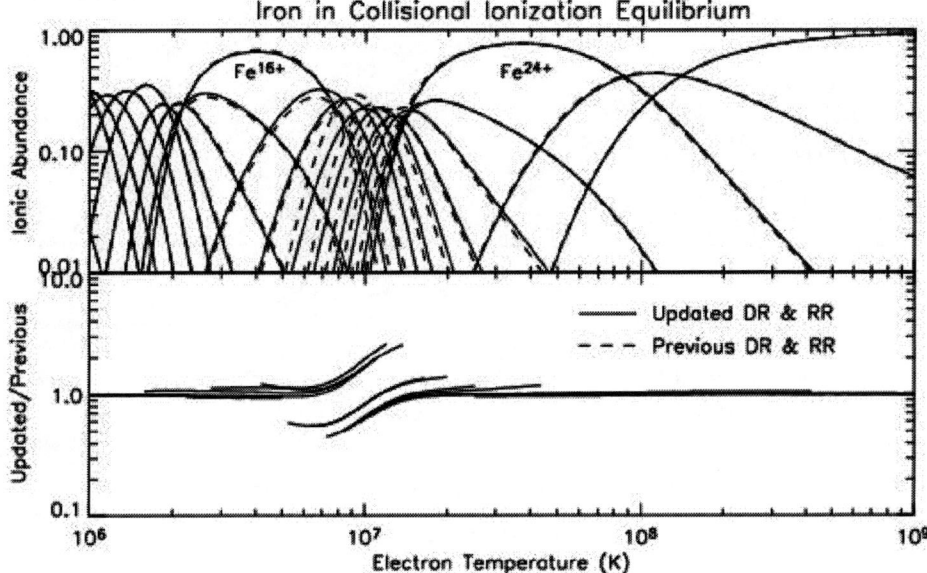

FIGURE 2. Calculated fractional ionic abundance for Fe in CIE [1]. The *solid curves* in the top plot shows the results using the state-of-the-art DR and RR data of Badnell et al. [19] and Badnell [38], respectively. The *dashed curves* use the previously recommended DR and RR data of Mazzotta et al. [39]. The bottom plot shows the ratio of the new to old fractional ionic abundances for abundances > 0.01. Little to no differences are seen in the ratio for charge states of Mg-like or less highly ionized as modern electron-ion recombination data for these isoelectronic sequences are only now becoming available (e.g., [35, 38]) and have yet to be incorporated into the models.

VI. REMAINING ATOMIC DATA NEEDS

The current oustanding data needs for calculating the ionization balance of collisionally-ionized and photoionized cosmic atomic plasmas can be summarized succinctly:

- EII data for K-, L-, and M-shell ions.
- DR for L- and M-shell ions.
- Improved atomic structure calculations for reliable low temperature DR data.
- RR for M-shell ions.
- Photoionization cross sections.
- Innershell ionization/Auger yields.
- CT data for H^{0+} and He^{0+} with systems ionized $\leq$ 4 times.

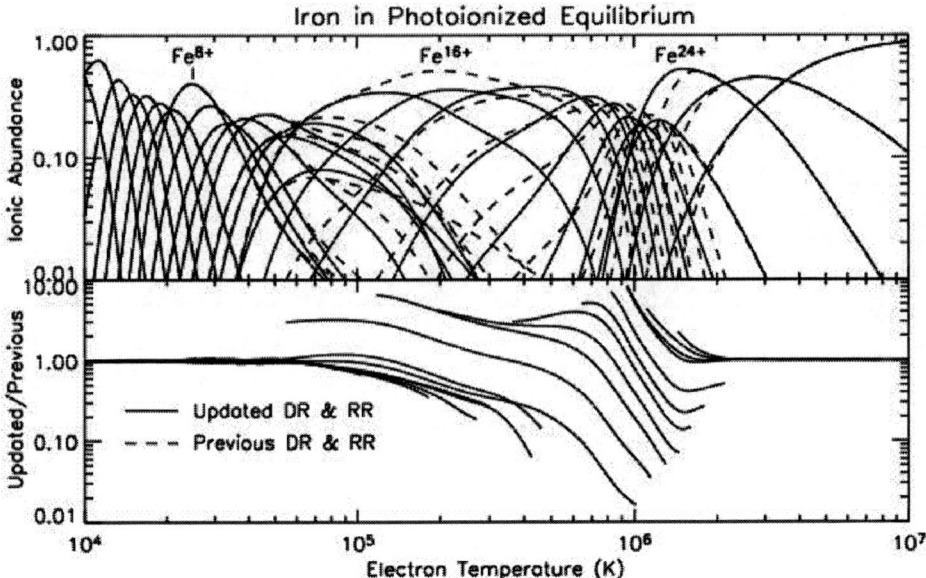

FIGURE 3. Calculated fractional ionic abundance for Fe in PIE as described in Fig. 1. The *solid curves* in the top plot shows the results using the state-of-the-art DR and RR data of Badnell et al. [19] and Badnell [38], respectively. The *dashed curves* use the previously recommended DR and RR data of Mazzotta et al. [39]. The bottom plot show the ratio of the new to old fractional ionic abundances for abundances > 0.01. Little to no differences are seen in the ratio for charge states Mg-like or less highly ionized as modern electron-ion recombination data for these isoelectronic sequences are only now becoming available (e.g., [35, 38]) and have yet to be incorporated into the models.

Here we have focused on the first three of these points. But significant improvements are still needed for much of the ionization and recombination data used in astrophysics. We propose that these data be generated with an aim for an accuracy of $\lesssim 35\%$ to match that of the best experimental and theoretical work.

Note added in manuscript.–After this paper was submitted, a re-evaluation and updating of the EII data base was published by Dere [40]. We plan to incorporate these new EII data into our CIE model [1] and will present the results elsewhere.

ACKNOWLEDGMENTS

The author thanks N. R. Badnell and M. Bannister for stimulating discussions, the anonymous referee for helpful comments, and P. Bryans and W. Mitthumsiri for making the figures. This work was funded in part by the NASA Astronomy and Physics Research and Analysis Program, the NASA Solar and Heliospheric Physics Supporting Research and Technology Program, and the NSF Astronomy and Astrophysics Research Program.

REFERENCES

1. P. Bryans, N. R. Badnell, T. W. Gorczyca, J. M. Laming, W. Mitthumsiri, and D. W. Savin, *Astrophys. J. Suppl. Ser.* **167**, 343–356 (2006).
2. M. Arnaud and R. Rothenflug, *Astron. Astrophys. Suppl. Ser.* **60**, 425–457 (1985).
3. J. S. Kaastra and R. Mewe, *Astron. Astrophys. Suppl. Ser.* **97**, 443–482 (1993).
4. G. J. Ferland, K. T. Korista, D. A. Verner, J. W. Ferguson, J. B. Kingdon, and E. M. Verner, *Publ. Astron. Soc. Pacific* **110**, 761-778 (1998); http://www.nublado.org/.
5. T. Kallman and M. Bautista, *Astrophys. J. Suppl. Ser.* **133**, 221–253 (2001).
6. J. B. Kingdon and G. J. Ferland, *Astrophys. J. Suppl. Ser.* **106**, 205–211 (1996).
7. M. A. Bautist, C. Mendoza, T. R. Kallman, and P. Palmeri *Astron. Astrophys.* **403**, 339–355 (2003).
8. P. Palmeri, C. Mendoza, T. R. Kallman, and M. A. Bautista *Astron. Astrophys.* **403**, 1175–1184 (2003).
9. P. Palmeri, C. Mendoza, T. R. Kallman, M. A. Bautista, and M. Meléndez, *Astron. Astrophys.* **410**, 359–364 (2003).
10. C. Mendoza, T. R. Kallman, M. A. Bautista, and P. Palmeri, *Astron. Astrophys.* **414**, 377–388 (2004).
11. M. Arnaud and J. Raymond, *Astrophys. J.* **398**, 394–406 (1992).
12. K. L. Bell, H. B. Gilbody, J. G. Hughes, A. E. Kingston, and F. J. Smith *J. Chem. Phys. Ref. Data* **12** 891–916 (1983).
13. M. A. Lennon, K. L. Bell, H. B. Gilbody, J. G. Hughes, A. E. Kingston, M. J. Murray, and F. J. Smith, *J. Chem. Phys. Ref. Data* **17**, 1285–1363 (1988).
14. T. Kato, K. Masai, and M. Arnaud, "Comparison of Ionization Rate Coefficients of Ions from Hydrogen through Nickel," *National Institute for Fusion Science Report, Nagoya, Japan, NIFS-DATA-14* (1991).
15. D. W. Savin, "Ionization and Recombination with Electrons: Laboratory Measurements and Observational Consequences," in *X-Ray Diagnostics of Astrophysical Plasmas: Theory, Experiment, and Observations*, edited by R. K. Smith, AIP Conference Proceedings Volume 774, American Institute of Physics, Melville, New York, 2005, pp. 297–303.
16. S. D. Loch, M. Witthoeft, M. S. Pindzola, I. Bray, D. V. Fursa, M. Fogle, R. Schuch, P. Glans, C. P. Ballance, and D. C. Griffin, *Phys. Rev. A* **71**, 012716 (2005).
17. M. Fogle, et al., in preparation.
18. D. W. Savin and J. M. Laming, *Astrophys. J.* **566**, 1166–1177 (2002).
19. N. R. Badnell, M. G. O'Mullane, H. P. Summers, Z. Altun, M. A. Bautista, J. Colgan, T. W. Gorczyca, D. M. Mitnik, M. S. Pindzola, O. Zatsarinny, *Astron. Astrophys.* **406**, 1151–1165 (2003); http://amdpp.phys.strath.ac.uk/tamoc/DR/
20. M. F. Gu, *Astrophys. J.* **590**, 1131–1140 (2003).
21. M. F.Gu, *Astrophy. J. Suppl. Ser.* **153**, 389–393 (2004).
22. D. W. Savin, et al. *Astrophys. J.* **576**, 1098–1107 (2002).
23. S. Schippers, M. Schmitt, C. Brandau, S. Kieslich, A. Müller, and A. Wolf, *Astron. Astrophys.* **421**, 1185–1191 (2004).
24. N. R. Badnell, *J. Phys. B* **39**, 4825–4852 (2006).
25. D. V. Lukić, M. Schnell, D. W. Savin, C. Brandau, E. W. Schmidt, S. Böhm, A. Müller, S. Schippers, M. Lestinsky, F. Sprenger, A. Wolf, Z. Altun, and N. R. Badnell, *Astrophys. J.*, submitted (arXiv:0704.0905).
26. M. Sako, S. M. Kahn, F. Paerels, and D. A. Liedahl *Astrophys. J.* **542**, 684–691 (2000).
27. S. Kaspi, et al., *Astrophys. J.* **574**, 643–662 (2002).
28. E. Behar, M. Sako, and S. M. Kahn, *Astrophys. J.* **563**, 497–504, (2001).
29. H. Netzer, et al., *Astrophys. J.* **599**, 933–948 (2003).
30. H. Netzer, *Astrophys. J.* **604**, 551–555 (2004).
31. S. B. Kraemer, G. J. Ferland, and J. R. Gabel *Astrophys. J.* **064** 556–561 (2004).
32. V. L. Jacobs, J. Davis, P. C. Kepple, and M. Blaha *Astrophys. J.* **211**, 605–616 (1977).
33. J. Linkemann, et al., *Nucl. Instrum. Methods B* **98**, 154–157 (1995).

34. A. Müller, *Int. J. Mass Spectrom.* **192**, 9–22 (1999).
35. E. W. Schmidt, S. Schippers, A. Müller, M. Lestinsky, F. Sprenger, M. Grieser, R. Repnow, A. Wolf, C. Brandau, D. Lukić, M. Schnell, and D. W. Savin, *Astrophys. J. Lett.* **641**, L157–L160 (2006).
36. N. R. Badnell, *Astrophys. J.* **651**, L73–L76 (2006).
37. M. F. Gu *Astrophys. J.* **589** 1085–1088 (2003).
38. N. R. Badnell, *Astrophys. J. Suppl. Ser.* **167**, 334–342 (2006); http://amdpp.phys.strath.ac.uk/tamoc/RR/
39. P. Mazzotta, G. Mazzitelli, S. Colafrancesco, and N. Vittorio, *Astron. Astrophys. Suppl. Ser.* **133**, 403–409 (1998).
40. K. P. Dere, *Astron. Astrophys.* **466**, 771–792 (2007).

SMALL LASER PLASMAS

Compact High Repetition Rate Soft X-Ray Lasers: A Doorway To High Intensity Coherent Soft X-Ray Science On A Table-Top

J. J. Rocca, Y. Wang, B. Luther, M. Berrill, M. Larotonda, D. Alessi, V.N. Shlyaptsev*, E. Granados, C.S. Menoni

NSF ERC for Extreme Ultraviolet Science and Technology
Department of Electrical and Computer Engineering and Department of Physics
Colorado State University, Fort Collins, CO
* University of California Davis

Abstract. We discuss recent advances in high repetition rate table-top soft x-ray lasers that allow the generation of laser beams in the 25-100 eV photon energy region with peak spectral brightness that surpasses by several orders of magnitude that of undulators in third generation synchrotrons, enabling new applications. These advances include the demonstration of 5 Hz repetition rate table-top soft x-ray lasers that operate in the gain-saturation regime to produce intense beams at wavelengths ranging from 13.2 to 32.6 nm, and the observation of lasing at wavelengths down to 10.9 nm. The results were obtained by collisional electron impact excitation of highly ionized atoms in dense plasmas efficiently heated with picosecond optical laser pulses of only 1 J energy. Further improvement in the brightness of these compact sources was obtained seeding the soft x-ray amplifiers with high harmonic pulses. We have demonstrated the saturated amplification of high harmonic seed pulses in a dense transient collisional soft x-ray laser amplifier medium created by heating a solid titanium target. Amplification of the seed pulses in the 32.6 nm line of Ne-like Ti generated laser pulses of sub-picosecond duration that were measured to approach full spatial coherence. The scheme is scalable to produce extremely bright lasers at very short wavelength with full temporal and spatial coherence. These new compact lasers are allowing the implementation of a variety of table-top experiments with intense soft x-ray laser light.

Keywords: Soft x-ray lasers , coherent soft x-ray light, seeded soft x-ray lasers
PACS: 42.55.Vc, 52.38.Ph

HIGH REPETITION RATE LASERS WITH WAVELENGTHS DOWN TO 10.9 NM

There is significant interest in the development of compact high average power soft x-ray lasers for a variety of science and technology applications. Fast discharge excitation of capillary plasmas has produced extremely compact saturated lasers emitting at 46.9 nm in Ne-like Ar at repetition rates up to 10 Hz [1]. Laser-driven optical field ionization lasers have produced saturated operation at 41.8 and 32.8 nm in Pd-like Xe and in Ni-like Kr respectively [2]. However, until recently gain-saturated laser operation at shorter wavelengths have required picosecond optical laser pump pulses with an energy of ~ 7 J, which limited their repetition rate to one shot every

CP926, *Atomic Processes in Plasmas—15th International Conference on Atomic Processes in Plasmas*
edited by J. D. Gillaspy, J. J. Curry, and W. L. Wiese
© 2007 American Institute of Physics 978-0-7354-0436-6/07/$23.00

several minutes [3,4]. Recently it was shown that the energy required to pump collisional soft x-ray lasers can be significantly decreased by directing the short pulse pump beam onto the target at a grazing angle, which increases the absorption efficiency of the pump pulse into the gain region of the plasma [5-9]. We have extended these results to shorter wavelengths, demonstrating for the first time high repetition rate saturated laser operation (5 Hz) of a table-top laser at wavelengths as short as 13.2 nm in Ni-like Cd [10], a wavelength that is within of the bandwidth of the Mo-Si multilayer coatings for EUV lithography. We also demonstrated the isoelectronic scaling of this high repetition rate laser scheme to shorter wavelengths with the observation of amplification at wavelengths of 11.4 and 10.9 nm in Ni-like Sb and Ni-like Te [9], respectively. The results were obtained heating a pre-created plasma with a 6-8 ps laser pulse of only 1 J energy impinging at a grazing incidence angle of 23 degrees. The experiments were conducted using a table-top 5 Hz repetition rate 800 nm Ti:Sapphire pump laser system, consisting of a mode-locked oscillator and three stages of chirped pulse amplification. The soft x-ray laser amplifier consisted of a line focus plasma of up to 4 mm in length generated by exciting polished slab targets with a sequence of an early prepulse of 120 ps duration and 10-15 mJ energy, followed after about 5 ns by a main prepulse of the same duration and ~350 mJ energy, which in turn was followed after a variable delay (typically 100 ps to a few 100s ps) by a 8 ps duration, ~ 1 J energy heating pulse that rapidly heats the plasma, creating a transient population inversion and gain in transitions of Ni-like ions. Overlapping line foci of 30 μm × 4.1 mm FWHM were generated for the pre-pulse beam and short pulse pump beams that impinged onto the target at near normal incidence and at a selected grazing incidence angle of 23 degrees, respectively. The on-axis plasma emission was spectrally resolved using a variably spaced spherical grating and recorded with a back-illuminated CCD detector.

Figure 1 shows on-axis spectra corresponding to 4 mm long targets of Ru (Z=44), Pd (Z=46), Ag (Z=47), Cd (Z=48), Sn (Z=50), Sb (Z=51) and Te (Z=52) that are dominated by the $3d^94d^1S_0 - 3d^94p^1P_1$ laser line of the corresponding Ni-like ions at wavelengths between 16.9 nm and 10.9 nm respectively. Gain measurements were performed on Ni-like Cd plasmas by monitoring the variation of the 13.2 nm laser line intensity as a function of target length. The results are shown in Fig. 2. The laser line intensity is observed to increase exponentially with an small signal gain coefficient of 69 cm^{-1}, until it rolls off into saturation. A fit of the data yields a gain-length product of g×l =17.6, value for which collisionally excited soft x-ray laser systems are normally in the gain-saturated regime. The laser pulse energy was estimated from the counts on the CCD taking into account the quantum efficiency of the detector and the losses. The energy of the most intense 13.2 nm shots obtained with a 4 mm target are estimated to exceed 430 nJ. The intensity of the Ni-like Cd laser is estimated to be about 1.6×10^{10} W/cm^2, a value that exceeds the computed saturation intensity of 0.6-0.8×10^{10} W/cm^2 for this laser line at the plasma conditions of this experiment.

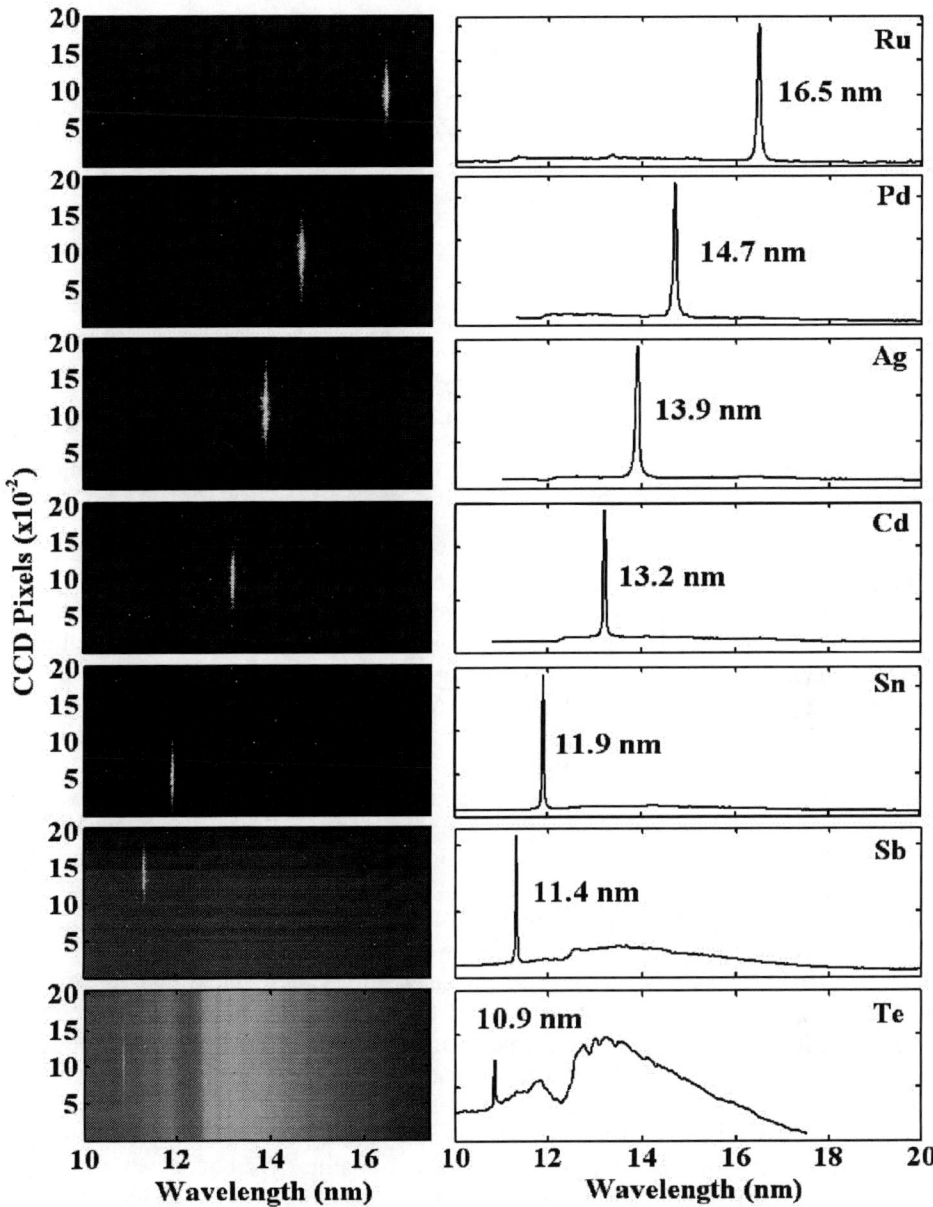

FIGURE 1. Single shot on-axis spectra of 4 mm line focus plasmas showing lasing in the $4d^1S_0$- $4p^1P_1$ transition of the Ni-like ions at wavelengths ranging from 16.5 to 10.9 nm. The laser lines completely dominate the spectrum except in the case of Ni-like Te. Lasers down to 13.2 nm are saturated.

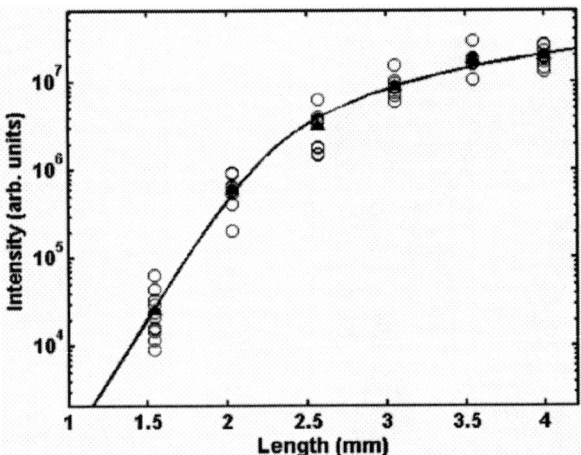

FIGURE 2. Intensity versus target length for the 13.2 nm line of Ni-like Cd. The solid triangles represent the average of the points shown. The fit corresponds to a small signal gain of 69 cm^{-1} and a gain-length product of 17.6.

Figure 3 illustrates continuous 5Hz repetition rate operation of the Ni-like 13.9 nm Ag laser for 250 shots. The data was obtained moving the target at a velocity of 0.2 mm/s. Lasing is observed in all shots, with an intensity variation characterized by a standard deviation of ~ 30% of the mean. The 13.9 nm laser average power approaches 2 µW. Measurements for Ni-like Ag at a grazing incidence pumping angle of 23 degrees for the optimum delay of 300 ps have yielded further increases in the laser output intensity, generating pulses with an energy up to 0.85 µJ. The shortest wavelength line for which amplification to gain-saturation levels was achieved was the 11.9 nm line of Ni-like Sn. The variation of the 11.9 nm laser line intensity as a function of plasma column length was measured using a main pre-pulse energy of ~ 350 mJ and a short pulse energy of ~ 1 J at a grazing incidence angle of 23 degrees. For an optimum delay of 125 ps the gain coefficient was measured to be 50 cm^{-1}, and a gain length product of g×l=14.3, value for which the onset of gain saturation effects was observed. The most intense 11.9 nm laser pulses were measured to have an energy of ~ 230 nJ. To demonstrate lasing in Ni-like Sb and Ni-like Te we reduced the FWHM length of the short pulse line focus to 3.5 mm, to increase the irradiation intensity up to 1.2×10^{14} W/cm^2. While the main focus of the work reported herein is the high repetition rate generation of gain-saturated soft x-ray laser radiation at wavelength shorter than 16.5 nm, we also re-visited the generation of 18.9 nm laser light from Ni-like Mo ions by grazing incidence pumping. We conducted a new Ni-like Mo laser experiment at a grazing incidence pumping angle of 20 degrees using 1 J of short pulse (~8 ps duration) excitation and the calibrated filters.

A series of shots made using a 4 mm long Mo slab target yielded an average 18.9 nm laser pulse energy of 1 µJ, with the most intense pulses reaching 1.4 µJ. Saturated laser operation at 5 Hz repetition rate with gain-length products of about 20 was also obtained for transition of Ne-like Ti and V with wavelength near 30 nm [11], with resulting average powers of about 1.5 to 2.5 µW.

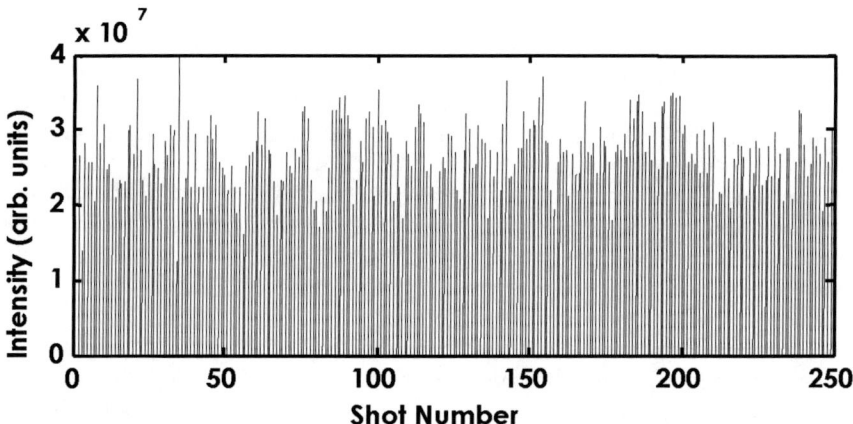

FIGURE 3. Shot-to-shot variation of the intensity of the 13.9 nm Ni-like Ag laser line at 5 Hz repetition rate. The data was obtained using a continuously rotating solid target.

DEMONSTRATION OF HIGH BRIGHTNESS SOFT X-RAY BEAM BY HIGH HARMONIC SEEDING OF A SOLID TARGET TRANSIENT COLLISIONAL AMPLIFIER

The output beam characteristics and peak spectral brightness of these lasers can be greatly improved by seeding of the amplifiers with high harmonic pulses. The high coherence, short pulse duration, low divergence, and defined polarization of the seed pulse can be imprinted in the amplifier output. In addition, the seeding can shorten the soft x-ray laser output pulse width to values approaching the limit set by the amplifier linewidth. This short pulse duration combined with the high energy stored in this type of soft x-ray amplifiers can result in table-top sources of unprecedented brightness. Saturated amplification of the 25th harmonic of a Ti:Sa laser was recently demonstrated in a 32.8 nm Ni-like Kr optical field ionization (OFI) soft x-ray laser amplifier [12]. However, the relatively low plasma density at which optimum lasing occurs in these OFI lasers results in a relatively low saturation intensity, that will ultimately limit the maximum brightness. The high electron density of soft x-ray laser amplifiers based on solid targets results in a higher saturation intensity and a broader laser linewidth, opening a route to higher soft x-ray laser pulse intensities and shorter pulsewidths. An early experiment performed in a Ne-like Ga plasma amplifier pumped by overlapping three 200 J optical laser beams onto a solid gallium target demonstrated the amplification of the harmonic seed, but only by a factor of about $3\times$ [13]. More recently, a Ne-like Mn soft x-ray laser amplifier was reported to amplify a high harmonics seed from 4.7 pJ to 3 nJ [14].

We report the first demonstration of saturated amplification of a high harmonic seed in the high density plasma of a transient collisional soft x-ray laser created by heating a solid target. A 32.6 nm table-top soft x-ray laser amplifier operating in the $3p^1S_0 \rightarrow 3s^1P_1$ line of Ne-like Ti at 5 Hz repetition rate [11] was used to amplify a seed pulse from the 25th harmonic of Ti:Sa laser into the gain saturation regime. The 27th harmonic of Ti:Sa was simultaneously amplified in the 30.1 nm 3d^{1}P$_1$→3p^{1}P$_1$ line of Ne-like Ti in the same plasma. This seeding is scalable to produce extremely bright lasers at very short wavelength. We demonstrated that the resulting soft x-ray beam is essentially fully spatially coherent. Below we discuss the measurements in comparison with modeling results that describe the dynamics of seeded amplification in a dense collisionally pumped soft x-ray laser amplifier. The model simulations indicate the soft x-ray laser pulses are sub-picosecond in duration. The results were obtained with a table-top laser operating at a repetition rate of 5 Hz, showing this is a practical scheme to produce extremely high brightness soft x-ray beams for applications in a small laboratory environment.

The experimental set up is schematically illustrated in Fig. 4. A single table-top Ti:Sa laser chirped pulse amplification laser system (λ = 800 nm) driver was used both to create the harmonic seed pulse, and the soft x-ray laser amplifier. High harmonic pulses of a few nJ energy are generated using 20 mJ drive pulses compressed to ~ 50 fs. These pulses were focused by a 5 m focal length lens into an 8.8 cm long gas cell filled with 5 Torr of argon. The center wavelength of the 25th harmonic was matched to the 32.6 nm wavelength of the highest gain laser line in the Ti amplifier by adjusting the compressor. The 0.5-1 nJ harmonic seed pulses exiting the gas cell were relay imaged onto a ~ 100 μm diameter spot at the entrance of the soft x-ray plasma amplifier using a toroidal mirror.

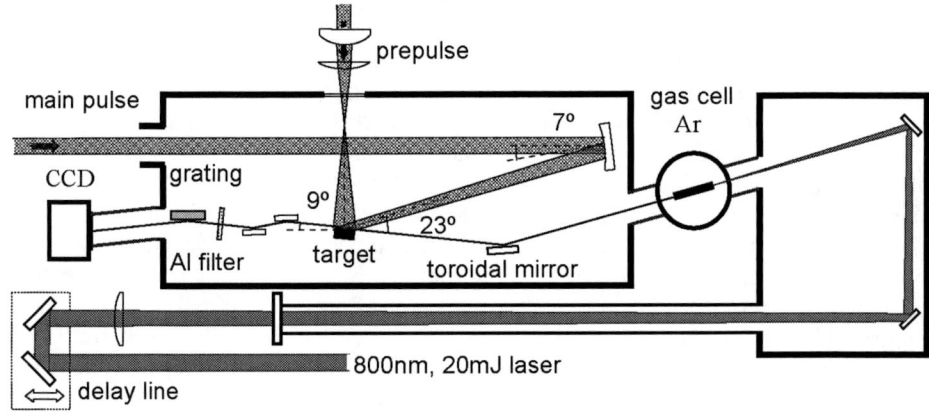

FIGURE 4. Schematic representation of the seeded soft x-ray laser amplifier based on a grazing incidence pumped plasma.

Figure 5 illustrates the dramatic increase in the output of a 3 mm long 32.6 nm Ne-like Ti amplifier and the large decrease in the beam divergence achieved by seeding the amplifier. The top frame (5a) shows the spectra of the un-seeded Ti soft x-ray laser amplifier, and the corresponding intensity distribution in the direction parallel to the target.

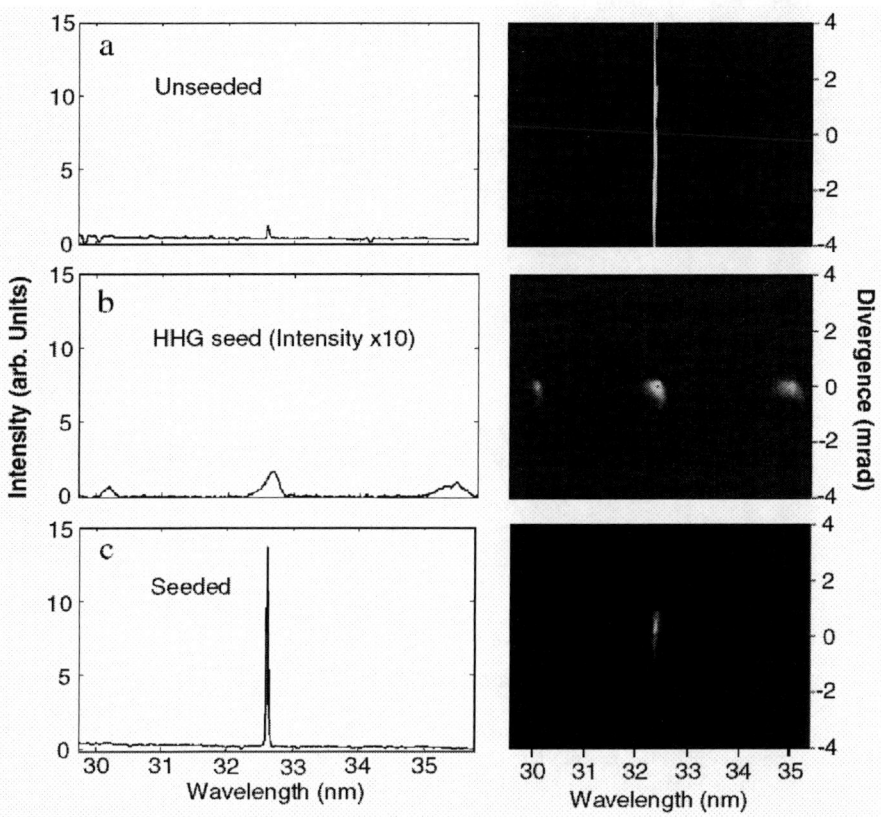

FIGURE 5. Spectra illustrating the relative intensity and beam divergence for; a) unseeded 32.6 nm soft x-ray laser amplifier; b) high harmonic seed pulse; c) seeded soft x-ray laser amplifier. The length of the plasma amplifier is 3 mm. The intensity scale of the seed pulse is magnified by 10 times.

The laser line at 32.6 nm dominates the spectrum and has a divergence of about 10 mrad. Fig. 5b shows the much lower divergence, about 1 mrad, but significantly broader spectra of the harmonic seed. The seeded amplifier output shown in Fig. 5c consists of a highly monochromatic spectral line with an energy that is ~ 64 times larger that of the seed pulse. The FWHM beam divergence is observed to be about 2.2 mrad, much smaller than that of the un-seeded laser. We also observed that it is possible to simultaneously inject seed the 32.6 nm and 30.1 nm lines of Ne-like Ti with the 25th and 27th harmonics.

Figure 6 illustrates the measured increase of the energy of the 32.6 nm seed pulse as a function of amplifier length for a time delay of 4 ps. The data was obtained by varying the length of the target between 0 and 4 mm while maintaining both the seed pulse and the amplifier pump excitation conditions constant. The measurements are compared with the results of model simulations in the same figure. The simulations were conducted using a 1½ dimension hydrodynamic/atomic physics code in which the hydrodynamics were solved in 1 dimension with lateral expansion taken into account using the self-similar solution for expansion into a vacuum. The code includes multi-cell radiation transport and computed the evolution of the gain and plasma density profile of the Ne-like Ti amplifier [16]. The propagation and amplification of a 0.7 nJ seed pulse was computed using a ray tracing post processor code taking into account the effects of gain narrowing and gain saturation. The experimentally measured amplification behavior is very similar to that predicted by the code, and can be divided into three distinct phases.

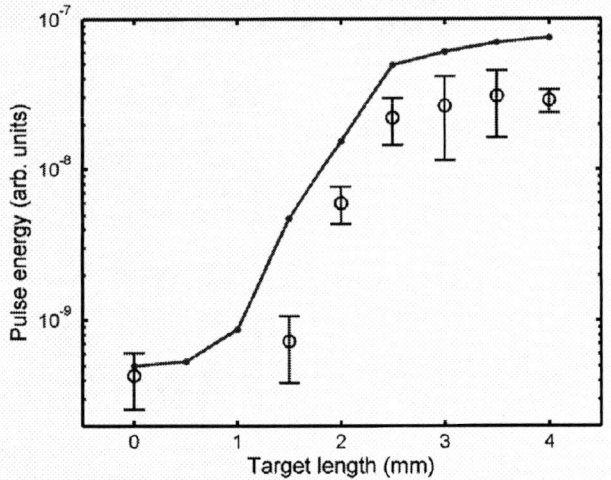

FIGURE 6. Measured and computed (continuous line) variation of the intensity of the amplified seed pulse as a function of plasma amplifier length for the 32.6 nm line of Ne-like Ti. The measured beam reaches the saturation intensity after ~ 2.5 mm into the plasma.

The first phase, which takes place in the first ~1 mm of the amplifier, is dominated by the gain narrowing of the seed pulse whose initial spectral bandwidth greatly exceeds that of the laser line (Fig. 7). This leads to the amplification of only a fraction of its bandwidth, resulting in the observed slow initial seed pulse energy increase. When the seed pulse bandwidth narrows sufficiently to approach the laser linewidth, a second amplification phase starts in which a quasi-exponential increase in the energy of the seed pulse takes place. This rapid increase ends after about 2.5 mm into the amplifier, when the amplified seed pulse reaches the saturation intensity of the 32.6 nm Ne-like Ti line. The third amplification phase corresponds to the gain saturated regime in which efficient energy extraction occurs. The maximum measured amplified seed pulse energy, 50-60 nJ, is similar to that predicted by the model. The model simulation of the pulse propagation in the amplifier predicts a pulse duration of 0.5-0.8 ps determined by the gain-narrowed bandwidth of the amplified laser transition, which is an order of magnitude shorter than the amplifier gain lifetime. We have also measured the variation of the intensity of the amplified seed pulse as a function of delay between the 6.7 ps pump pulse and the arrival of the harmonic seed pulse. The time span during which the seed pulse is amplified is determined by the duration of the gain in the amplifier [15] and is about 5 ps. The maximum amplification is observed at delays between 2 and 4 ps.

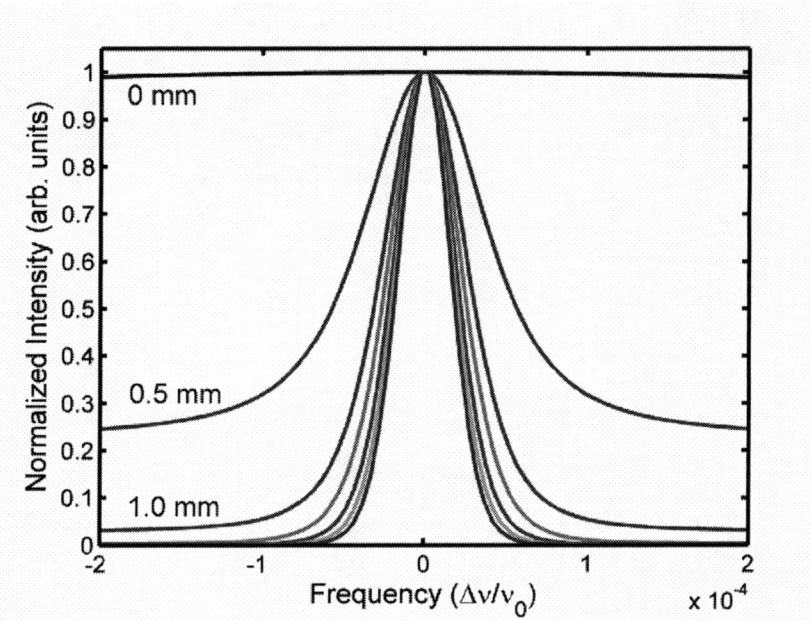

FIGURE 7. Computed variation of the normalized spectral intensity distribution of the injected seed pulse as a function of Ti plasma amplifier length. Most to the spectral narrowing occurs during the first 1mm.

Measurements in grazing incidence transient collisional lasers have shown that the spatial coherence length is nearly an order of magnitude smaller than the beam diameter [8]. The results of a Young's slit interference experiment illustrated in Fig. 8 show the degree of spatial coherence improves dramatically when the amplifier is seeded. Pairs of 5 µm wide slits separated by 30, 75, 150 and 200 µm were placed at 10 cm from the exit of the amplifier, a location at which the FWHM beam diameter is about 200 µm. The measured fringe visibility as a function of slit separation is illustrated in Fig. 8e. A fit of the data with a Gaussian profile yields a coherence length of L_c = 150 µm. The measurement shows the beam approaches full spatial coherence as the majority of the beam energy falls within a coherent length. The equivalent incoherent source diameter of this laser is about 6.9 µm. In conclusion, we have demonstrated a seeded solid target laser amplifier with high spatial and temporal coherence that is readily scalable to significantly shorter wavelengths and higher plasma densities, leading towards table-top soft x-ray sources of extremely high peak brightness. These compact lasers should allow a large number of applications of intense soft x-ray light using table-top set ups. For example, we have recently used the 13.2 nm Ni-like Cd to demonstrate broad area imaging with sub-38 nm resolution [17].

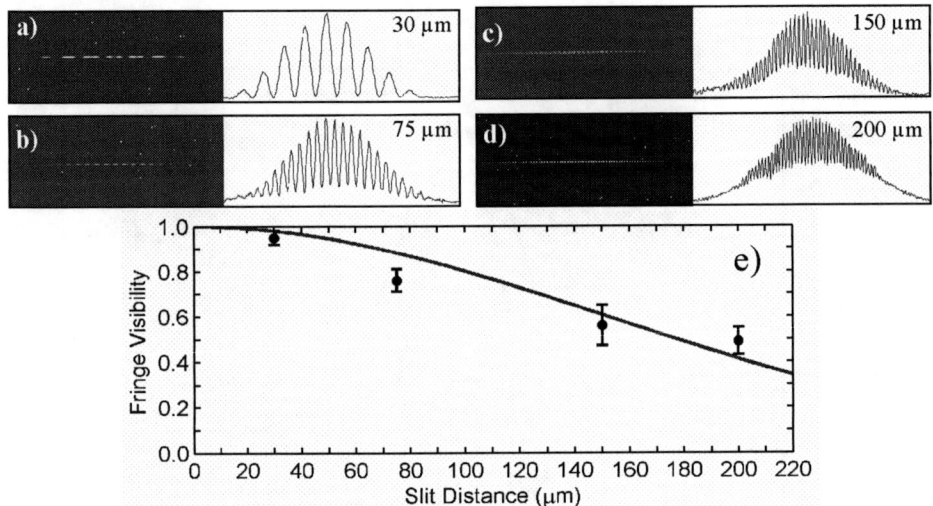

FIGURE 8. Results of Young's slit pair interference experiment for the output of the seeded 32.6 nm laser amplifier. a-d) Interferograms and their lineout for the slit separation indicated. e) Plot of the degree of coherence as a function of the slit separation.

ACKNOWLEDGMENTS

We would like to thank David Attwood, Henry Kapteyn, Margaret Murnane and Yanwei Liu for their contributions. This work was supported by the NSF ERC for Extreme Ultraviolet Science and Technology under NSF Award EEC-0310717 and by the US Department of Energy.

REFERENCES

1. B.R. Benware et al., Phys. Rev. Lett. **81**. 5804 (1998)
2. S. Sebban et al., Phys. Rev. Lett. **86**, 3004 (2001), and S. Sebban, et al., Phys. Rev. Lett. **89**, 253901 (2002).Brown, M. P., and Austin, K., *Appl. Phys. Letters* **65**, 2503-2504 (1994)
3. P.V. Nickles et al., Phys. Rev. Lett. **78**, 2748 (1997).
4. J. Dunn et al., Phys. Rev. Lett. **84**, 4834 (2000).
5. V.N. Shlyaptsev et al., Proc. of SPIE **5197**, 221 (2003).
6. R. Keenan et al., Phys. Rev. Lett. **94**, 103901, (2005)
7. B.M. Luther et al., Opt. Lett. **30**, 165 (2005).
8. M. A. Larotonda et al., IEEE J. Sel. Top. Quantum Electr. **10**, 1363, (2004); Y.Liu et al, unpublished.
9. Y. Wang et al., Physical Review A **72**, 053807, (2005).
10. J.J. Rocca et al., Optics Letters **30**, 2581, (2005).
11. D. Alessi et al, Opt. Express **13**, 2093, (2005).
12. P. Zeitoun et al., Nature **431**, 426, (2004).
13. T. Ditmire et al., Phys. Rev. A **51**, R4337, (1995).
14. T. Kawachi et al., SPIE Vol. **5919**, 155, (2005).
15. T. Mocek et al., Phys. Rev. Lett. **95**, 173902, (2005).
16. M. Berrill and J. J. Rocca, unpublished
17. G. Vaschenko et al., Optics Lett. **31**, 1214, (2006)

Attosecond Nonlinear Optics in Plasmas for Coherent X-ray Generation

Xiaoshi Zhang, Amy L. Lytle, Oren Cohen, David M. Gaudiosi, Tenio Popmintchev, Ariel Paul, Margaret M. Murnane, and Henry C. Kapteyn

Department of Physics and JILA, and NSF Engineering Research Center in Extreme Ultraviolet Science and Technology, University of Colorado and NIST, Boulder, CO 80309-0440
kapteyn@jila.colorado.edu

Brendan Reagan, Mike Grisham, and Jorge J. Rocca

Department of Electrical and Computer Engineering, and NSF Engineering Research Center in Extreme Ultraviolet Science and Technology, Colorado State University, Ft. Collins, CO

Abstract. The process of high-order harmonic generation (HHG) can be used to generate bright, coherent beams of light in the extreme-ultraviolet and soft x-ray region of the spectrum by upconverting intense femtosecond pulses to very short wavelengths. These high-order harmonics result from ionization of the gas used as a nonlinear medium; thus, a full understanding of the process involves incorporating atomic physics, quantum dynamics, and plasma physics into an already non-trivial nonlinear optics problem. In the past several years, we have developed a new technology of "extreme" nonlinear optics that uses the rich, attosecond time-scale physics of the process in novel ways to manipulate the characteristics of this source, improving both the flux and the spectral characteristics. Most recently, we have (1) demonstrated that quasi phase matching of the high-order harmonic conversion process can be accomplished by the use of weak counterpropagating pulse trains that modulate the conversion process, constituting a nonlinear-optical "crystal" made of light; and (2) we have demonstrated that high-order harmonics can be generated by ionization of ions in a guided-wave geometry, using a discharge-created plasma waveguide to pre-ionize the gas and form a guiding electron density profile.

High-order harmonic generation (HHG) driven by ultrashort laser pulses is a source of extreme-ulraviolet and soft x-ray light with the unique properties of ultrashort pulse duration and high spatial and temporal coherence.[1] This source has made possible a number of new ultrafast spectroscopic probes of atoms, molecules and materials. To date, however, most applications use relatively long wavelengths (>10nm), due to the fact that the efficiency of the HHG process decreases rapidly at shorter wavelengths. This decrease in efficiency is *not* due primarily to the very high-order nonlinearity of the process, but as a result of the phase mismatch between the fundamental laser field and high order harmonic field. Dispersion of the free-electron plasma produced by ionization creates a phase mismatch that speeds up the phase velocity of the driving laser with respect to the generated harmonics, and that cannot be compensated using conventional phase matching techniques. This limits efficient harmonic generation to

CP926, *Atomic Processes in Plasmas—15th International Conference on Atomic Processes in Plasmas*
edited by J. D. Gillaspy, J. J. Curry, and W. L. Wiese
© 2007 American Institute of Physics 978-0-7354-0436-6/07/$23.00

relatively low levels of ionization, below a "critical" ionization level of $\approx$ 5% for argon, or $\approx$ 0.5% for helium, corresponding to photon energies of $\approx$ 50eV and 130eV, respectively.[1,3]

Recent work as shown that a weak counterpropagating pulse can disrupt harmonic emission in a gas cell.[2] Although a number of reasons for this counterpropagating pulse interaction could be contemplated, the work presented herein definitively confirms that the interaction is due to field interferences that interact with the high order harmonic generation process itself. The superposition of a pump beam and a counterpropagating pulse results in a standing wave field that amplitude and phase modulates the driving laser with a periodicity corresponding to half the laser wavelength. These rapid variations prevent the coherent buildup of the harmonic field, suppressing the generation process in regions where the counterpropagating pulses intersect.[4] Therefore, if the laser beams collide in a region where the harmonic polarization is out of phase with the pump, then the harmonic signal is enhanced.

Our work shows for the first time that this interaction can be used to increase the brightness of high-order harmonic generation. We observe a substantial enhancement (300x) using this all-optical quasi phase matching (QPM) implemented using a train of counterpropagating pulses.[4] In our first experiment, we used harmonic generation in a 6 cm long hollow waveguide filled with 7 torr of argon.[5] The driving laser pulse, with an energy of 0.5 mJ and a pulse duration of ~25 fs, drives the generation process. The length of each counterpropagating pulse (sent through the waveguide in the opposite direction) was 0.34 mm, with a separation of 1.1 mm, and with energy 0.12mJ. Figures 1a and c show the measured enhancement in the output EUV signal as a function of harmonic order, using one, two or three counterpropagating pulses. A large, 300x enhancement of the 41st harmonic is shown in Fig. 1c. To benchmark the output flux, we compare the brightness of harmonics emitted from 3-pulse QPM with conventionally phase-matched harmonic generation in helium at the same energy (Fig. 1b). Helium is the brightest emitter that can be phase matched at this photon energy using conventional approaches. We find that the brightness of the 39th, 41st and 43rd harmonics are comparable to or brighter than those from phase matched HHG in helium, showing that this technique can result in brighter emission than any other geometry in this spectral range. This demonstrates that all-optical QPM does significantly enhance the harmonic flux harmonic flux in regimes that cannot normally be phase-matched, such as those involving highly ionized plasmas.[3] Since the three-pulse sequence is not yet near the maximum enhancement possible in this wavelength range, further improvements should be possible.

In further work, we enhanced harmonic emission at energies up to 150 eV, also using a 6cm-long hollow waveguide filled with 200 torr of helium. In this experiment, a pulse train with 3 pulses (each 0.16 mm long) and a separation of 0.45 mm is used. The pulse energy of the pump pulse and each of the counterpropagating pulses was 1.24 mJ and 0.1 mJ, respectively. Figure 2a shows these data. An enhancement of 60x is observed for the 95th harmonic around 150eV.

Finally, we demonstrate selective enhancement of the two different quantum trajectories that generate a given harmonic (the long and short trajectories) by varying the counterpropagating pulse separation. In this experiment, using three counterpropagating pulses, each separated by 0.45 mm, the long trajectory or high frequency peak of a harmonic is selectively enhanced (Fig. 2c, red curve); When blocking the middle of the three counterpropagating pulses to increase the separation to 0.9 mm, the short trajectory (low frequency peak of the harmonic order) is selectively enhanced (Fig. 2c. blue curve). Since the two quantum trajectories spend slightly different times in the field (slightly less and slightly greater than 1fs) and since the quantum phase of the two trajectories is different, they respond differently to the counterpropagating field. This allows for coherent control over the radiating electron wavefunction on attosecond timescales.

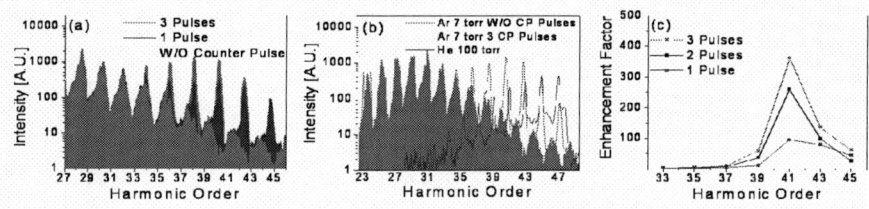

Fig. 1: Quasi phase matching of high harmonic generation using either one, two, or three counterpropagating pulses in a 6cm hollow waveguide filled with 7 torr of Ar. a) HHG emission for no (grey), one (blue), and three (red) counterpropagating pulses, each of width 0.34 mm, spaced by 1.1 mm. b) Comparison of brightness between 3-pulse QPM (red), no counterpropagating pulses, (grey) in argon at 7 torr, and the phase matched emission from 100 torr of He (blue) under same conditions; c) Enhancement factor as a function of harmonic order, for one (blue), two (black) and three (red) counterpropagating pulses.

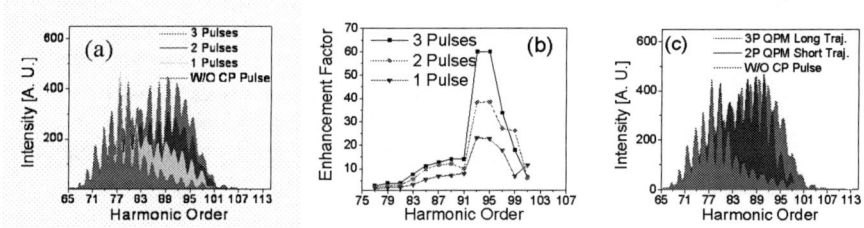

Fig. 2: Quasi phase matching of high harmonic generation using either one, two or three counterpropagating pulses in a long hollow waveguide (6 cm) filled with 200 torr of He. a) Observed HHG emission for no (grey), one (green), two (blue) and three (red) counterpropagating pulses, each of width 0.16mm, spaced by 0.45mm. b) Enhancement factor as a function of harmonic order, for one (blue), two (black) and three (red) counterpropagating pulses. c) Quasi phase matching of different quantum trajectories: long trajectories using 3 counterpropagating pulses with a separation of 0.45 mm (red); short trajectories using 2 counterpropagating pulses with a separation of 0.9 mm (blue); Non-phasematched harmonic spectrum without counterpropagating pulses (grey).

This work demonstrates that phase-matching in high-order harmonic generation can be obtained even for very high-energy harmonics where the gas is highly ionized and plasma dispersion prevents conventional phase matching. It is thus of interest to push these novel phase matching techniques to higher energies in the soft x-ray region of the spectrum. In other recent work,[6] we showed that it is possible to generate very high energy harmonics from further ionization of a plasma pre-ionised using a capillary discharge. This allows one to tailor the electron density radial profile to guide the pulse. This guiding avoids plasma-induced defocusing that would otherwise limit the peak intensity. Figure 3 shows that, using such a discharge, significantly higher photon energies can be obtained than without the discharge. Other experimental data show that the effects of ionization self-phase-modulation are suppressed in a discharge preionized plasma.[7] The next step in these experiments is to implement the quasi phase matching techniques introduced earlier to the case of the plasma waveguide.

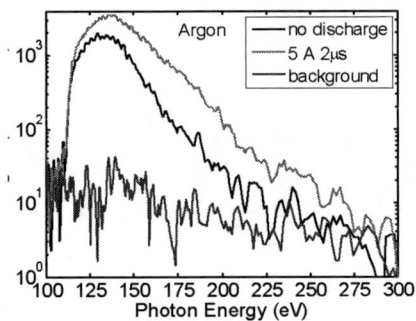

Fig. 3. High harmonic spectrum obtained from 6.0 Torr argon when the discharge is pulsed with a 5 A current pulse (red). The harmonic spectrum obtained in a hollow core waveguide with no discharge is also shown (black).

In summary, this work represents an important advance in overcoming a critical challenge in laser science, for implementing useful coherent EUV sources at wavelengths of interest to advanced lithography, and for high-resolution imaging. Promising future directions include the application of this technique to very high-order harmonics, the use of multiple pulse trains and pulse sequences that are shaped in time, amplitude, and polarization to optimize EUV efficiency, as well as attosecond pulse generation by enhancing selected quantum paths while applying gated phase matching techniques.

REFERENCES

1   H.C. Kapteyn, M.M. Murnane, and I.P. Christov, Physics Today (March), March Issue (2005).
2   S. Voronov et al., Phys. Rev. Lett. 87, 133902 (2001).
3   E. A. Gibson, A. Paul, N. Wagner et al., Science 302 (5642), 95 (2003).
4   X. Zhang, A. L. Lytle, H. C. Kapteyn, M. M. Murnane, and O. Cohen, Nature Physics 3, 270-275 (2007).
5   A. Rundquist, C.G. Durfee III, S. Backus et al., Science 280 (5368), 1412 (1998).
6   D. M. Gaudiosi, B. Reagan, T. Popmintchev et al., Physical Review Letters 96 (20), 203001 (2006).
7   B. A. Reagan, T. Popmintchev, M. E. Grisham, D. M. Gaudiosi, M. Berrill, O. Cohen, B. C. Walker, M. M. Murnane, J. J. Rocca, and H. C. Kapteyn, Physical Review A TBP(2007).

Particle Manipulation with Nonadiabatic Ponderomotive Forces

I. Y. Dodin and N. J. Fisch

Department of Astrophysical Sciences, Princeton University, Princeton, New Jersey 08544

Abstract. Average, or ponderomotive potentials effectively seen by particles in oscillating fields allow advanced techniques of particle manipulation inaccessible with static potentials. In strongly inhomogeneous fields the ponderomotive force is phase-dependent, and the particle dynamics resembles that of a quantum object in a conservative barrier. Probabilistic transmission through a ponderomotive potential is possible then and can be used for particle beam slicing. Resonant fields can also cool and trap particles exhibiting natural oscillations (e.g., Larmor rotation), as well as transmit them asymmetrically, hence acting as one-way walls. An approximate integral of particle motion is found for this case and a new ponderomotive potential is introduced accordingly.

Keywords: nonadiabatic ponderomotive forces, Hamiltonian dynamics
PACS: 52.35.Mw, 45.20.Jj, 05.45.-a

Numerous methods of manipulating atoms with laser light have been proposed in the literature and have enjoyed experimental verification [1, 2, 3]. Similar capabilities apply also to other objects ranging from molecules to micron-sized particles and permit one to selectively and stably trap particles, levitate them against gravity, channel particles along laser beams, and use them as sensitive probes for measuring optical, electric, magnetic, viscous drag, and gravity forces. The new tools of particle control yield present and potential applications in a variety of subjects such as light scattering, cloud physics, quantum optics, isotope separation, and high-resolution spectroscopy [4]. What we offer here is the extension of these techniques, which is based on classical interactions of undamped particles with intense electromagnetic fields, extrapolation being possible to quantum objects such as atoms and molecules in metastable energy states [5].

Even without a bias, an oscillatory field can exert a nonzero time-averaged force on a particle [6, 7]. This so-called ponderomotive force consists of two components: the dipole force due to the inhomogeneity of the field envelope and the light pressure due to the radiation scattering off the particle. Often, the light pressure is negligible (the particle is "undamped"), and the induced dipole moment of the particle p is nearly a local, or "adiabatic" function of the field E. In the linear approximation, the two conditions yield $p = \hat{\alpha} \cdot E$, where the polarizability tensor $\hat{\alpha}$ is Hermitian. The force on the particle can then be described in terms of the ponderomotive potential Φ, equal to the average energy of the dipole-field interaction:

$$\Phi = -\frac{1}{4}\left(E^* \cdot \hat{\alpha} \cdot E\right). \tag{1}$$

We assume a stationary field ($k = 0$, $\partial_t|E| \equiv 0$); hence Φ is a function of space only. For example, in the absence of additional forces, an elementary particle with charge

CP926, *Atomic Processes in Plasmas—15th International Conference on Atomic Processes in Plasmas*
edited by J. D. Gillaspy, J. J. Curry, and W. L. Wiese
© 2007 American Institute of Physics 978-0-7354-0436-6/07/$23.00

e and mass m has $\hat{\alpha} = -(e^2/m\omega^2)\hat{I}$, so one recovers the well known formula $\Phi = e^2|E|^2/4m\omega^2$ [6, 8].

Under the above assumptions oscillatory fields act like static potentials, which yields numerous applications including atomic traps, rf plugs, low-frequency mode stabilization, and edge control in fusion plasmas [9]. Violating the approximation (1) though renders extra flexibility allowing for more advanced and otherwise inaccessible techniques of particle manipulation. Refs. [1, 2, 3] illustrate how the additional capabilities result from dissipation. What we contemplate here is how similar effects can be practiced on undamped particles – via breaking the adiabatic approximation.

To offer new tools for manipulating classical particles with ac fields and advance the analytical treatment of nonadiabatic ponderomotive forces is the purpose of this work (also reviewed in Ref. [9]), which re-examines our selected results reported in Refs. [10, 11, 12, 13, 14, 15, 16, 17, 18, 19, 20] from a unifying standpoint. We show that in strongly inhomogeneous fields the ponderomotive force is phase-dependent, and the particle dynamics resembles that of a quantum object in a conservative barrier [13]. Probabilistic transmission through a ponderomotive potential is possible then [11] and can be used for particle beam slicing. In the ultra-relativistic regime, attosecond electron bunches can be produced in this way and are suitable for coherent x-ray generation via Thomson backscattering of the laser light [10]. We demonstrate that resonant fields can also cool and trap particles exhibiting natural oscillations (e.g., Larmor rotation) [12], as well as transmit them asymmetrically, hence acting as one-way walls [5, 14, 16, 17]. We find an approximate integral of particle motion for this case and introduce a new ponderomotive potential accordingly [12].

ACKNOWLEDGMENTS

This work was supported by U. S. DOE Contract No. DE-FG02-05ER54838, by the U. S. NNSA under the SSAA Program through DOE Research Grant No. DE-FG52-04NA00139, and by the U. S. National Science Foundation under Grant No. PHY99-07949.

REFERENCES

1. S. Chu, *Rev. Mod. Phys.* **70**, 685 (1998).
2. C. N. Cohen-Tannoudji, *Rev. Mod. Phys.* **70**, 707 (1998).
3. W. D. Phillips, *Rev. Mod. Phys.* **70**, 721 (1998).
4. A. Ashkin, *Proc. Natl. Acad. Sci. USA* **94**, 4853 (1997).
5. M. G. Raizen, A. M. Dudarev, Q. Niu, and N. J. Fisch, *Phys. Rev. Lett.* **94**, 053003 (2005).
6. A. V. Gaponov and M. A. Miller, *Zh. Eksp. Teor. Fiz.* **34**, 242 (1958) [*Sov. Phys. JETP* **7**, 168 (1958)].
7. V. G. Minogin and V. S. Letokhov, *Laser light pressure on atoms*, Gordon and Breach, New York, 1987.
8. J. R. Cary and A. N. Kaufman, *Phys. Rev. Lett.* **39**, 402 (1977).
9. I. Y. Dodin and N. J. Fisch, *Phys. Plasmas* **14**, 055901 (2007).
10. I. Y. Dodin and N. J. Fisch, *Phys. Rev. Lett.* **98**, 234801 (2007).
11. I. Y. Dodin and N. J. Fisch, *Phys. Rev. E* **74**, 056404 (2006).
12. I. Y. Dodin and N. J. Fisch, *Phys. Lett. A* **349**, 356 (2006).
13. I. Y. Dodin and N. J. Fisch, *Phys. Rev. Lett.* **95**, 115001 (2005).

14. I. Y. Dodin and N. J. Fisch, *Phys. Rev. E* **72**, 046602 (2005).
15. I. Y. Dodin and N. J. Fisch, *Phys. Lett. A* **341**, 187 (2005).
16. I. Y. Dodin, N. J. Fisch, and J. M. Rax, *Phys. Plasmas* **11**, 5046 (2004).
17. N. J. Fisch, J. M. Rax, and I. Y. Dodin, *Phys. Rev. Lett.* **91**, 205004 (2003); Erratum: *Phys. Rev. Lett.* **93**, 059902(E) (2004).
18. I. Y. Dodin and N. J. Fisch, *J. Plasma Phys.* **71**, 289 (2005).
19. I. Y. Dodin, N. J. Fisch, and G. M. Fraiman, *JETP Lett.* **78**, 202 (2003).
20. I. Y. Dodin and N. J. Fisch, *Phys. Rev. E* **68**, 056402 (2003).

Ultra-high Intensity Optical Slow Wave Structure and Applications

A. York, B.D. Layer, T.M. Antonsen, S. Varma, Y.-H. Chen, and H.M. Milchberg

Institute for Physical Science and Technology, Dept. of Electrical and Computer Engineering, and Dept. of Physics, University of Maryland, College Park, MD 20742

Abstract. We report the development of corrugated 'slow wave' plasma guiding structures and discuss their potential application to quasi-phase-matched direct laser acceleration of charged particles and generation of a wide spectrum of electromagnetic radiation. These structures support guided propagation at intensities up to 2×10^{17} W/cm^2, limited by our current laser energy and side leakage. Hydrogen and argon plasma waveguides up to 1.5 cm in length with corrugation period as short as 35μm are generated in extended cryogenic cluster jet flows. Side-imaged scattering of guided intense pulses shows evidence of periodic modulations of the laser pulse intensity

PACS: 52.38.Hb, 36.40.Gk, 36.40.Vz, 52.50.Jm

INTRODUCTION

In the rapidly developing field of ultra-intense laser-plasma interactions, applications including electron acceleration by laser-driven wakefields [1], x-ray lasers, and coherent electromagnetic wave generation [2] must rely for their fullest realization on the extended diffraction-suppressed propagation of extreme intensity laser pulses in plasma optical guiding structures. Plasma is unavoidable at laser intensities well in excess of the ionization threshold of any constituent atom of interest. Early results demonstrated CO_2 laser refraction by He plasmas with on-axis density minima [3]. Plasma waveguides for intense optical pulses were first generated through the radial hydrodynamic shock expansion of picosecond laser-heated gas plasma [4]. Later, waveguides were demonstrated in electrical discharge capillaries [5], in variations of the hydrodynamic shock technique [6], and, most recently using laser-driven hydrodynamic shocks in cluster jet targets, both end-pumped [7] and side pumped [8]. The optical mode structure and dispersion properties of plasma waveguides have been discussed in detail [9].

Beyond material and modal dispersion and the confinement of beam intensity, waveguide control of beams can be augmented by additional 'branches' to the ω versus k dispersion diagram, where $k = k(\omega)$ is the axial wavenumber of the guide and ω is the angular frequency. Adding an axial (z) modulation of period d through variations in geometry, materials, or both gives rise to a wider class of solutions $u(\mathbf{r}_\perp, z, \omega) \exp(ikz)$ where $u(\mathbf{r}_\perp, z+d, \omega) = u(\mathbf{r}_\perp, z, \omega)$ and $k = k_c(\omega) + 2\pi m/d$ (from the Floquet-Bloch theorem), where u is an electromagnetic field component, $\mathbf{r}_\perp$ is transverse position, m is an integer, and k_c is the fundamental axial wavenumber [10].

CP926, *Atomic Processes in Plasmas—15th International Conference on Atomic Processes in Plasmas*
edited by J. D. Gillaspy, J. J. Curry, and W. L. Wiese

The extra dispersion branches arise from the multiplicity of k values assigned to each ω for each transverse mode.

For low intensity beams, two examples of modulated guiding structures are illustrative of applications. In Radio Frequency (RF) accelerators (for example, SLAC [11]), charged particles are accelerated by the E_z field of a TM wave guided in an axially modulated copper waveguide. In an ordinary unmodulated metal waveguide, guided wave phase velocity $v_p = \omega/k$ is greater than c, so a sub-light speed particle would be accelerated and then decelerated as the wave oscillations pass the particle, giving zero net acceleration. However, a modulated structure can have an effective k, as discussed above, sufficiently larger than k_0 to give $v_p < c$ so that the particle and wave speeds can be matched. Such a guide is called a slow wave structure, and finds wide application in both accelerators and microwave sources [10]. Another view of this matched interaction—called *quasi phase-matching* – is that the wave speeds up and slows down in successive half-periods of the modulation in such a way that the partial acceleration in the first half is not completely canceled by deceleration in the second half. We note that for a number of applications, quasi-phase matching is enabled not only by the modulations in linear dispersion, but also by the accompanying laser intensity modulations. In low-intensity guided wave optics, periodic modulation in the material's optical nonlinearity or refractive index makes possible the quasi-phase matched generation of low-order harmonics of the fundamental pump wave [12]. At higher intensity ($\sim 10^{14}$-10^{15} W/cm^2), this process has been demonstrated in modulated, gas-filled hollow core fibres for the generation of extreme ultraviolet high harmonics [13]. Recent work [14] has shown that the propagation of an intense pulse over several plasma segments with total extent <1 mm could assist the phase matching of third harmonic generation. However, quasi-phase matching becomes important only after diffraction is removed as a limiting factor. Without the direct measurement of plasma density profiles, energy throughputs, guided laser spot sizes, or guiding itself, it is unclear if that experiment effectively extended the nonlinear interaction beyond a Rayleigh length.

Here, we demonstrate for the first time high intensity optical guiding in an extended corrugated plasma slow wave structure. Such plasma structures were first discussed in reference [15]. Spontaneous, but uncontrolled modulated channels were reported in reference [16]. Here we have produced exceptionally stable plasma waveguides with adjustable axial modulation periods as short as 35 μm, where the period can be significantly smaller than the waveguide diameter. We have measured guided propagation at intensities up to 2×10^{17} W/cm^2, limited only by our current laser energy.

EXPERIMENTS

Figure 1 shows the experimental setup. Initially, unmodulated waveguides were generated in cluster jets [8] using lowest order (J_0) Bessel beam pulses produced by an axicon-focused beam from a 10Hz, 1064 nm, 100ps Nd:YAG laser with pulse energy of 200-500 mJ. The cluster source in these experiments was a cryogenically cooled supersonic gas jet with a 1.5 cm long by 1 mm wide nozzle exit orifice. Clusters form when a highly pressurized gas expands through the nozzle into vacuum, with the atoms or molecules attracted to one another by van der Waals forces. Aggregates form at solid density with diameter 1-50 nm ($\sim 10^2$ to 10^7 atoms), depending on nozzle geometry, gas species, and jet backing temperature and pressure, which were controlled in the range 115 to 295 K and 100 to 1000 psi.

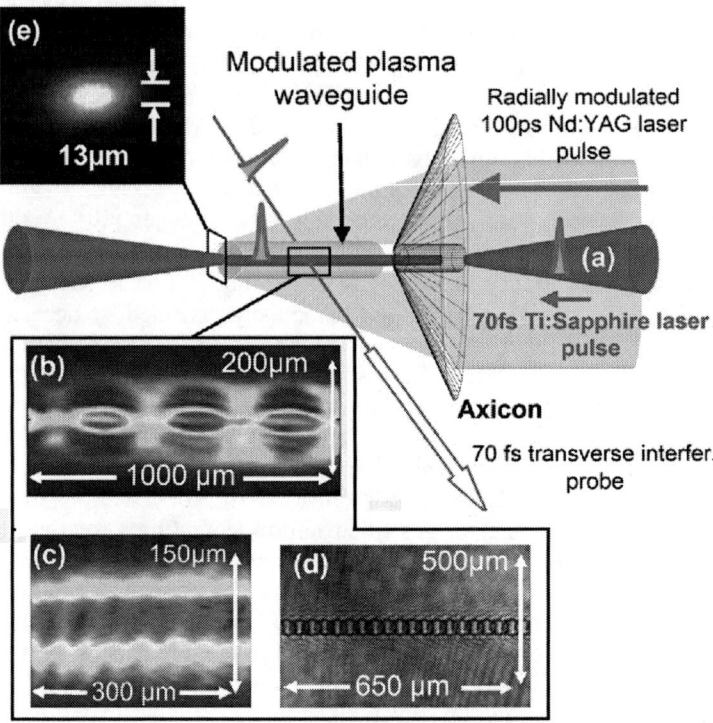

FIGURE 1. This Schematic of experimental geometry showing radially modulated Nd:YAG laser pulse (200-500mJ, 100ps, 1064nm) from ring grating imaging system (not shown) focused by an axicon onto a liquid nitrogen-cooled elongated cluster jet target, generating a 1.5cm long corrugated plasma channel. (a) Ti: Sapphire laser pulse (70mJ, 70fs, 800nm) guided down the channel at an adjustable delay with respect to the Nd: YAG pulse. A 1mJ portion of the 70fs, 800 nm pulse was directed transversely through the corrugated guide and into a folded wavefront Michelson interferometer for time-resolved interferometric/shadowgraphic images. Examples are phase images of channels with (b) 300μm and (c) 35μm modulation periods in argon cluster targets and (d) a 35μm period in air. (e) Lowest order exit mode from the guide of Fig. 3(b)(1ii), with w_{fwhm}=13μm.

The 25 mm line-focus length of the Bessel beam overlapped the 1.5 cm length of the cluster jet, resulting in 1.5 cm long plasma channels. Longer channels can be obtained using longer orifices, or using several contiguous gas jets. Waveguides were injected at f/10 through a hole in the axicon (Fig. 1(a)) with a 70 mJ, 70 fs, 800nm Ti: Sapphire laser pulse synchronized [17] and delayed with respect to the channel-generating Nd:YAG pulse. A small portion of this pulse (~ 1 mJ) was split off into a delay line, directed transversely through the waveguide, and then imaged through a femtosecond folded wavefront Michelson interferometer onto a CCD camera. Electron density profile images of the evolving corrugated waveguide were obtained by phase extraction of the time-resolved interferograms, followed by Abel inversion. In addition, shadowgrams were obtained by blocking one arm of the interferometer, and images of the guided pulse side-scattering from the waveguide were obtained by blocking both the transverse probe beam and one arm of the interferometer, and placing an 800nm interference filter in front of the CCD camera.

We have shown previously [8] that the use of clusters can increase the 100 ps Bessel beam absorption efficiency by an order of magnitude compared to unclustered gas targets of the same volume average density. This high absorption efficiency (measured up to ~35%) occurs in spite of the fact that typical clusters in this experiment (size ~5-30 nm) explosively disassemble and expand below the plasma critical density on a subpicosecond timescale [18]. With cluster targets the far leading edge of the 100 ps pulse prepares a highly ionized, locally uniform, and cool plasma from the expanded and merged individual cluster plasmas, which the remainder of the pulse can heat efficiently [8].

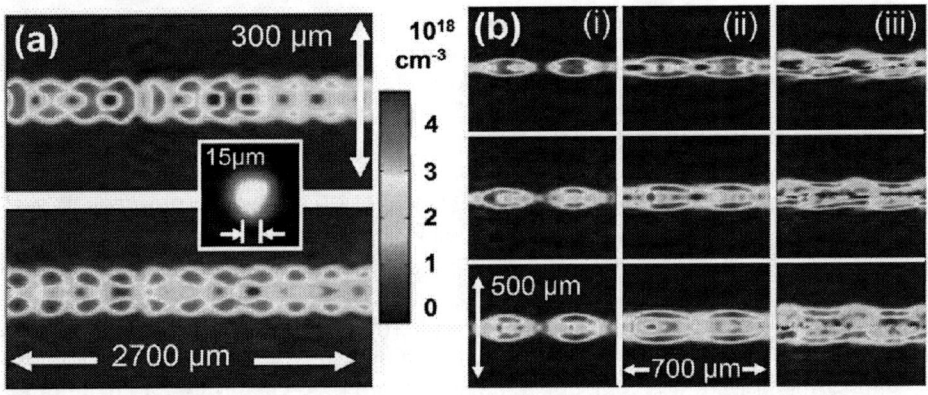

FIGURE 2. Corrugated channels produced in a hydrogen cluster jet at 800psi and -145°C backing pressure and temperature. **(a)** ~3 mm section of channel without (top panel) and with (bottom panel) injection at 1 ns delay of a 70 mJ, 70 fs, 800 nm laser pulse, viewed 100ps after pulse passage. The channel exit mode of $w_{hwhm}=15\mu m$ is shown in the inset. **(b)** Higher magnification Abel-inverted electron density profiles of 2 corrugation periods for channel creation pulse energies of (i) 200 mJ , (ii) 300 mJ, and (iii) 500 mJ (with RG/axicon misalignment) at interferometer probe delays of, top to bottom, 0.5ns, 1ns, and 2 ns.

To impose axial modulations upon the plasma channel, a transmissive 'ring grating' (RG) and associated imaging optics were centered in the path of the Nd:YAG laser pulse. The RG used in these experiments was a 1" diameter, lithographically etched fused silica disc with variable groove period, groove structure, and duty cycle. The axicon projects the diffraction pattern produced by the RG onto the optical axis, leading to axial intensity modulations of the Bessel beam. The dominant axial modulation of the central spot intensity imposes axial modulations in the heating and plasma generation in the cluster jet. For the present work, we imposed a periodic modulation, but arbitrary RG patterns are possible. Phase matching direct ion acceleration, for example, would require a graded modulation period.

The 100ps heater pulse is essentially an impulse on the hydrodynamic timescale (~0.1-0.5 ns) of the heated bulk plasma (formed from merged cluster explosions) that remains after the pulse [8]. This plasma then undergoes radial hydrodynamic shock expansion, producing a corrugated plasma waveguide. Examples of two phase images of the magnified central waveguide region are shown in Fig. 1(b) and 1(c), with modulation periods of 300μm and 35μm, using two different RGs and an argon cluster jet. Corrugated guides can also be generated in backfill gases: Fig. 1(d) shows a shadowgram of a modulated channel produced in air with a period of 35μm. A typical guided exit mode from a modulated cluster plasma channel (half-width-at half-maximum mean radius 13μm) is shown in Fig 1(e). Note that the cluster-generated channels are highly stable and reproducible: all density profiles shown in this paper are extracted from the average phase of 200 consecutive interferograms. The shot-to-shot extracted density variation is less than 5%.

Figure 2 shows results for a corrugated hydrogen plasma waveguide. Hydrogen plasma waveguides are attractive for laser-plasma acceleration [1] because they are easily fully ionized during their formation, making impossible further ionization by guided intense pulses, which can lead to distortion and ionization-induced refractive defocusing. The modulation period d~300μm was chosen to ensure clearly observable periodic oscillations in laser intensity. Figure 2(a) shows the electron density $N_e(r,z)$ of a 3 mm section near the entrance of a 1.5 cm hydrogen waveguide, 1 ns after generation. The bottom panel shows the guide ~100ps after passage of an injected 70 mJ, 70 fs Ti: Sapphire laser pulse injected from the right at 1 ns delay after waveguide formation. The top panel shows the uninjected guide. The density profiles are very similar, showing that little change to the guide was produced by the guided pulse. The inset in 2(a) shows the exit mode of the guided pulse. With modulated hydrogen guides, energy throughput is ~10%, yielding output intensity of 10^{17} W/cm^2. The low throughput is due to leakage and side scattering out of the guide due to the modulations. This leakage is directly seen in argon results. Figure 2(b) shows higher magnification profiles of two modulation periods as a function of interferometer probe pulse delay (0.5 ,1, and 2 ns), for channel creation pulse energies of (i) 200 mJ and (ii) 300 mJ. It is seen that lower pulse energy ((i)) can produce periodic 'beads' of plasma, separated by zones of neutral clusters/atoms, while higher energy ((ii)) results in a more continuous channel. The beads act as plasma lenslets, collecting the light emerging from a neutral gap and focusing the beam into the next gap. Figure 2(b)(iii) shows the result of an intentional misalignment at 500 mJ of the Bessel beam axis and the RG optical axis: a continuous plasma fibre is generated with angular fluting. In this case, owing to the top-bottom asymmetry in the extracted phase image, separate Abel inversions were performed above and below the optical axis.

Channels with higher ionization Z were generated in argon cluster jets. Figure 3(a) shows an extended region of channel at 1.5 ns delay with and without guided pulse injection (bottom and top panels respectively) for a probe delay ~10 ps after guided pulse passage. Separate Abel inversions were done above and below the optical axis to highlight the cylindrical symmetry. It is seen that the channel itself is little affected by the guided pulse, but in contrast to the hydrogen results, there is a significant electron density 'halo' located at a radial distance ~100 μm from the channel. Short interval sequences of probe images show that the halo propagates right-to-left at the speed of light with the guided pulse. The halo radial position remains constant over the full 1.5 cm length of the corrugated channel but decays in density, suggesting that it originates from additional cluster ionization from channel side leakage [9] rather than from uncoupled entrance light skimming the outside of the channel. The leakage light moves radially across a zone of low density plasma and neutrals until it reaches the layer of argon clusters that was unperturbed during channel formation. With hydrogen, the clusters are smaller and more fragile, so it is unlikely they survive so close to the channel after its formation, and hence there is no observed halo in Fig. 2.

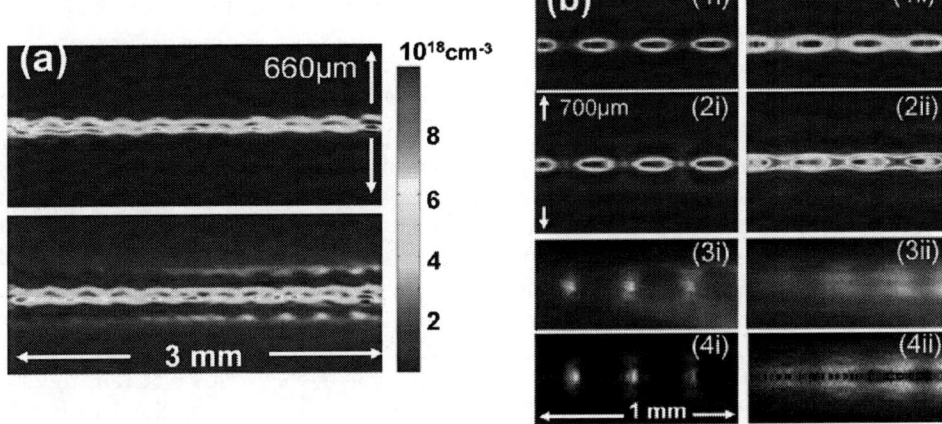

Figure 3. Corrugated channels produced in an argon cluster jet at 800psi backing pressure and room temperature. **(a)** 3mm section of channel without (top panel) and with (bottom panel) injection at 1.5 ns delay of a 70 mJ, 70 fs, 800 nm pulse. **(b)** Magnified images at 2 ns delay of beaded (300 mJ pump, left column) and more continuous (500 mJ pump, right column) channel modulations. Left and right columns: **(1)** density profile of uninjected guide, **(2)** density profile of guide injected with 70 mJ, 70 fs, 800 nm pulse, **(3)** scattering image at 800 nm corresponding to (2), **(4)** Abel-inversion of (3).

Higher resolution images of 3 periods of modulation near the argon channel centre are shown in Fig. 3(b), for cases of beaded (300 mJ pump, left column) and continuous (500 mJ pump, right column) modulations, without guided pulse injection (panels (1)) and with injection (panels (2)-(4)). In the case of the beaded guide, Fig. 3(b)(2i) shows strong additional ionization by the guided pulse in the neutral gaps as the beam is focused by each plasma lenslet and collected by the next. Remarkably in this case, the overall channel coupled energy throughput is still 10%, showing that there is significant injection/capture of light by successive lenslets over the full 1.5 cm length of the channel. Throughput for the continuous modulation case is 20%, yielding 2×10^{17} W/cm^2 peak intensity at a beam waist, using the fact that the beam exits the channel at a guide bulge. By comparison, throughput for this injection delay in unmodulated waveguides is 60%. Panels (2) both show in more detail the ionization halo induced by side-leakage of the guided pulse. Side-imaged Thomson/Rayleigh scattering of guided 800 nm light shows scattering strongly localized at the neutral gaps in the beaded guide (Fig. 3(b)(3i)), but more smoothly modulated in the more continuous guide (Fig. 3(b)(3ii)). In both cases, the strongest scattering originates from the highest intensity zones, either during ionization of clusters in the neutral gaps (panel (3i)) or from leakage at the beam waists of the continuous modulation (panel (3ii)). Panels (4i) and (4ii), which show the scattering source (r,z) profiles, are Abel inversions of panels (3i) and (3ii). These images make clear that the dominant scatterers are likely clusters, either those surviving in the gaps between beads, or those external to the continuously modulated guide. The relative intensity modulation of the guided pulse is obtained from the bulge-to-waist area ratio of the guide from the density and scattering profiles, and from propagation simulation. To better understand leakage and mode dynamics, we simulated the propagation of short pulses in these waveguides using our intense laser pulse propagation code WAKE [19], and found results consistent with our measurements. We refer readers to reference [27] for further discussion of simulations.

SELECTED APPLICATIONS

A novel and promising application of the corrugated plasma waveguide is direct laser acceleration of charged particles. Since the Woodward-Lawson theorem forbids the exchange of energy between radiation and a relativistic charged particle in vacuum [21], direct laser acceleration requires some nearby structure to phase-match the interaction. It is ultimately the damage threshold of this structure which limits laser intensity and therefore acceleration gradients, making the corrugated plasma waveguide an ideal structure for this application.

Kimura et al. demonstrated direct acceleration of relativistic electrons by laser light using the inverse Cherenkov effect [22]. Serafim et al. proposed enhancing the direct laser acceleration efficiency by guiding a radially polarized laser mode in a plasma channel [23]. The dominant radial component E_r guides as a hollow mode ($m=0$, $p=1$, where m and p are azimuthal and radial mode indices) with peak intensity at $r=w_{ch}/\sqrt{2}$, where w_{ch} is the $1/e^2$ intensity radius of the channel's fundamental ($m=0$, $p=0$) Gaussian mode and is given by $w_{ch} = (1/\pi r_e \Delta N_e)^{1/2}$, where r_e is the classical electron radius and ΔN_e is the electron density difference between $r = 0$ and $r = w_{ch}$. The accelerating field is the associated axial component E_z, which peaks at $r=0$ and passes through zero at $r = w_{ch}$. Following reference [23], the axial acceleration gradient from hollow beam guiding in a plasma channel is given by

$$E_z = \frac{4\lambda}{\pi w_{ch}^2} \sqrt{\frac{2P}{c}}$$

(1)

where P is the peak laser power (erg cm^{-2} s^{-1}).

Serafim et al. calculate [23] that 17 MW of peak laser power (at laser wavelength $\lambda \sim 1$ μm) guided in $p=1$, $m=0$ hollow mode in a channel supporting $w_{ch}=$ 20 μm gives an on-axis acceleration gradient of $E_z \sim 1$ MV/cm over a distance limited only by pump beam depletion and the length of the plasma channel. It is assumed that there is sufficient neutral gas in the channel to slow the wave phase velocity to c for phase matching to the relativistic electron speed.

If we now consider a 1 TW laser pulse with $\lambda = 0.8$ μm in a channel supporting w_{ch}=12 μm, Eq. (1) yields a gradient of $E_z \sim 0.55$ GV/cm. If there were no phase slippage issue, this would compare very favorably to wakefield acceleration: Malka et al. used a 30 TW laser at $\lambda = 0.8$ μm to produce an acceleration gradient of ~ 0.66 GV/cm (200 MV over 3 mm) [24]. Also, unlike wakefield acceleration, direct laser acceleration is a linear process that scales down to lower pulse energies. Even a single regenerative amplifier can easily produce 20 GW peak power, giving a 77 MV/cm gradient.

Of course, phase slippage is a major issue: a means must be found to slow the wave phase velocity to c or less to match the relativistic electron velocity. Neutral gas as used in [23] will not survive at the laser intensities essential for high values of accelerating field E_z. Even pulses well below the terawatt level will propagate in fully ionized waveguides, making conventional phase matching impossible. Here we show that imposing axial modulations along the plasma waveguide provides a solution to the above phase matching problem.

The dispersion relation for a unmodulated plasma waveguide with a parabolic electron density profile is [25]:

$$\beta^2 = k_0^2 - \omega_{p0}^2 / c^2 - 4(2p + m + 1) / w_{ch}^2 \qquad (2)$$

where β is the guide parallel wavenumber, $k_0 = \omega/c$ is the vacuum wavenumber, $\omega_{p0} = (4\pi N_{e0} e^2/m)^{1/2}$ is the plasma frequency on the waveguide axis (and N_{e0} is the electron density there), and the other parameters were defined earlier. Clearly, the phase velocity $v_p = \omega/\beta$ is superluminal, so the matching condition $v_{electron} = v_p$ can never be satisfied. A relativistic electron ($v_e \approx c$) initially co-propagating in phase with a single mode would experience some acceleration, but after propagating a distance

$$L_d = \frac{\pi}{(k_0 - \beta)} \qquad (3)$$

the electron will move out of phase with the accelerating field and will experience an equal and opposite deceleration over the next L_d. Combining Eqs. (2) and (3), and keeping terms only to first order in N_{e0}/N_{cr} and $(k_\perp/k_0)^2$, the dephasing length becomes

$$L_d = \lambda \left(\frac{N_{e0}}{N_{cr}} + \frac{4(2p + m + 1)}{k_0^2 w_{ch}^2} \right)^{-1} \qquad (4)$$

where N_{cr} is the critical density at the vacuum wavelength λ. A simple model shows how the technique of quasi-phase matching can overcome dephasing (See Fig. 4).

Consider a corrugated plasma waveguide consisting of alternating segments with the same spot size but different central electron densities N_{e01} and N_{e02}. The length of each region is chosen to match the dephasing length L_d for that region's lowest order hollow mode (m=0, p=1). A properly phased relativistic electron will gain energy

159

$$\Delta U = \int_0^{L_{d1}} eE_{z1} \sin(\pi z / L_{d1}) dz - \int_0^{L_{d2}} eE_{z2} \sin(\pi z / L_{d2}) dz$$

$$= \frac{2}{\pi} e(E_{z1} L_{d1} - E_{z2} L_{d2})$$

(5)

Using $E=E_0/\varepsilon = E_0/(1-N_e/N_{cr}) \approx E_0$ where E_0 is the peak waveguide input field gives

$$E_{z,eff} = \frac{2}{\pi} \frac{\eta_1 - \eta_2}{(\eta_1 + \eta_2 + \frac{8(2p+m+1)}{k_0^2 w_{ch}^2})} E_0$$

(6)

for the effective (quasi-phase matched) axial accelerating field, where $\eta_i = N_{e0i}/N_{cr}$.

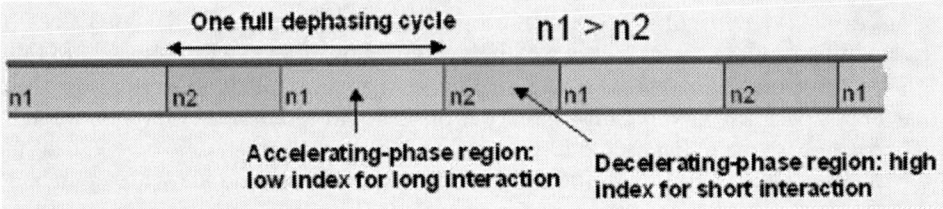

FIGURE 4. Simple alternating index waveguide model. By matching the length of each region to the local dephasing length, acceleration is emphasized and deceleration is diminished, giving net gain over each modulation period.

Suppose the modulated channel supports $w_{ch}=12$ μm, with $N_{e01} = 3 \times 10^{18}$ cm^{-3} and $N_{e02} = 6 \times 10^{18}$ cm^{-3}. At $\lambda=0.8$ μm, $N_{cr}=1.7 \times 10^{21}$ cm^{-3}, so that Eq. (7) gives for the m=0, p=1 mode $E_{z,eff} = 0.14 E_0$. A 1 TW laser pulse ($\lambda = 0.8$ μm) guided in this channel therefore gives $E_{z,eff} \sim 77$ MV/cm, with increasing gradients scaling as the square root of laser power. Even a 20 GW pulse from a regenerative amplifier would still give a respectable gradient of ~ 11 MV/cm.

Note that this simple model is far from optimized. Annular plasma waveguides have been demonstrated [26] which guide annular modes with zero on-axis intensity. A corrugated waveguide consisting of alternating annular-mode and Gaussian-mode regions could eliminate the decelerating regions. Finally, we note that guiding radially polarized light in the corrugated plasma waveguide could also benefit high-harmonic generation. In addition to phase matching, which the corrugated plasma waveguide could easily improve, one of the fundamental limitations of high harmonic generation is that at sufficient laser intensity, $\mathbf{v} \times \mathbf{B}$ forces can no longer be ignored. These forces generally deflect electron trajectories away from recollision with their parent ion, suppressing harmonic generation and giving a fundamental intensity limit beyond which high harmonic generation is impossible. However, perfectly radially polarized light is a TM wave with a vanishing on-axis magnetic field, and a diminished magnetic field in the vicinity of the propagation axis. In theory, radially polarized light should allow harmonic generation to occur at a relativistic intensity that would be impossible with linearly polarized light.

CONCLUSIONS

In conclusion, we have demonstrated guiding and dispersive control of extremely intense pulses in miniature plasma slow wave guiding structures, making possible the direct laser acceleration of charged particles and quasi-phase matched generation of intense electromagnetic waves. Future experiments will explore reduced leakage guides. However, we note that even with the ~80% leakage shown here, the guided laser field at the channel exit is reduced by only ~50% from that at the entrance.

ACKNOWLEDGEMENTS

This work was supported by the US Department of Energy and the National Science Foundation. The authors are grateful to Y. Leng and C. Pesto for technical assistance.

REFERENCES

1. T. Tajima and J. M. Dawson, Phys. Rev. Lett. **43**, 267 (1979); E. Esarey *et al.*, IEEE Trans. Plasma Sci. **24**, 252 (1996).
2. A. Butler *et al.*, Phys. Rev. Lett. **91**, 205001 (2003); D. V. Korobkin *et al.*, *ibid.* **77**, 5206 (1996); H.M. Milchberg *et al.*, J. Opt. Soc. Am. B **12**, 731 (1995); H.M. Milchberg *et al.* Phys. Rev. Lett. **75**, 2494 (1995).
3. L.C. Johnson and T.K. Chu, Phys. Rev. Lett. **32**, 517 (1974); T.K. Chu and L.C. Johnson, Phys. Fluids **18**, 1460 (1975).
4. C.G. Durfee J. Lynch, and H.M. Milchberg, Phys. Rev. E **51**, 2368 (1995).
5. Y. Ehrlich *et al.*, Phys. Rev. Lett. **77**, 4186 (1996); A. Butler *et al.*, Phys. Rev. Lett. **89**, 185003 (2002).
6. P. Volfbeyn *et al.*, Phys. Plasmas **6**, 2269 (1999); E.W. Gaul *et al.*, Appl. Phys. Lett. **77**, 4112 (2000).
7. V. Kumarappan, K.Y. Kim, and H. M. Milchberg, Phys. Rev. Lett. **94**, 205004 (2005).
8. H. Sheng *et al.*, Phys. Rev. E **72**, 036411(2005).
9. T.R. Clark and H.M. Milchberg, Phys. Rev. E **61**, 1954 (2000).
10. L. Schachter, *Beam-Wave Interaction in Periodic and Quasi-Periodic Structures*, Springer (Berlin, 1997).
11. http://www2.slac.stanford.edu/vvc/accelerators/structure.html
12. R. H. Stolen and H. W. K. Tom, Opt. Lett. **12**, 585 (1987); R. Kashyap, J. Opt. Soc. Am. B **6**, 313 (1989); M. Fejer et al., IEEE J. Quant. Elec. **28**, 2631 (1992); S. Chao et al., Opt. Expr. **13**, 7091 (2005).
13. A. Paul *et al.*, Nature **421**, 51 (2003).
14. C.-C. Kuo *et al.*, Phys. Rev Lett. **98**, 033901 (2007); M.-W. Lin *et al.*, Phys. Plasmas **13**, 110701 (2006).
15. H.M. Milchberg *et al.*, Phys. Plasmas **3**, 2149 (1996).
16. J. Cooley *et al.*, Bull Am. Phys. Soc, V. 45, No. 7, 237 (2000); J. Cooley *et al.*, Phys. Rev. E **73**, 036404 (2006).
17. S. Nikitin *et al.*, Phys. Rev. E **59**, R3839 (1999).
18. K.Y. Kim *et al.*, Phys. Rev. Lett. **90**, 023401 (2003).
19. P. Mora and T. M. Antonsen Jr., Phys. Plasmas **4**, 217 (1997).
20. A. York, B.D. Layer, and H.M. Milchberg, in *Advanced Accelerator Concepts*, AIP Conf. Proc. **877**, 807 (2006).
21. P.M.Woodward, J. Inst. Electr. Eng. **93**, 1554 (1947).
22. W. D. Kimura et al., Phys. Rev. Lett. **74**, 546 (1995)
23. P. Serafim, P. Sprangle, and B. Hafizi, IEEE Trans. Plasma Sci **28**,1190 (2000).
24. V. Malka et al., Science **298**, 1596 (2002).
25. T.R. Clark and H.M. Milchberg, Phys. Rev. E **61**, 1994 (2000).
26. C. G. Durfee, T. R. Clark, and H. M. Milchberg, J. Opt. Soc. Am. B **13**, 59 (1996); T. R. Clark and H. M. Milchberg, Phys. Rev. E **57**, 3417 (1998).
27. B.D. Layer, A. York, T.M. Antonsen, S. Varma, Y.-H. Chen, and H.M. Milchberg, submitted for publication

Laser Wakefield Acceleration: A path to creating 100 GeV electron beams on a tabletop

W. B. Mori

Departments of Physics and Astronomy and of Electrical Engineering
University of California, Los Angeles, Los Angeles, CA 90095

Abstract. The extraordinary ability of space-charge waves in plasmas to accelerate charged particles at gradients that are orders of magnitude greater than in current accelerators has been well documented. We show here that 100 TW to 2000 TW class lasers can excite large amplitude wakefields and be stably self-guided in very underdense plasmas to produce 1 to 10 GeV mono-energetic, self-injected electron beams with nC of charge [1]. For such powers the plasma wakes can be excited by the nearly complete blowout, i.e., expulsion, of plasma electrons by the radiation pressure of a short pulse laser [2]. We also show that these wakefields are ideal for accelerating externally injected beams of electrons [1]. The proposed regime is distinct from the "bubble regime" [3,4] in that it advocates using lower densities and wider spot sizes while keeping the intensity relatively constant in order to increase the output electron beam energy and keep the efficiency high. We discuss what laser parameters would be needed to generate 100 GeV beams with more than a nC of charge in a single stage in this LWFA regime [1]. Our theoretical results are verified by three-dimensional particle-in-cell simulations.

This work was done in collaboration with W. Lu, M. Tzoufras, F.S. Tsung, C. Huang, C. Joshi, L.O. Silva, and J. Vieira, and R.A. Fonseca. Work supported by DOE and NSF.

REFERENCES

1. Wei Lu, et al., "Generating multi-GeV electron bunches using single stage laser wakefield acceleration in a 3D nonlinear regime", Phys. Rev. STAB, **10**, 061301 (2007).
2. Wei Lu, et al., "Nonlinear theory for relativistic plasma wakefields in the blowout regime", Phys. Rev. Lett. **96**, 165002 1:4 (2006); Wei Lu et al, "A nonlinear theory for multi-dimensional relativistic plasma wakefields", Phys. Plasmas, **13**, 056709 (2006): and references therein.
3. A. Puhkov and J. Meyer-ter-vehn, "Laser wakefield acceleration: the highly non-linear broken-wake regime", Appl. Phys. B: Lasers Opt. **74**, 355 (2002).
4. S. Gordienko and A. Puhkov, "Scalings for ultra-relativistic laser plasmas and quasi-monoenergetic electrons", Phys. Plasma **12**, 043109 (2005).

CP926, *Atomic Processes in Plasmas—15th International Conference on Atomic Processes in Plasmas*
edited by J. D. Gillaspy, J. J. Curry, and W. L. Wiese
© 2007 American Institute of Physics 978-0-7354-0436-6/07/$23.00

FUNDAMENTAL DATA AND MODELING

Yong-Ki Kim – His Life and Recent Work

Philip M. Stone

National Institute of Standards and Technology
Gaithersburg, MD 20899, USA

Abstract. Dr. Kim made internationally recognized contributions in many areas of atomic physics research and applications, and was still very active when he was killed in an automobile accident. He joined NIST in 1983 after 17 years at the Argonne National Laboratory following his Ph.D. work at the University of Chicago. Much of his early work at Argonne and especially at NIST was the elucidation and detailed analysis of the structure of highly charged ions. He developed a sophisticated, fully relativistic atomic structure theory that accurately predicts atomic energy levels, transition wavelengths, lifetimes, and transition probabilities for a large number of ions. This information has been vital to model the properties of the hot interior of fusion research plasmas, where atomic ions must be described with relativistic atomic structure calculations. In recent years, Dr. Kim worked on the precise calculation of ionization and excitation cross sections of numerous atoms, ions, and molecules that are important in fusion research and in plasma processing for manufacturing semiconductor chips. Dr. Kim greatly advanced the state-of-the-art of calculations for these cross sections through development and implementation of highly innovative methods, including his Binary-Encounter-Bethe (BEB) theory and a scaled plane wave Born (scaled PWB) theory. His methods, using closed quantum mechanical formulas and no adjustable parameters, avoid tedious large-scale computations with main-frame computers. His calculations closely reproduce the results of benchmark experiments as well as large-scale calculations requiring hours of computer time. This recent work on BEB and scaled PWB is reviewed and examples of its capabilities are shown.

Keywords: Yong-Ki Kim, Ionization, Excitation, BEB theory, scaled PWB
PACS: 34.80.Dp, 34.80.Gs, 34.80.Kw

INTRODUCTION

Many scientists knew Yong-Ki Kim. Some knew him well, as a good friend or close colleague. Perhaps visiting with him at home or joining him in various cities, looking for the best Asian restaurants and the best calamari. Others maybe knew Yong-Ki more casually. Perhaps exchanging telephone and email messages, visiting him at NIST (the National Institute of Standards and Technology) or, earlier, at Argonne, and maybe working with him on joint papers. Maybe it was just chatting in hallways at meetings, joining him at lunches, or exchanging occasional emails. Nearly all know of his published articles. There were more than 125 of them. Everyone knows then what a great loss it has been to the atomic physics community, to NIST, and especially to his family, for Yong-Ki to have been killed so tragically this past September in a senseless automobile accident.

CP926, *Atomic Processes in Plasmas—15th International Conference on Atomic Processes in Plasmas*
edited by J. D. Gillaspy, J. J. Curry, and W. L. Wiese
© 2007 American Institute of Physics 978-0-7354-0436-6/07/$23.00

FIGURE 1. Yong-Ki Kim at his official retirement in 2002

Korea

Yong-Ki was born in 1932 in Korea. At that time, Korea was annexed by Japan, and Japanese was the language used in the schools. The curriculum was based on the Japanese system. Yong-Ki came from a distinguished academic family of scholars, teachers, and translators. His is the 22nd generation of this distinguished family, starting with a senior government minister about 1300 AD. He did very well in school, learning English along with Japanese, and was selected as one of only two students from throughout the country to attend a special high school, a well recognized and prized honor in Korea. Some courses were done in English. The high school was poor and text books were often in short supply. They had one copy of an American calculus text book for the whole high school class.

Yong-Ki and a few others spent many extra hours laboriously copying out each page of the book and making copies for each class member on an old mimeograph machine. He was admitted to Seoul National University, the premier university in the country. Unfortunately, in June 1950 during his first week of classes, the Korean war began. Classes were suspended and he, along with other university students, was sent south to the Pusan area, walking almost all the way and with little food and no medical care. Some died. Many families and the general population also fled south, and through a sister's family living in the south Yong-Ki was eventually able to board with them and escape the tent cities that sprung up.

The situation was chaotic. Knowing English well, Yong-Ki applied for work as a translator with the U.S. army. They required competency in typing as well as English and a test was to be given shortly. So he borrowed an old typewriter and slaved away for several days, teaching himself to type. He passed the test and got the job. Payment was in rice and food, and in this way, Yong-Ki helped his whole family survive the early years of the war.

The war ended in 1953 and Yong-Ki was back at Seoul National University. Graduating in 1957, he found a position as an Instructor in the Physics Department of the Korean Air Force Academy. After two years at the Academy, he set off for the

U.S. and the University of Delaware. To keep his costs low, he arranged for passage to San Francisco on a cargo ship that was a converted World War II Liberty ship.

The ship was caught in a violent storm at sea. Everyone was sea sick, and the ship ended up finally at Seattle, weeks behind schedule. The immigration people were suspicious of Yong-Ki. Here was a rather young Korean, speaking fluent English, on a third rate cargo ship, asking for entry. It was unusual at that time for a young Korean on his own to be admitted. But students were admitted, and he had legitimate papers. After much interrogation he was allowed to proceed to San Francisco to get a plane for Delaware. The U.S. has benefited ever since.

Education and Research in the U. S.

In 1961, Yong-Ki received a Masters degree from the University of Delaware and moved to the University of Chicago for further studies. Working under Prof. Roothaan, he received a Ph.D. in 1966 for pioneering calculations of closed-shell atoms with the relativistic Hartree-Fock method. His work was the first relativistic atomic calculation using basis sets. The work was published in the Phys. Rev. 154, 17 (1967), his first published article.

After graduation, he joined the Argonne National Laboratory working with Drs. Inokuti and Platzman, initially as a Research Associate. He remained at Argonne for 17 years, rising eventually to the rank of Senior Physicist. His early work at Argonne was with Inokuti in the general area of inelastic scattering by charged particles. Over some 13 publications, 11 of them with Inokuti, he developed and explored a sum rule for the Bethe cross section for scattering and explored properties of the generalized oscillator strength. Applications were to the helium atom and the negative helium ion, among other small atoms and ions and molecules.

Starting about 1972, he began to look at collisional ionization, especially the energy distribution of secondary electrons ejected by fast charged particles, a subject central to studies of radiation damage. He was able to develop consistent relationships and theoretical constraints on differential and total ionization cross sections. He did this by relating collisions of charged particles with atoms to photoionization, a very fruitful and important connection. The work contributed to a very helpful understanding of the interpretation and extrapolation of experimental data. He maintained an active interest in this subject through at least 1995, publishing some 18 articles, many with Gene Rudd as a coauthor. Two Reviews of Modern Physics articles in 1985 and 1992 with Gene Rudd collected and analyzed all the available data on differential and total cross sections for proton impact on gases.

At the same time, he continued interest in his original Ph.D. thesis area of relativistic effects and calculations. Working often with Kwok-tsang Cheng and Jean-Paul Desclaux, first at Argonne and later at the National Bureau of Standards, he studied the relativistic theory of atomic structure and calculated many relativistic energy levels, oscillator strengths, and transition probabilities. With these and other authors he published five major data papers in Atomic Data and Nuclear Data Tables. The one in 1979 contains many thousands of oscillator strengths for atomic ions isoelectronic with the first row atoms Lithium through Fluorine, and has accumulated more than 335 citations. In all, he published some 45 articles on relativistic effects and it was his continuing interest throughout his career.

It was these relativistic calculations that eventually brought attention of the magnetic fusion community to Kim and indirectly led to his joining the National Bureau of Standards in 1983 (renamed NIST in 1988).

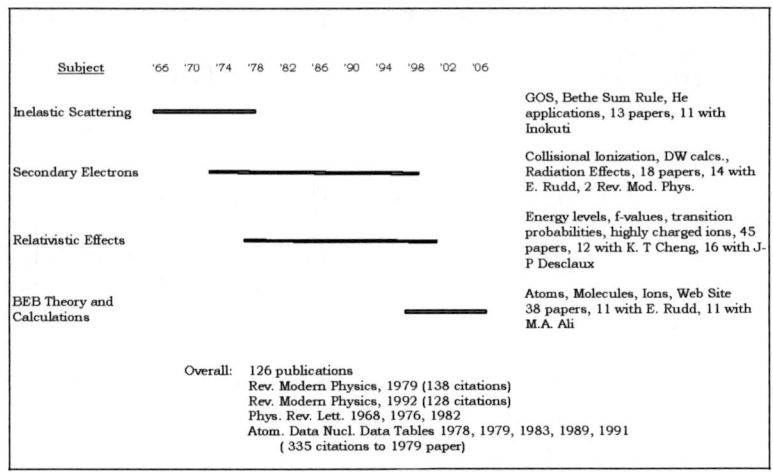

FIGURE 2. A summary of publications by Yong-Ki Kim

A brief summary of Kim's publications is shown in Fig. 2. The last area, encompassing the development with Gene Rudd of the Binary-Encounter-Bethe (BEB) method, has been one of Yong-Ki Kim's latest interests. I believe it has much to offer the plasma community and is not yet fully appreciated.

BED AND BEB THEORY

It was 1994 when Yong-Ki and Gene Rudd published their paper on the Binary-Encounter-Dipole (BED) theory of electron impact ionization [1]. Earlier, Gene Rudd had developed a quasi-empirical equation to calculate the cross section, but in 1994 Yong-Ki made a crucial addition to the derivation that gave a full-blown theoretical solution to the problem.

This theory combines the Mott cross section for electron impact at low energies with the leading dipole part of the Bethe cross section at high impact energies. There have been many attempts to use the Mott cross section, or its variations under the general category of the binary encounter theory, and the first Born approximation. These attempts differ in the way the two theories are joined at some intermediate impact energy. Some of the attempts introduced empirical parameters into the theoretical models, such as work by Khare and colleagues [2]. Others have developed additivity rules for molecular ionization cross sections by introducing parameters derived from ionization of constituent atoms of the molecule, such as work by Märk and colleagues [3].

The BED theory of Kim and Rudd differs in that the combination of the low and high energy approximations is done without introducing arbitrary parameters. The key idea is to combine the two approximations in such a way that both the ionization cross section and the stopping cross section have the correct asymptotic form at high energy predicted by the Bethe theory. No adjustable or empirical parameters are introduced. The theory can be applied to molecules, atoms, or ions. In fact, it is remarkably accurate, compared to experimental measurements, and especially so for molecules.

The Mott cross section for two free electrons with exchange is

$$\frac{d\sigma}{dW} = \frac{4\pi a_0^2 R^2}{T}\left[\frac{1}{W^2} - \frac{1}{W(T-W)} + \frac{1}{(T-W)^2}\right] \tag{1}$$

where T is the incident electron energy, W is the target energy after the collision, R is the Rydberg energy (13.6057 eV), and a_0 is the Bohr radius (0.592 Å). The target electron is initially at rest. After the collision the two electrons are indistinguishable with energies T and T-W. The equation is symmetric with respect to the kinetic energy of the two electrons. The first term, $1/W^2$, is the classical Rutherford cross section for a direct collision and the last term is the exchange collision term. The middle term is the interference between the direct and the exchange terms.

For a real atom, where the target electron is bound with binding energy B, the energy transferred is $E = W + B$, and the Mott equation becomes

$$\frac{d\sigma}{dW} = \frac{4\pi a_0^2 R^2}{T}\left[\frac{1}{E^2} - \frac{1}{E(T-W)} + \frac{1}{(T-W)^2}\right] \tag{2}$$

The denominators in this equation are never zero because the energy transferred, E, and the scattered energy, T-W, always differ by at least the binding energy B.

Including the kinetic energy of the bound electron further alters the cross section equation and in the usual binary encounter theory [4] becomes

$$\frac{d\sigma}{dW} = \frac{4\pi a_0^2 R^2}{T+U+B}\left[\frac{1}{E^2} - \frac{1}{E(T-W)} + \frac{1}{(T-W)^2} + \frac{4U}{3}\left(\frac{1}{E^3} + \frac{1}{(T-W)^3}\right)\right] \tag{3}$$

where the orbital kinetic energy of the bound electron is U and is given by the expectation value of the square of its momentum, $<p^2>/2m$, determined from the wave function of the target electron. This is the version of the Mott cross section used by Kim and Rudd in their theory.

The Born approximation cross section for scattering of an electron at high impact energies from a target atom or molecule is given by

$$\frac{d\sigma_n}{d\ln(Ka_0)^2} = \frac{4\pi a_0^2}{T/R}\frac{f_n(K)}{E_n/R} \tag{4}$$

169

where E_n is the excitation energy to state n (bound or free), K is the magnitude of the momentum change (in units of a_0^{-1}) in the collision, and f_n is the generalized oscillator strength for transitions from the ground state, labeled $|0>$, to the state $|n>$.

$$f_n(K) = \left(\frac{E_n}{R}\right)\frac{1}{(Ka_0)^2}\left|\sum_j \langle n|\exp(i\mathbf{K}\bullet\mathbf{r}_j)|0\rangle\right|^2 \tag{5}$$

The summation in the generalized oscillator strength is over the target electrons with position vectors $\mathbf{r}_j$.

The Bethe cross section is the asymptotic behavior of the first Born approximation and in terms of the energy transferred can be written as

$$\frac{d\sigma}{dW} = \frac{4\pi a_0^2 R^2 B}{T}\ln\left(\frac{T}{B}\right)\frac{1}{W+B}\frac{df(W)}{dW} \tag{6}$$

where $df(W)/dW$ is the continuum oscillator strength [5].

A key idea of Kim and Rudd's theory is to require both the ionization cross section and the stopping cross section – that is, the product of the ionization cross section and the energy transferred from the incident electron to the target – to have the correct asymptotic form required by the Bethe theory. This involves some additional small approximations to the above equations but does not introduce any new or adjustable parameters.

Finally, after integrating the combined equations over W, the total ionization cross section in the Binary-Encounter-Dipole theory (BED) is

$$\sigma_{BED}(t) = \frac{4\pi a_0^2 N \left(R/B\right)^2}{t+u+1}\left[D(t)\ln t + \left(2 - \frac{N_i}{N}\right)\left(\frac{t-1}{t} - \frac{\ln t}{t+1}\right)\right] \tag{7}$$

where the energies are all expressed in reduced dimensionless units, that is, energies divided by the binding energy, or $t = T/B$, $u = U/B$, and $w = W/B$.

The expression for D(t) is

$$D(t) \equiv N^{-1}\int_0^{(t-1)/2}\frac{1}{w+1}\frac{df(w)}{dw}dw \tag{8}$$

where the constant N_i is defined by the integration over the continuum oscillator strength

$$N_i \equiv \int_0^\infty \frac{df(w)}{dw}dw \tag{9}$$

The number of electrons, N, in each target subshell has been added so that the equation gives the cross section for the orbital shell. The total cross section for the atom or molecule is obtained by summing equation (7) for each shell over all the occupied orbital shells.

The BED model requires knowledge of the differential oscillator strength df/dw, a quantity known sometimes from theory (for hydrogen for instance) or from experimental measurements. Most of the time, there is not a good source for the differential oscillator strength, especially for individual subshells.

Because of the difficulty of obtaining the differential oscillator strength in most cases, a simplified version of BED theory, referred to as binary-encounter-Bethe theory, or BEB, was developed. The BEB model assumes a simple analytic form for df/dw which is similar in shape to the df/dw of the ground state of the hydrogen atom. Using this assumed df/dw and integrating over w to get the total ionization cross section from a subshell gives the BEB cross section

$$\sigma_{\text{BEB}} = \frac{4\pi a_0^2 N (R/B)^2}{t+u+1}\left[\frac{\ln t}{2}\left(1-\frac{1}{t^2}\right)+1-\frac{1}{t}-\frac{\ln t}{1+t}\right] \tag{10}$$

As before, summing over the subshells gives the total cross section for the target.

Some representative results of the BEB theory for atoms and molecules are shown in Figs. 3-6 for helium, carbon, H_2, and CH_3.

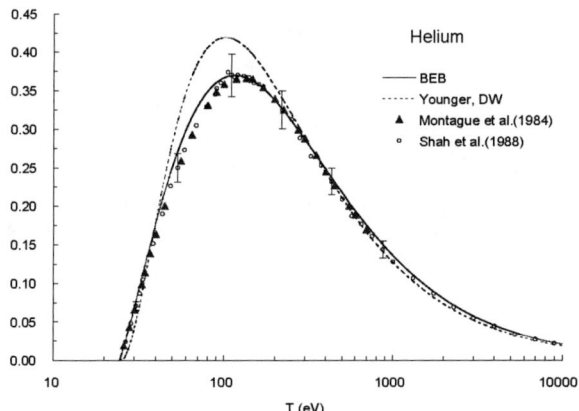

FIGURE 3. BEB ionization cross section for helium (solid line). Measurements by Montague et al. [6] (solid triangles), and Shah et al. [7] (open circles). The distorted wave calculation of Younger [8] is the dotted line.

171

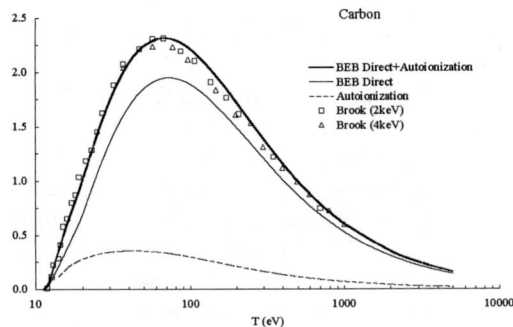

FIGURE 4. BEB direct ionization cross section for carbon (solid line) and with autoionization (heavy line) [9]. The autoionization contribution (dashed line) is calculated with the scaled PWB theory. Measurements by Brook et al. [10] are for ion beams of 2 keV (open squares) and 4 keV (open triangles).

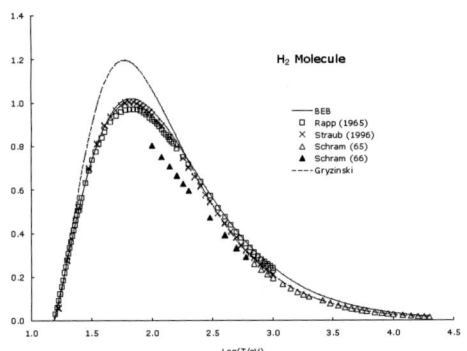

FIGURE 5. BEB ionization cross section for the H_2 molecule (solid line) . Measurements by Schram in 1965 [11] and 1966 [12] are open and closed triangles. Measurements of Rapp et al. [13] are open squares. The classical calculation of Gryzinsky theory [14] is also shown (dashed line).

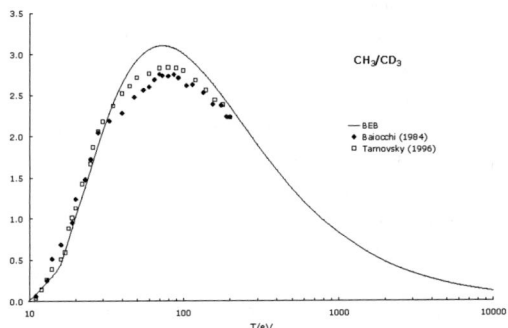

FIGURE 6. BEB ionization cross section for the CH_3 molecule (solid line). The measurements were done on CD_3 by Baiocchi et al. [15] (solid diamonds) and by Tarnovsky et al. [16] (open squares). The BEB calculation is within the uncertainty ranges of the experiments at all energies.

There are many more results presented on a NIST web site. When accessing the site, at http://physics.nist.gov/PhysRefData/Ionization/index.html, a page is presented that directs the viewer to an introduction to the theory and to pages to select a target. The molecular targets are listed as shown on the web page in figure 7.

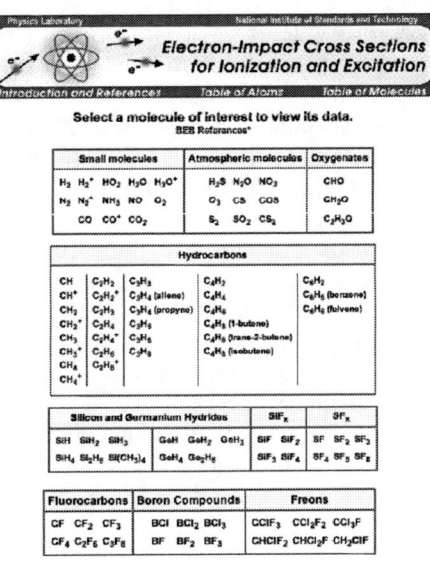

FIGURE 7. Molecule selection page in the NIST website as of 2006 (see text). Clicking on a molecule brings up the BEB cross section and references.

173

There are a number of points to make about the BEB formula, equation (10). First, the equation is quite simple. There are no arbitrary parameters. All that is needed are the values of B, U and N for each target orbital. These values are obtained from experiments or calculated from an electron structure computer program, and once calculated can be stored in any application that uses the cross sections, such as a program to model a gas or plasma discharge. The cross sections can be calculated as needed very quickly using the formula.

More advanced quantum mechanical theories such as close coupling, convergent close coupling, and R-matrix methods are of course a better treatment but require large computers and large amounts of computer time. Because of this, they have not been done for many of the atoms, ions, and molecules needed in most plasma and gas discharge applications. The quick BEB calculation is ideal for these applications.

Second, there is no dependence on continuum wavefunctions. This is why it is so simple and quickly calculated. The continuum is implicitly included through the use of the Mott cross section and the approximate expression for the differential continuum oscillator strength.

Third, the denominator $(T+U+B)$, or in reduced energies $(t+u+1)$, has a strong effect on the cross section at low energies. At high impact energies, the denominator goes to the initial energy T which is the denominator in usual quantum mechanical theory, such as the first Born theory or distorted wave theory. The denominator is not based on as rigorous an argument as the other terms in the BEB equation but comes in via the binary encounter formula. The energy of the incident electron is assumed to gain kinetic energy before it interacts with the atomic electron. This idea originated with Burgess in 1964 [4,17] and is often used in binary encounter theories. The additional kinetic energy of $B+U$ comes from the form of the BED model. The denominator reduces the calculated cross section at low energies where the Born approximation is known to be too large. The denominator also shifts the peak of the cross section to slightly higher energy and usually brings the results into better agreement with measurements.

Fourth, I should emphasize limitations of the BEB model. Because it is based on the Bethe cross section at high energies, which treats collisions through the dipole interaction between the incident and target electrons, the calculation is appropriate only for electric-dipole ionizations. In addition, the theory is a binary impact theory and cannot account for multiple ionization in a single collision. Also, without explicit consideration of the continuum wavefunctions, the calculation cannot include such effects as resonances in the scattering that are included in the more advanced quantum mechanical calculations. In applications to plasmas and discharges, however, the resonances are smoothed out and the BEB method is very appropriate.

SCALED PLANE WAVE BORN (SCALED PWB)

Because of the success of the denominator in the BEB theory for ionization, it was natural to use a similar denominator in first order Born approximation calculations of excitation cross sections. The plane wave Born cross section is proportional to the collision strength and the usual $1/T$ energy dependence and can be written as

$$\sigma_{\text{PWB}} = \frac{4\pi a_0^2 R}{T} F_{\text{PWB}}(T) \qquad (11)$$

where F_{PWB} is the usual collision strength except for a multiplicative constant. Kim simply replaces the T denominator with $T+B+E$, where E is the excitation energy, and refers to the cross section as BE scaled [17].

$$\sigma_{\text{BE}} = \sigma_{\text{PWB}}\left[\frac{T}{T+B+E}\right] \qquad (12)$$

The addition of $B+E$ to T in the denominator is an estimate of the change of incident energy during the collision in the spirit of Burgess [4,17] and the BEB theory. It is not necessarily the best modification and, indeed, in some cases a somewhat larger constant gives better results.

This kind of scaling of the Plane-Wave Born (PWB) cross section for excitation has not been introduced before. Usually, the collision strength F is modified in some way. Kim uses the BEB collision strength and scales only by changing the incident energy. As with BEB ionization, the denominator decreases the PWB cross section at low energies and shifts the peak to slightly higher energy, and gives good agreement with measurements in most cases.

Further scaling is sometimes useful when an accurate value of the dipole oscillator strength is available from experiment or a sophisticated calculation. In such a situation the scaled PWB cross section is

$$\sigma_{\text{scaledPWB}} = \left(\frac{f_{\text{accurate}}}{f_{\text{approximate}}}\right)\frac{T}{T+B+E}\sigma_{\text{PWB}}(T) \qquad (13)$$

where $f_{\text{approximate}}$ is the f-value calculated with the same wave function used to obtain the PWB cross section.

An example of this scaling for calcium is shown in figure 8, from Kim's first article on the subject [18]. There are many other examples in that paper and in subsequent publications.

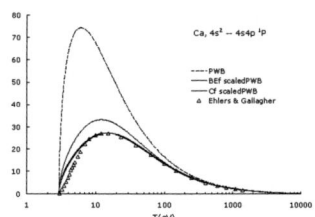

FIGURE 8. Ca excitation cross section [18]. The scaled PWB calculation, with a constant $C = 14$ eV rather than $B+E$ in the denominator, is the heavy solid line. The normal solid line is scaling with $B+E$ = 9 eV in the denominator. The original PWB cross section is the dotted line and is clearly much too large below 100 eV. Measurements by Ehlers and Gallagher [19] are the open triangles.

The scaling of Born excitation cross sections has been used by Kim and others to estimate the magnitude of excitation-autoionization; that is, excitation from an inner shell to a bound state in the continuum followed by autoionization.

The case of aluminum is a good example, because excitation-autoionization is a large contributor to the ionization cross section.

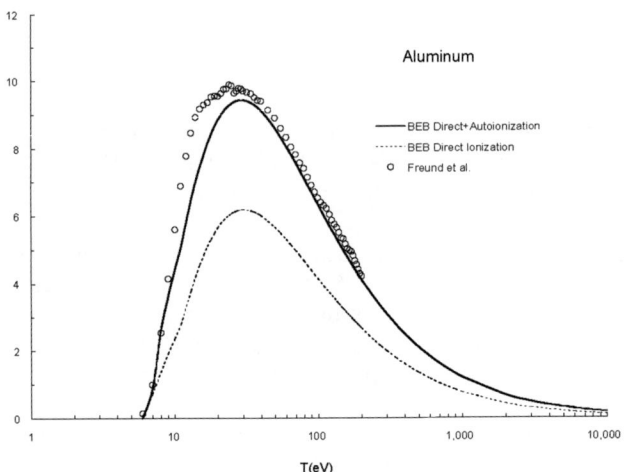

FIGURE 9. Ionization of Al [20]. Direct ionization by the BEB theory is the dotted line. Excitation to autoionizing levels, calculated by the scaled PWB method, has been added to direct ionization to give total ionization (solid line). Autoionization nearly doubles the direct cross section and must be included to get good agreement with the measurements of Freund et al. [21]. Autoionization has been assumed to follow immediately after excitation with 100 % probability.

Ionization of single ions of molybdenum, Mo^+, is a more complicated situation. There are dozens of metastable levels for Mo^+ and hundreds of autoionizing levels. The calculation of Kim includes the two lowest metastable levels and autoionizing levels are included only if their calculated dipole oscillator strengths are greater than 0.05. The scaling for excitation is done with the Coulomb-Born cross section rather than the Born cross section, as appropriate for an ion.

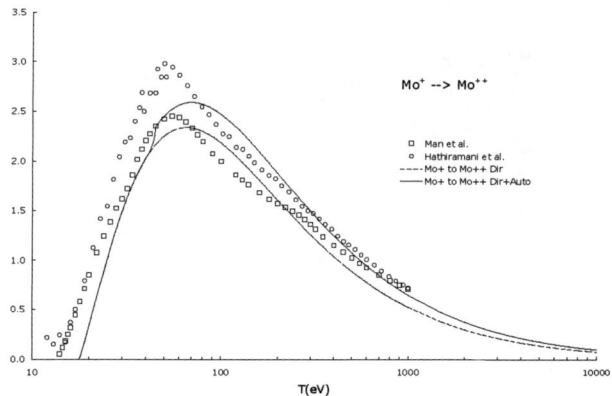

FIGURE 10. Ionization of Mo^+ from D-H Kwon et al. [22]. Direct ionization by the BEB method (dashed line) and with excitation-autoionization (solid line)are shown. Measurements are by Man et al. [23] (open squares) and by Hathiramani et al. [24] (open circles).

The ionization measurements for Mo^+ shown in Fig. 10 are higher than the calculation at the lower energies and show an ionization threshold at a lower energy. These results suggest that more metastable levels of Mo^+ were present in the experimental measurements than the two metastables included in the calculation by Kwon et al. [22].

CLOSING

Yong-Ki's style of doing research was very open and collaborative. He communicated regularly with scientists around the world in meetings, by telephone, and by email, sharing ideas and insights. He was very thorough and exact in calculations but also had great physics insight, especially into electronic structure and orbital wavefunctions. He participated on numerous committees, such as the International Advisory Committee on Radiation Units and Measurements. He was particularly active in supporting this conference series, serving on its Program Committee since 1987 and actively working to organize the conference at NIST in Gaithersburg in 1989. He was a faithful and loyal friend to so many scientists and colleagues.

Yong-Ki was athletic in a quiet way. He was a rock climber, a very accomplished table tennis player, a good automobile mechanic, played tennis regularly, and was recently taking up golf. While serious about his work, he had a good sense of humor and was not above clowning around.

Though officially retired, he was still fully engaged in research. He was full of energy and had many ideas that he wished to pursue. He was always a perfect gentleman, quick to give credit to others and to help with ideas and constructive advice. He seemed to have endless patience with coworkers, visitors, and students. We have truly lost a warm and valued personality.

FIGURE 11. Yong-Ki Kim in China on the border with North Korea. The mountains and lake of North Korea are in the background.

Yong-Ki was greatly helped throughout his career by the support of his wife and family, and was immensely proud of his two children and their own wonderful accomplishments. I believe he would give them credit for much of his success as a scientist and as a very decent person.

In the end, it is the legacy of Yong-Ki's work that will continue to enlighten us.

ACKNOWLEDGMENTS

I want to express my appreciation to many others who have helped me to prepare this paper, and to understand the importance of Yong-Ki's work. I especially want to mention Gene Rudd, Mitio Inokuti and Mahamed Asgar Ali for their thoughts, and Yong-Ki's wife and family for sharing remembrances of the years with Yong-Ki.

REFERENCES

1. Yong-Ki Kim and M. Eugene Rudd, Phys. Rev. A **50**, 3954 (1994).
2. S.P. Khare and W.J.Meath, J. Phys. B **20**. 2101 (1987).
3. H. Deutsch, K. Becker, and T.D. Märk, Int. J. Mass Spectrom. **197**, 37 (2000).
4. L. Vriens, *Case Studies in Atomic Physics*, ed. E.W. McDaniel and M.R.C. McDowell, North Holland, Amsterdam, 1969, vol. 1, p. 335.
5. See, for instance, M. Inokuti, Rev. Mod. Phys. **43**, 297 (1971).
6. R.G. Montague, M.F.A. Harrison, and A.C. Smith, J. Phys. B **17**, 3295 (1984).
7. M.B. Shah, D.S. Elliot, P. McCallion, and H.B. Gilbody, J. Phys. B **21**, 2751 (1988).
8. S.M. Younger, J. Quant. Spectrosc. Radiat. Transfer **26**, 329 (1981).
9. Yong-Ki Kim and Jean-Paul Desclaux, Phys. Rev. A **66**,012708 (2002).
10. E. Brook, M.F.A. Harrison, and A.C.H. Smith, J. Phys. B **11**, 3115 (1978).
11. B.L. Schram, F.J. de Heer, M.J. van der Wiel, and J. Kistenmaker, Physica **31**, 94 (1965).
12. B.L. Schram, H.R. Moustafa, J. Schutten, and F.J. de Heer, Physica **32**, 734 (1966).
13. D. Rapp and P. Englander-Golden, J. Chem. Phys. **43**, 1464 (1965).
14. M. Gryzinsky, Phys. Rev. **138**, A305 (1965); **138**, A322 (1965); **138**, A336 (1965).
15. F.A. Baiocchi, R.C. Wetzel, and R.S. Freund, Phys. Rev. Lett. **53**, 771 (1984).
16. V. Tarnovsky, A Levin, H Deutsch and K Becker, J. Phys. B **29**, 139 (1996).
17. A. Burgess, *Proceedings of the Symposium on Atomic Collision Processes in Plasmas*, Culham, 1964, p. 63.
18. Y.-K. Kim, Phys. Rev. A **64**, 032713 (2001).
19. V.J. Ehlers and A. Gallagher, Phys. Rev. A 7, 1573 (1973).

20. Y.-K. Kim and P.M. Stone, Phys. Rev. A **64**, 052707 (2001).
21. R.S. Freund, R.C. Wetzel, R.J. Shul, and T.R. Hayes, Phys. Rev. A **41**, 3575 (1990).
22. D.-H. Kwon, Y.-J. Rhee, and Y.-K. Kim, Int. J. Mass Spectrom. **245**, 26 (2005).
23. K.F. Man, A.C.H. Smith, and M.F.A. Harrison, J. Phys. B **20**, 1351 (1987).
24. D. Hathiramani, K. Aichele, G. Hofmann, M. Steidl, M. Stenke, R. Völpel, E. Salzborn, M.S. Pindzola, J.A. Shaw, D.C. Griffin, and N.R. Badnell, Phys. Rev. A 54, 587 (1996).

Recent developments in the modeling of dense plasmas

J. Colgan*, C.J. Fontes†, J. Abdallah, Jr.,* and B. Streufert**

*Theoretical Division, Los Alamos National Laboratory, Los Alamos, NM 87545
†Applied Theory Division, Los Alamos National Laboratory, Los Alamos, NM 87545
**University of Virginia, Charlottesville, VA 22192

Abstract. Recent experiments using intense laser pulses on thin targets have produced spectra in which it has been speculated that certain features are due to multiple ionization or recombination events [1, 2]. To explore this possibility, the rate coefficients for collisional double ionization and its inverse process, four-body recombination, have been added to the collisional rate matrix computed within the Los Alamos plasma kinetics code ATOMIC. The collisional double ionization cross sections are obtained from semi-empirical fits to experimental measurements, and the corresponding four-body recombination rates are derived from detailed-balance considerations. We have examined emission spectra produced from solving the coupled rate equations, including the double ionization and four-body recombination rate coefficients, for an Ar plasma in which various fractions of hot electrons are present. We have also explored the sensitivity of our results to the approximations made for the ionization cross sections used in our calculations. We find that inclusion of these multiple-electron effects can make appreciable differences to the average ionization stage of the plasma and the resulting emission spectra at moderately high electron densities, but is strongly dependent on the form of the differential cross sections used in our model.

INTRODUCTION

Most plasma kinetic models designed to calculate emissivities and/or opacities generated under a wide variety of plasma conditions include the various single-electron-impact processes which can distribute population among the various ion species present in the plasma. For example, the Los Alamos plasma kinetics code ATOMIC [3] includes electron-impact collisional excitation and ionization processes, as well as their inverse processes, collisional deexcitation, and three-body recombination. Autoionization and dielectronic recombination are also considered, as well as the radiative processes of photo-excitation and ionization along with their inverse processes photo-deexcitation and radiative recombination.

However, more exotic multi-electron-impact processes are typically not included in such a plasma model. At first glance, one would expect such processes to have little bearing on plasma kinetics, since the cross sections for, say, electron-impact double ionization, are usually much smaller than the cross sections for electron-impact single-ionization from the same ionic species under typical conditions. The threshold energies for such multi-electron processes are also much higher than for single ionization. Since the single-electron processes can allow most ionic species to be reached via a series of stepwise ionization/recombination processes under typical conditions, all multi-electron

CP926, *Atomic Processes in Plasmas—15th International Conference on Atomic Processes in Plasmas*
edited by J. D. Gillaspy, J. J. Curry, and W. L. Wiese
© 2007 American Institute of Physics 978-0-7354-0436-6/07/$23.00

processes are usually ignored. However, if the free-electron energies in the plasma are sufficiently large, and the electron densities are such that the plasma is collisionally dominated, it is interesting to examine whether the effect of such multi-electron processes can be detected. The purpose of this paper is to explore the importance of this effect and is motivated by recent sets of theoretical and experimental comparisons for an Ar plasma produced from a high-intensity ultrashort laser [1, 2]. It has been speculated [4] that differences in the theoretical modeling and the measurements shown in [1, 2] could potentially be explained by further collisional ionization and recombination processes beyond the standard single-electron considerations.

We have added the rate coefficients for electron-impact double ionization, and its inverse process, four-body recombination, to the Los Alamos plasma kinetics code ATOMIC. In the following section we give a short outline of the theoretical methods used, and then we show calculations in which double ionization and recombination processes are included in Ar plasma modeling. We end with a short conclusion.

THEORETICAL BACKGROUND

Collisional double ionization occurs when an electron of energy E collides with an ion A^{q+} in state il and doubly ionizes the target to a state jm,

$$e^- + A^{q+}_{il} \rightarrow e^- + e^- + e^- + A^{(q+2)+}_{jm} . \tag{1}$$

If the post-collision electrons have energies denoted by E_1, E_2, and E_3, then the energy balance can be written as

$$
\begin{aligned}
E &= E_{jm} - E_{il} + E_1 + E_2 + E_3 \\
&= E_0 + E_1 + E_2 + E_3
\end{aligned} \tag{2}
$$

where E_0 is the threshold for double ionization. In this work, we use the empirical fit formulae given by Fisher $et\ al$ [5] to estimate the collisional double ionization cross sections. Although accurate time-dependent close-coupling calculations exist for electron-impact double ionization of helium [6], these methods are computationally intensive and difficult to extend to an arbitrary ion. Furthermore, little or no distorted-wave calculations exist for collisional double ionization. We do note that several models exist for multiple ionization as the result of ion-impact, based on independent-electron models [7] which have recently been extended to include post-collision contributions [8]. However, the extension of such methods to electron-impact has not yet been made. The empirical formula for the total collisional double ionization cross sections used in the present work is given by

$$Q_{il\,jm}(E) = \frac{4\pi a_0^2}{17} \left(\frac{I_H}{I_2} \right)^2 \zeta \frac{\ln x}{x^{1.4}} \left(1 - 2e^{-(\ln 2)x} \right) , \tag{3}$$

where here I_2 is the double ionization threshold, x denotes the incoming electron energy in threshold units, and ζ is a scaling factor which depends on the subshells of the ionized electrons. I_H and a_0 are the ionization potential of hydrogen and the Bohr radius, respectively. This scaling formula produces cross sections which are in reasonable agreement with experiment. For example, Fig. 1 shows the double ionization cross sections for Ar^{7+} and Ar^{8+}. In both cases the scaling formula of Eq. 3 is within a factor of two of the measure1

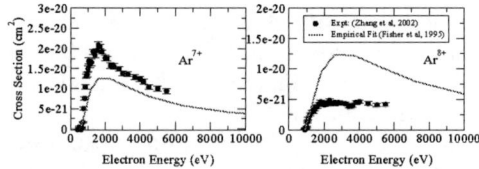

FIGURE 1. Electron-impact double ionization cross sections for Ar^{7+} and Ar^{8+}. In both cases calculations from the scaling formula given by Eq. 3 are compared with measurements of [9].

The (energy) differential cross section for collisional double ionization depends not only on the initial electron energy, but also on the energy of two of the ejected electrons. The energy of the third ejected electron is fixed by energy conservation. This cross section will be denoted as $\sigma_{il\,jm}(E,E_2,E_3)$, and it is useful to express this quantity as the product of the total double ionization cross section $Q_{il\,jm}(E)$ and a 'doubly differential' cross section $\Omega(E,E_2,E_3)$:

$$\sigma_{il\,jm}(E,E_2,E_3) = Q_{il\,jm}(E)\Omega(E,E_2,E_3) . \tag{4}$$

As will be shown below, a knowledge of Ω is required to obtain four-body recombination rate coefficients for non-Maxwell-Boltzmann electron distributions. Expressions for such doubly differential cross sections in energy are not available. In this work, we consider various types of expressions for Ω. The form of this cross section should satisfy two criteria: (1) the doubly differential cross section Ω should be symmetric with respect to the interchange of $E_1 \leftrightarrow E_2$, $E_1 \leftrightarrow E_3$, and $E_2 \leftrightarrow E_3$, and (2) it should be normalized so that integration of the doubly differential cross section Ω gives unity:

$$\int_0^{E-E_0} \int_0^{(E-E_0-E_2)/2} \Omega(E,E_2,E_3)dE_2dE_3 = 1 . \tag{5}$$

However, since no other theoretical or experimental results are available for this differential cross section, we have no information to guide us as to the correct behavior of Ω as a function of the initial electron energy E. In this paper we consider two forms of Ω. The first form is to use an expression which is an extension of a singly differential cross section in energy given by Clark *et al* [10]:

$$\Omega(E,E_2,E_3) = \frac{4}{(E-E_0)^2} - \frac{3}{80} + \left(\frac{1}{2} - \frac{E_2}{E-E_0} - \frac{E_3}{E-E_0}\right)^4$$
$$+ \left(\frac{E_2}{E-E_0} - \frac{1}{2}\right)^4 + \left(\frac{E_3}{E-E_0} - \frac{1}{2}\right)^4 . \tag{6}$$

We also consider an alternative form of the doubly differential cross section in energy which is similar to Eq. 6 but has a more explicit dependence on the incoming electron energy:

$$\Omega(E,E_2,E_3) = \frac{4}{(E-E_0)^2}\left[1-\frac{3(u^2-1)}{2(14.4+u^2)}\right] + \frac{160(u^2-1)}{(14.4+u^2)(E-E_0)^2} \tag{7}$$

$$\times \left\{\left(\frac{1}{2}-\frac{E_2}{E-E_0}-\frac{E_3}{E-E_0}\right)^4 + \left(\frac{E_2}{E-E_0}-\frac{1}{2}\right)^4 + \left(\frac{E_3}{E-E_0}-\frac{1}{2}\right)^4\right\},$$

where $u = E/E_0$. The expressions for the singly differential cross section given by Clark et al [10] are based on fits to distorted-wave calculations and, in the absence of such distorted-wave calculations for doubly differential cross sections, we extend the fitting method of Clark et al [10] to the current case involving three outgoing electrons. These forms of Ω clearly possess the appropriate symmetry about interchange of $E_1 \leftrightarrow E_2$, $E_1 \leftrightarrow E_3$, and $E_2 \leftrightarrow E_3$ and satisfy the normalization condition given by Eq. 5. The dependence on the outgoing energies E_2 and E_3 is plotted in Fig. 2. We see that any slices through the doubly differential cross section parallel to the E_2 or E_3 axis recover a symmetric singly differential cross section, as required. These forms of the cross section Ω describe the probability of ionization producing the two ejected electrons of energies E_2 and E_3 for a given impact energy E. (Again, the energy of the other outgoing electron, E_1, is determined from energy conservation).

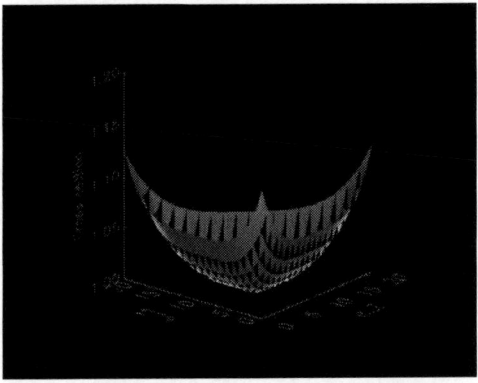

FIGURE 2. Doubly differential energy cross section Ω given by Eq. 6 as a function of the energy of two of the outgoing electrons, for a fixed initial electron energy.

The Los Alamos multipurpose ionization code, GIPPER, has been modified to provide the total double ionization cross sections Q as given by Eq. 3, in units of cm^2. The number of double ionizations per unit volume per unit time in a given energy width is given by

$$dR = N_{il}N_eF(E,T)v(E)Q_{iljm}(E)\Omega(E,E_2,E_3)\,dE\,dE_2\,dE_3. \tag{8}$$

In Eq. 8 N_e is the electron density, N_{il} is the ion density of the initial state il, and $v(E)$ is the electron velocity given by $v(E) = \left(\frac{2E}{m_e}\right)^{1/2}$, with m_e the electron mass. This equation can be integrated over E, E_2, and E_3 (using Eq. 5) to obtain the total double ionization rate as

$$R = N_e N_{il} c(il, jm; T) , \tag{9}$$

where $c(il, jm; T)$, the rate coefficient for collisional double ionization, is given by

$$c(il, jm; T) = \int_{E_0}^{\infty} F(E, T) v(E) Q_{iljm}(E) dE \tag{10}$$

where $F(E, T)$ is the electron energy distribution. For the specific case of a Maxwell-Boltzmann distribution

$$F(E, T) = \frac{2}{\sqrt{\pi}} \frac{\sqrt{E}}{(kT)^{3/2}} e^{-E/kT} , \tag{11}$$

where T is the electron temperature and k is the Boltzmann constant.

The four-body recombination rate coefficient can be derived using the principle of detailed balance. The number of recombinations per unit volume per unit time in a given energy width can be written as

$$dR' = N_{jm} N_e^3 F(E_1, T) F(E_2, T) F(E_3, T) \sigma'_{iljm}(E_1, E_2, E_3) \, dE_1 \, dE_2 \, dE_3 \tag{12}$$

In Eq. 12 σ' is a quantity that depends on the energy of all three post-collision electrons and has CGS units of $cm^9 \, sec^{-1}$. Here N_{jm} is the ion density of the final jm state. Invoking detailed balance, the rates given by Eq. 8 and Eq. 12 are equated and, using Eq. 11, we obtain

$$\sigma'_{iljm}(E_1, E_2, E_3) = \frac{N_{il}}{N_{jm} N_e^2} \frac{1}{2} \frac{\pi}{\sqrt{2m_e}} (kT)^3 \frac{E}{\sqrt{E_1 E_2 E_3}} e^{-E_0/kT} \sigma_{iljm}(E, E_2, E_3) \tag{13}$$

where $dE = dE_1$ when E_2 and E_3 are fixed. Eq. 13 can be further reduced by using the definition of the partition function

$$Z_e = 2 \left(\frac{2\pi m_e kT}{h^2} \right)^{3/2} , \tag{14}$$

where h is the Planck constant, and the Saha-Boltzmann relation that is applicable for populations belonging to ion stages that are separated by two charge states

$$\frac{N_{il}}{N_{jm}} = \frac{g_{il}}{g_{jm}} \left(\frac{N_e}{Z_e} \right)^2 e^{E_0/kT} , \tag{15}$$

to obtain

$$\sigma'_{iljm}(E_1, E_2, E_3) = \frac{g_{il}}{g_{jm}} \frac{h^6}{64\sqrt{2}\pi^2 m_e^{7/2}} \frac{E}{\sqrt{E_1 E_2 E_3}} \sigma_{iljm}(E, E_2, E_3) . \tag{16}$$

Here g_{il} denotes the statistical weight of the initial state il. This expression can be substituted into Eq. 12 and, after integration over E_1, E_2, and E_3, one obtains the total four-body recombination rate as

$$R' = N_{jm} N_e^3 b(il, jm; T) , \qquad (17)$$

where $b(il, jm; T)$ is the four-body recombination rate coefficient given by

$$
\begin{aligned}
b(il, jm; T) &= \frac{g_{il}}{g_{jm}} \frac{h^6}{64\sqrt{2}\pi^2 m_e^{7/2}} \int_0^\infty \int_0^\infty \int_0^\infty F(E_1, T)F(E_2, T)F(E_3, T) \\
&\times \frac{E}{\sqrt{E_1 E_2 E_3}} \sigma_{il\,jm}(E, E_2, E_3)\, dE_1\, dE_2\, dE_3 .
\end{aligned} \qquad (18)
$$

This is the general expression for the four-body recombination rate coefficient for an arbitrary electron distribution and involves a triple integration.

If F is Maxwell-Boltzmann and Eq. 2 is inverted, we obtain the convenient relationship

$$b(il, jm; T) = \frac{g_{il}}{g_{jm}} \frac{e^{E_0/kT}}{Z_e^2} c(il, jm; T) . \qquad (19)$$

Remembering Eq. 14, we see that, for the Maxwell-Boltzmann case, the four-body recombination rate has the explicit temperature dependence T^{-3} as compared to the three-body recombination rate which contains only a single factor of $1/Z_e$ and hence possesses a $T^{-3/2}$ dependence. If the electron distribution is non-Maxwellian, we can rewrite Eq. 18 in the computationally more convenient form

$$
\begin{aligned}
b(il, jm; T) &= \frac{g_{il}}{g_{jm}} \frac{h^6}{64\sqrt{2}\pi^2 m_e^{7/2}} \int_{E_0}^\infty dE\, Q_{il\,jm}(E) E \\
&\times \int_0^{E-E_0} \frac{dE_2 F(E_2, T)}{\sqrt{E_2}} \\
&\times \int_0^{(E-E_0-E_2)/2} \frac{dE_3 F(E_3, T)F(E - E_0 - E_2 - E_3, T)\Omega(E, E_2, E_3)}{\sqrt{E_3}\sqrt{E - E_0 - E_2 - E_3}} .
\end{aligned} \qquad (20)
$$

The rate coefficients for collisional double ionization and four-body recombination have been included in the Los Alamos plasma kinetics code ATOMIC, for cases where the electron distribution is either Maxwellian or non-Maxwellian. The following section details the results of our calculations.

RESULTS

Our initial exploration of the effect of collisional double ionization and four-body recombination (CDI/4BR) considered an Ar plasma for various electron temperatures and

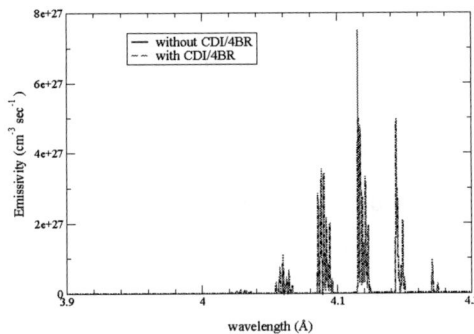

FIGURE 3. Ar emissivity at an electron density of 10^{23} cm^{-3}. A Maxwellian electron distribution is used at an electron temperature of 120 eV. The black dashed line shows calculations without the effects of CDI/4BR; the red solid line shows calculations including CDI/4BR.

densities. In these calculations, a relatively simple model was used at a configuration-average level of approximation, for ion stages ranging from Ar^{9+} to Ar^{18+}. We explored the effect of CDI/4BR in the same regions as the measurements made by Sherrill *et al* [2], although our calculations are not directly comparable, as we do not solve the Boltzmann equation to obtain the electron energy distribution, and our calculations are not in the fine-structure approximation. However, this study should provide a qualitative guide as to the importance of CDI/4BR for these plasma conditions.

In Fig. 3 we show the emissivity from an Ar plasma at an electron temperature of 120 eV and a density of 10^{23} cm^{-3}. A Maxwellian electron distribution has been used. Here it is clear that the effect of CDI/4BR (shown in the solid red line) are minimal, compared with calculations in which no CDI/4BR have been included (dashed black line).

In Fig. 4 we show similar sets of calculations to those displayed in Fig. 3, but in this case a bi-Maxwellian electron distribution is used, where 50% of the electrons are at a temperature of 10 keV. The two figures show the emissivities of the Ar plasma when CDI/4BR are included (red lines) for the two different forms of the doubly differential cross section Ω discussed in the previous section. The calculations without CDI/4BR are shown as the dashed black lines. In this case the inclusion of CDI/4BR makes a noticeable difference to the emissivity when the first form of Ω is used. In particular, the emissivity is enhanced for the lower ion stages. However, for the second form of Ω, no difference is observed in the emissivity when CDI/4BR is included.

After performing a series of kinetics calculations over a broad range of the available parameter space (e.g. N_e, T_e, fraction of hot electrons, temperature of hot electrons, etc.), several conclusions may be drawn from this study. If the form of Ω is such that the four-body recombination rate coefficients become comparable to, or larger than, the three-body recombination rate coefficients, for high electron densities, then inclusion of CDI/4BR makes a difference to the emissivity. We remember that the four-body recombination rates have a N_e^3 dependence (Eq. 17). Further investigation shows that the four-body recombination is the dominant factor, with the collisional double ionization rates making little difference for this case. Large four-body recombination rates drive the plasma towards a lower average ionization stage, which results in higher emission from charge states with more electrons. Further calculations show that an electron density above 10^{22} cm^{-3} is necessary before the effects of the four-body recombination rate coefficients become noticeable in the average ionization stage and in the resulting emissivity. Also, a hot-electron fraction of over 10%, where the hot electrons are at a temperature of at least 1 keV, is also required.

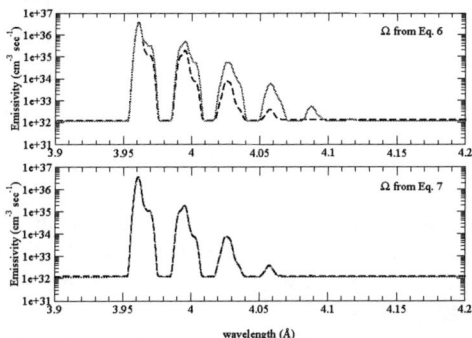

FIGURE 4. Ar emissivity at an electron density of 10^{23} cm^{-3} and electron temperature of 120 eV. A bi-Maxwellian electron distribution is used, with 50% of the electrons at a temperature of 10 keV. The black dashed line shows calculations without the effects of CDI/4BR; the red solid line shows calculations including CDI/4BR. The two figures show the emissivity produced when the two forms of the doubly differential cross section Ω discussed in the previous section are used.

The four-body recombination rate coefficients are quite sensitive to the form of the doubly differential cross section Ω, used in Eq. 20. In particular, the recombination rate coefficients are sensitive to the behavior of Ω as a function of the initial electron energy E. The first form of Ω considered has a clear dependence on the initial electron energy E but also contains some terms which are independent of E. This choice allows the four-body recombination rate coefficients to have a somewhat similar behavior to the collisional double ionization cross section and, in particular, allows the possibility that the four-body recombination rate coefficient can increase as a function of temperature, for low electron temperatures. (We remember that, near threshold, the collisional double ionization cross section follows the Wannier power law and increases as E^2 [5]). These rate coefficients can then approach (and eventually exceed) the three-body recombination rate coefficients for this high electron density as the temperature increases, resulting in noticeable differences in the emissivity whenever CDI/4BR is included.

The final form of Ω considered, Eq. 7, has an overall initial electron energy dependence due to the $(E - E_0)^2$ terms appearing in the denominator of each term. This choice produces differential cross sections that rise sharply as the initial electron energy is decreased, producing behavior that is similar to the singly differential cross section expression used in three-body recombination as discussed by Clark *et al* [10]. It also constrains the four-body recombination rate coefficients to always decrease as a function of temperature. Thus, even for the high electron densities considered in these calculations, use of this form of Ω results in relatively small four-body recombination rate coefficients, which in turn make little difference to the calculated emissivity, as shown in the bottom plot of Fig. 4.

It could be argued that this final form of Ω considered is perhaps the closest to what the actual physical form of Ω must be, since it is the closest analogue with the expression used for singly differential cross sections by Clark *et al* [10]. The singly differential cross section defined by [10] has been used in three-body recombination rate coefficients for non-Maxwellian distributions in the Los Alamos suite of kinetics codes for many years. However, without a derivation of the form of the energy doubly differential cross section for collisional double ionization, or an experimental measurement of such a quantity, the final effect of four-body recombination on dense, non-Maxwellian plasmas must remain an open question.

SUMMARY

The multi-electron processes of collisional double ionization and four-body recombination have been added to the Los Alamos suite of atomic physics codes and the effects of these processes have been studied for both Maxwellian and non-Maxwellian electron distributions. We have found that, for non-Maxwellian distributions, the four-body recombination rates are sensitive to the form of the doubly differential (in energy) cross sections for collisional double ionization. Unfortunately, these differential cross sections have not been studied in any detail, either theoretically or experimentally, to the best of our knowledge. Although we have shown that a particular form of this differential cross section implies relatively large four-body recombination rate coefficients which can lower the average charge state and modify the resulting emission spectra of the Ar plasma under investigation, it appears that the most likely, physically intuitive, form of the doubly differential cross section implies four-body recombination rate coefficients which do not significantly alter the characteristics of the plasma.

ACKNOWLEDGMENTS

This work was performed under the auspices of the Department of Energy by Los Alamos National Laboratory. We thank Prof. R. C. Mancini for useful discussions.

REFERENCES

1. J. Abdallah, Jr., G. Csanak, Y. Fukuda, Y. Akahane, M. Aoyama, N. Inoue, H. Ueda, K. Yamakawa, A. Ya. Faenov, A. I. Magunov, T. A. Pikuz, and I. Yu. Skobelev, *Phys. Rev. A* **68**, 063201 (2003).
2. M.E. Sherrill, J. Abdallah, Jr., G. Csanak, E.S. Dodd, Y. Fukuda, Y. Akahane, M. Aoyama, N. Inoue, H. Ueda, K. Yamakawa, A. Ya. Faenov, A. I. Magunov, T. A. Pikuz, and I. Yu. Skobelev, *Phys. Rev. E* **73**, 066404 (2006).
3. N.H. Magee, J. Abdallah, J. Colgan, P. Hakel, D.P. Kilcrease, S. Mazevet, M. Sherrill, C.J. Fontes, and H.L. Zhang, *14th APS Topical Conference on Atomic Processes in Plasmas*, (Eds: J.S. Cohen, S. Mazevet, and D.P. Kilcrease) AIP Conference Proceedings, Melville, New York, 2004, p168.
4. R. C. Mancini, private communication.
5. V. Fisher, Y. Ralchenko, A. Goldgirsch, D. Fisher, and Y. Maron, *J. Phys. B* **28**, 3027 (1995).

6. M.S. Pindzola, F. Robicheaux, J.P. Colgan, M.C. Witthoeft, and J.A. Ludlow, *Phys. Rev. A* **70**, 032705 (2004).
7. W.E. Meyerhof, R. Anholt, X.-Y. Xu, H. Gould, B. Feinberg, R.J. McDonald, H.E. Wegner, and P. Thieberger, *Phys. Rev. A* **35**, 1967 (1987).
8. C.D. Archubi, C.C. Montanari, and J.E.Miraglia, *J. Phys. B* **40**, 943 (2007).
9. H. Zhang, S. Cherkani-Hassani, C. Belenger, M. Duponchelle, M. Khouilid, E.M. Oualim, and P. Defrance, *J. Phys. B* **35**, 3829 (2002).
10. R.E.H. Clark, J. Abdallah, Jr., and J.B. Mann, *Ap. J.* **381**, 597 (2000).

Atomic Data For Determining Abundances In Interstellar Clouds

S. R. Federman

Department of Physics and Astronomy, University of Toledo, Toledo, OH 43606 USA

Abstract. Chemical abundances in interstellar clouds are used to examine processes taking place in this environment, to extract gas density and pressure, and to infer astronomical sources for the synthesis of nuclei. The derivation of abundances from absorption lines at ultraviolet and visible wavelengths relies on knowledge of oscillator strengths. I will discuss our multi-pronged approach involving laboratory measurements, interstellar observations, and theoretical computations for determining oscillator strengths. Results for atoms and ions of chlorine, magnesium, carbon, and phosphorus will be presented. Comparisons with other results will highlight where consensus has been achieved and where further work is needed.

Keywords: Interstellar atoms; interstellar abundances; atomic lifetimes and transition probabilities; atomic oscillator strengths.
PACS: 31.15.Ct, 32.30-r, 32.30.Jc, 32.70.Cs, 32.70.Fw, 39.90+d, 98.38.–j

INTRODUCTION

The interstellar medium within the disks of galaxies like our own is a mixture of neutral and ionized gas. The physical conditions span an enormous range, with temperatures from 10 to 10^6 K and gas densities from 0.001 atoms per cm^3 in highly ionized material to 10^6 molecules per cm^3 in sites of star formation. Ionization arises from a combination of absorbing (stellar) ultraviolet radiation permeating space, high-speed collisions involving electrons and ions, and collisions involving relativistic Galactic cosmic rays. The focus of this presentation is on interstellar clouds, which occupy about 1% of the volume but contain most of the mass. The emphasis is on the outer portions of a cloud, traditionally called diffuse clouds, where the gas is neutral and is a mixture of atomic and molecular hydrogen. The main means of probing this material is through measurements of absorption lines at visible and ultraviolet (UV) wavelengths, using a background hot star as the continuum source.

The amount of absorption yields abundances, from which information about the processes taking place in the gas as well as the pertinent physical conditions is obtained. The inferred abundances are also used to improve our understanding of the synthesis of elements. A key ingredient in such analyses is the conversion of the amount of absorption into abundance. The UV spectra, acquired with the *Hubble Space Telescope* and the *Far Ultraviolet Spectroscopic Explorer*, are of such high quality that the precision in the atomic and molecular data is now often the limitation.

In order to address the need for more precise atomic and molecular data at UV wavelengths, our group at the University of Toledo has conducted a series of studies

CP926, *Atomic Processes in Plasmas—15th International Conference on Atomic Processes in Plasmas*
edited by J. D. Gillaspy, J. J. Curry, and W. L. Wiese
© 2007 American Institute of Physics 978-0-7354-0436-6/07/$23.00

involving laboratory measurements, astronomical observations, and theoretical calculations. Beam-foil spectroscopy is used to determine lifetimes, branching fractions, and the resulting oscillator strengths. The effects of systematic errors, such as foil thickening and beam divergence, are inferred through measurements taken at two energies and by acquisition of forward and reverse decay curves [1]. Whenever possible, cascades are treated via the method of Arbitrarily Normalized Decay Curves [2]. Two techniques are used to derive oscillator strengths from interstellar spectra; both rely on the availability of well-determined oscillator strengths for a subset of transitions so that relative values can be converted to absolute ones. One method involves the acquisition of curves of growth, which are plots of the amount of absorption versus abundance × oscillator strength. There are three distinct portions of curves of growth: (1) the linear portion; (2) the flat portion that is controlled by thermal and turbulent broadening in the core of a line; and (3) the square-root portion that is controlled by the lifetime of the upper state involved in the transition. This method is best when numerous lines, spanning a significant range in optical depth at line center and thereby sampling much of a curve of growth, are included in the analysis. An example is the study of lines in neutral sulfur [3]. Profile synthesis can also be used to extract oscillator strengths. Here the amount of absorption is reproduced by a synthetic line, whose profile contains a thermally-broadened (Doppler) core and Lorentzian wings, convolved with a Gaussian instrumental function, in a least-squares manner. The work on neutral carbon [4] that is described below utilized this method. Our theoretical efforts are based on a suite of techniques, including Coulomb approximation with a central potential, relativistic Hartree Fock, and Multi-configuration Hartree Fock, as well as semi-empirical methods.

The remainder of the paper presents results on several atomic systems studied by the techniques outlined above. In particular, work on neutral chlorine, singly-ionized magnesium, neutral carbon, and singly-ionized phosphorus is highlighted. The final section includes a summary and future directions for our program.

ILLUSTRATIVE EXAMPLES

Neutral Chlorine

The ionization balance for chlorine in diffuse clouds differs from that of all other abundant elements. For most elements, the dominant charge state is determined by the ionization potential (IP) for the neutral atom. If the IP is less than 13.6 eV, the IP for hydrogen, singly-charged ions dominate, and neutrals dominate when the IP is greater than 13.6 eV. In the former situation, neutral atoms arise from electron recombination, a relatively slow process. However, reactions between Cl^+ and H_2 lead to a significant abundance of Cl^0 in molecular-rich gas. The chemical coupling produces comparable abundances for neutral and singly-ionized chlorine in this environment [5, 6].

TABLE 1. Oscillator Strengths for Lines of Neutral Chlorine. The numbers in parentheses indicate uncertainties in the last digit(s). Labels L and V refer to calculations performed in the length or velocity formalism.

Reference	λ1088	λ1097
Schectman et al. [10]	0.081(7)	0.0088(13)
Federman [7]	0.094(26)	...
Jura and York [5]	...	0.019(5)
Ojha and Hibbert [8]	0.016 (L); 0.006(V)	0.042 (L); 0.017(V)
Biémont et al. [11]	0.00495 (L); 0.00450(V)	0.0803 (L); 0.0730(V)
	0.0688 (L); 0.0618(V)	0.0122 (L); 0.0111(V)
Froese Fischer et al. [12]	0.2099	0.00293

The first studies of interstellar chlorine involved spectra acquired with the *Copernicus* satellite. Because the Cl I line at 1347 Å is relatively strong, weaker lines yield more precise abundance determinations, but their *f*-values were not known at the time. Astronomically-derived values were inferred for the Cl I lines at 1097 Å [5] and 1088 Å [7] from the amount of absorption seen at 1347 Å (see Table 1). The first large-scale computations [8], however, indicated that λ1097 was stronger than λ1088, contrary to astronomical observations [see also 9]. In order to resolve which set of *f*-values to use, we [10] performed beam-foil measurements that confirmed the astronomical results. As part of this effort, exploratory relativistic Hartree-Fock calculations revealed significant configuration interaction between the $3p^4 5d$ and $3p^4 5s$ configurations; this led us to suggest that the identifications for the upper states in the two interstellar lines needed to be reversed. Subsequent theoretical efforts [11, 12] are consistent with this hypothesis, as shown in Table 1. When the upper-state designation was switched [11] (as in lines 6 and 7 of Table I), the agreement with empirical results was greatly improved. *Ab initio* multi-configuration Hartree-Fock (MCHF) calculations [12] indeed suggest that the $3p^4 5d\,^2D$ levels have higher energy.

Recently, astronomical *f*-values for lines at 1005, 1079, 1091, and 1095 Å were published [13]. The comparison with theoretical results [11] for the three transitions with the longest wavelength shows factor-of-2 differences for λλ1091,1095. All three transitions involve $3p^4 5s$ as the upper state. Further theoretical effort is required to discern whether these discrepancies have a similar origin to those described above.

Singly-ionized Magnesium

Magnesium is readily incorporated into interstellar dust grains because it is very refractory. The mineralogy of interstellar dust is inferred through comparisons of gas phase abundances with abundances found in the Solar System [e.g., 14]. The gas phase abundance of Mg is deduced from measurements of Mg II absorption from doublets near 2800 and 1240 Å. The lines near 2800 Å are very strong and usually have zero intensity at line center; placement of the stellar continuum can lead to imprecise measures of absorption in the line wings. Those near 1240 Å are much weaker, the result of accidental cancellation in the radial dipole matrix elements [e.g., 15]. Analysis of the weak doublet is preferred when extracting the Mg^+ abundance for studies on grain mineralogy as well as for deriving electron densities from the Mg^0/Mg^+ ratio [e.g., 16].

TABLE 2. Results for the Doublet of Singly-ionized Magnesium at 1240 Å. The numbers in parentheses indicate uncertainties in the last digit(s).

Reference	Multiplet f-Value	Branching Ratio
Theodosiou and Federman [15]	$9.88(7) \times 10^{-4}$	1.78(3)
Fitzpatrick [17]	$9.5(6) \times 10^{-4}$	1.82(8)
Sofia et al. [18]	$9.71(32) \times 10^{-4}$	1.78(3)
Majumber et al. 2002 [19]	9.3×10^{-4}	1.80
Froese Fischer et al. [12]	7.014×10^{-4}	1.72

Higher order terms in the Hamiltonian that are usually insignificant become important when accidental cancellation of the matrix elements arises. As a result, the branching ratio for the two transitions can differ from the 2:1 ratio expected from LS coupling rules [15]. As shown in Table 2, subsequent analyses of interstellar spectra [17, 18] and theoretical calculations [12, 19] confirmed our results for multiplet oscillator strength and branching ratio. The astronomical studies compared the abundance derived from line wings seen in the doublet at 2800 Å with the abundance extracted from the weaker lines. Our semi-empirical work [15] was based on the Coulomb approximation with a central potential method, which is especially suitable for alkali-like systems. The more recent *ab initio* calculations utilized relativistic coupled cluster theory [19] and MCHF theory [12].

Neutral Carbon

The three fine structure levels of the ground state in neutral carbon can be populated under the densities (50 cm^{-3}) and temperatures (50 K) prevalent in diffuse clouds. The relative populations obtained from UV observations, therefore, can be used to extract the conditions for individual clouds [e.g., 20, 21]. Accurate determinations require knowledge of oscillator strengths for the numerous lines of C I seen in interstellar spectra. Laboratory measurements and modern theoretical computations for f-values of dipole-allowed transitions above 1200 Å are in very good agreement [e.g., 22]. However, the situation is different for spin-changing transitions above 1200 Å and all transitions below this wavelength. Below 1200 Å the only comparisons are between astronomical determinations and theoretical results.

TABLE 3. Oscillator Strengths for Neutral Carbon Lines near 1158 Å in units of 10^{-3}. The numbers in parentheses indicate uncertainties in the last digit(s).

Transition	Wiese et al. [23]	Federman and Zsargó [4]	Zatsarinny and Froese Fischer [24]
$^3P_1 - 2p5d\,^3F_2^{\circ}$	...	2.23(0.29)	1.71
$^3P_1 - 2p6s\,^3P_0^{\circ}$	1.86	1.90(0.27)	1.55
$^3P_2 - 2p6s\,^3P_2^{\circ}$	4.18	5.91(1.64)	1.74
$^3P_0 - 2p6s\,^3P_1^{\circ}$	5.57	13.7(2.8)	6.80
$^3P_2 - 2p5d\,^3D_2^{\circ}$	3.66	3.53(1.84)	5.28
$^3P_1 - 2p5d\,^3D_1^{\circ}$	6.09	6.97(0.79)	3.98
$^3P_1 - 2p6s\,^3P_2^{\circ}$	2.32	17.8(1.8)	10.5
$^3P_2 - 2p5d\,^3D_3^{\circ}$	20.5	34.0(4.0)	15.6
$^3P_0 - 2p5d\,^3D_1^{\circ}$	24.4	40.5(5.6)	21.2
$^3P_1 - 2p5d\,^3D_2^{\circ}$	18.3	9.36(1.01)	8.64

Table 3 highlights results for the many transitions occurring near 1158 Å, where astronomical f-values [4] are compared with those in a NIST compilation [23] and a large-scale MCHF calculation [24]. The interstellar results are based on lines above 1200 Å. The compilation is based on R-matrix calculations with branching fractions derived from LS coupling rules. For transitions to the $2p6s$ $^3P^o$ levels, the results presented in the compilation and the astronomical determinations differ greatly. Exploratory Hartree-Fock calculations [25] suggest that the $2p6s$ $^3P^o$ levels are affected by configuration interaction with the $2p5d$ $^3P^o$ levels and spin-orbit mixing with the $2p5d$ $^3D^o$ levels. The comparison with the MCHF results [24] is similar, however, with differences seen here as well. Sometimes, the compilation and the theoretical results are in better agreement. A further complication arises when considering the f-values inferred from another interstellar study [26] because in this study, the differences between their values and those in [23] gets larger as weaker lines are considered. Clearly, more work, both theoretical and empirical, is needed. On a more encouraging note, a recent theoretical effort [27] that builds on earlier work [24] shows improved correspondences with interstellar analyses [4, 25] for spin-forbidden transitions near 1270 Å.

Singly-ionized Phosphorus

Singly-ionized phosphorus represents the dominant charge state for this element in diffuse clouds. Its abundance is used to examine the amount of phosphorus incorporated into interstellar grains [28-30] and to study the metal content and nucleosynthetic history in distant galaxies [e.g., 31-33]. The resonance line at 1153 Å is commonly used for these studies because it is relatively strong and in a clean portion of the spectrum.

In the course of a comprehensive experimental investigation yielding f-values for all lines in the multiplet containing $\lambda 1153$ [34], precise branching fractions were obtained. The branching fractions were compared with theoretical and semi-empirical results [35] to ascertain the usefulness of semi-empirical branching fractions for calibrating UV instrumentation via in-beam methods. A synopsis of the comparison appears in Table 4. A variety of calculations are presented here: configuration interaction [36]; MCHF [12, 37]; and semi-empirical [38]. The agreement among

TABLE 4. Branching Fractions for Transitions in the Singly-ionized Phosphorus Multiplet at 1154 Å. The numbers in parentheses indicate uncertainties in the last digit(s). The primes indicate levels that are affected by spin-orbit mixing.

Transition	Hibbert [36]	Tayal [37]	Froese Fischer et al. [12]	Curtis [38]	Federman et al. [34]
$^3P_0{}' \leftarrow {}^3P_1{}^{o\prime}$	33.4	32.9	33.3	33.1	35.9(2.7)
$^3P_1 \leftarrow$	24.8	24.3	24.6	24.5	25.4(1.4)
$^3P_2{}' \leftarrow$	41.8	40.9	41.3	41.0	38.7(2.6)
$^1D_2{}' \leftarrow$	...	2.0	0.8	1.3	...
$^1S_0{}' \leftarrow$	...	...	0.04	0.12	...
$^3P_1 \leftarrow {}^3P_2{}^o$	25.3	25.3	25.3	25.2	26.7(1.6)
$^3P_2{}' \leftarrow$	74.7	74.7	74.7	74.8	73.3(3.3)
$^1D_2{}' \leftarrow$	...	0.0003	0.025	0.005	...

determinations is excellent for dipole-allowed transitions. This is not unexpected because there exists little configuration interaction among the levels involved in the multiplet and because intermediate coupling allows one to derive branching fractions from spin-orbit mixing angles. Unfortunately, the fractions for spin-changing transitions were too weak to measure and could not be used to help clarify the differences seen in the Table. Overall, the comparison in Table 4 suggests that semi-empirical results for other similar systems, such as the Sn I isoelectronic sequence, can be used to calibrate instruments for experiments involving high-speed beams over the wavelength range 400 to 1500 Å [35]. We will pursue this over the course of the coming years. We will also study other P II multiplets ($\lambda\lambda$964, 967, 1308) where the correspondence between theoretical branching fractions is less satisfactory, with a goal of discerning the best theoretical model for the ion.

CONCLUDING REMARKS

This review presented our group's ongoing activities that provide precise oscillator strengths for use in determining interstellar abundances. Examples involving laboratory measurements, astronomical observations, and theoretical calculations for transitions in Cl I, Mg II, C I, and P II were highlighted. In many instances, theoretical and empirical (laboratory and astronomical) are converging (e.g., some f-values for UV transitions in Cl I, Mg II f-values and branching ratios, and P II branching fractions and the resulting f-values). In other cases, suggestions for additional work are given (e.g., other transitions in Cl I and P II and lines below 1200 Å in C I). This will surely keep providers of atomic data for the astrophysics community busy for some time to come.

ACKNOWLEDGMENTS

I want to thank the many colleagues and students who participated in this research. Special thanks go to Larry Curtis, Dave Ellis, and Dick Schectman for their insight into the underlying physics. The work described here was supported by grants from NASA and the Space Telescope Science Institute. Additional support from NSF under its Research Experience for Undergraduates (REU) Program is gratefully acknowledged.

REFERENCES

1. S. R. Federman, D. J. Beideck, R. M. Schectman and D. G. York, *Astrophys. J.* **401**, 367-370 (1992).
2. L. J. Curtis, H. G. Berry and J. Bromander, *Physica Scripta* **2**, 216-220 (1970).
3. S. R. Federman and J. A. Cardelli, *Astrophys. J.* **452**, 269-274 (1995).
4. S. R. Federman and J. Zsargó, *Astrophys. J.* **555**, 1020-1026 (2001).
5. M. Jura and D. G. York, *Astrophys. J.* **219**, 861-869 (1978).
6. P. Sonnentrucker, S. D. Friedman, D. E. Welty, D. G. York and T. P. Snow, *Astrophys. J.* **576**, 241-254 (2002).
7. S. R. Federman, *Astrophys. J.* **309**, 306-310 (1986).
8. P. C. Ojha and A. Hibbert, *Physica Scripta* **42**, 424-430 (1990).

9. S. R. Federman, Y. Sheffer, D. L. Lambert and V. V. Smith, *Astrophys. J.* **619**, 884-890 (2005).
10. R. M. Schectman, S. R. Federman,, D. J. Beideck and D. G. Ellis, *Astrophys. J.* **406**, 735-738 (1993).
11. E. Biémont,, R. Gebarowski and C. J. Zeippen,, *Astron. Astrophys.* **287**, 290-296 (1994).
12. C. Froese Fischer, G. Tachiev and A. Irimia, *At. Data Nucl. Data Tables* **92**, 607-812 (2006).
13. P. Sonnentrucker, S. D. Friedman and D. G. York, *Astrophys. J.* **650**, L115-L118 (2006).
14. U. J. Sofia, J. A. Cardelli and B. D. Savage, *Astrophys. J.* **430**, 650-666 (1994).
15. C. E. Theodosiou and S. R. Federman, *Astrophys. J.* **527**, 470-473 (1999).
16. D. E. Welty, L. M. Hobbs, J. T. Lauroesch, D. C. Morton, L. Spitzer and D. G. York, *Astrophys. J. Suppl.* **124**, 465-501 (1999).
17. E. Fitzpatrick, private communication.
18. U. J. Sofia, D. Fabian and J. C. Howk, *Astrophys. J.* **531**, 384-390 (2000).
19. S. Majumder, H. Merlitz, G. Gopakumar, B. P. Das, U. S. Mahapatra and D. Mukherjee, *Astrophys. J.* **574**, 513-517 (2002).
20. E. B. Jenkins, M. Jura and M. Loewenstein, *Astrophys. J.* **270**, 88-104 (1983).
21. J. Zsargó and S. R. Federman, *Astrophys. J.* **589**, 319-337 (2003).
22. D. C. Morton, *Astrophys. J. Suppl.* **149**, 205-238 (2003).
23. W. Wiese, J. Fuhr and T. M. Deters, "Atomic Transition Probabilities of Carbon, Nitrogen, and Oxygen: A Critical Data Compilation" Amer. Chem. Soc., Washington, DC 1996.
24. O. Zatsarinny and C. J. Froese Fischer, *J. Phys. B* **35** 4669-4683 (2002).
25. J. Zsargó, S. R. Federman and J. A. Cardelli, *Astrophys. J.* **484**, 820-827 (1997).
26. E. B. Jenkins and T. M. Tripp, *Astrophys. J. Suppl.* **137**, 297-340 (2001).
27. C. Froese Fischer, *J. Phys. B* **39**, 2159-2167 (2006).
28. P. L. Dufton, F. P. Keenan, and A. Hibbert, *Astron. Astrophys.* **164**, 179-183 (1986).
29. E. B. Jenkins, B. D. Savage, L. Spitzer Jr., *Astrophys. J.* **301**, 355-379 (1986).
30. C. Mallouris, et al., *Astrophys. J.* **558**, 133-144 (2001).
31. P. Molaro, S. A. Levshakov, S. D'Odorico, P. Bonifacio and M. Centurión, *Astrophys. J.* **549**, 90-99 (2001).
32. S. A. Levshakov, M. Dessauges-Zavadsky, S. D'Odorico and P. Molaro, *Astrophys. J.* **565**, 696-719 (2002).
33. M. Pettini,, S. A. Rix, C. C. Steidel, K. L. Adelberger, M. P. Hunt and A. E. Shapley, *Astrophys. J.* **569**, 742-757 (2002).
34. S. R. Federman, M. Brown, S. Torok, S. Cheng, R. E. Irving, R. M. Schectman and L. J. Curtis, *Astrophys. J.* **660**, 919-921 (2007).
35. L. J. Curtis, S. R. Federman, S. Torok, M. Brown, S. Cheng, R. E. Irving and R. M. Schectman, *Physica Scripta* **75**, C1-C7 (2007).
36. A. Hibbert, *Physica Scripta* **38**, 37-44 (1988).
37. S. S. Tayal, *Astrophys. J. Suppl.* **146**, 459-465 (2003).
38. L. J. Curtis, *J. Phys. B*, **33**, L259-L263 (2000).

Experiments on Interactions of Electrons with Molecular Ions in Fusion and Astrophysical Plasmas

M. E. Bannister, H. Aliabadi, E. M. Bahati, M. R. Fogle, P. Krstić, and C. R. Vane* and A. Ehlerding, W. Geppert, F. Hellberg, V. Zhaunerchyk, M. Larsson, and R. D. Thomas†

*Physics Division, Oak Ridge National Laboratory, Oak Ridge, TN 37831-6372, USA
†Department of Physics, Stockholm University, S106 91 Stockholm, Sweden

Abstract. Through beam-beam experiments at the Multicharged Ion Research Facility (MIRF) at Oak Ridge National Laboratory (ORNL) and at the CRYRING heavy ion storage ring at Stockholm University, we are seeking to formulate a more complete picture of electron-impact dissociation of molecular ions. These inelastic collisions play important roles in many low temperature plasmas such as in divertors of fusion devices and in astrophysical environments. An electron-ion crossed beams experiment at ORNL investigates the dissociative excitation and dissociative ionization of molecular ions from a few eV up to 100 eV. Measurements on dissociative recombination (DR) experiments are made at CRYRING, where chemical branching fractions and fragmentation dynamics are studied. Taking advantage of a 250-kV acceleration platform at the MIRF, a merged electron-ion beams energy loss apparatus is employed to study DR down to zero energy. Recent results on the dissociation of molecular ions of importance in fusion and astrophysics are presented.

Keywords: dissociation, ion recombination
PACS: 34.80.Ht

INTRODUCTION

In low-temperature, dense plasmas, collisions of electrons with molecular ions play a major role in determining dynamics such as energy and particle balance and transport. In fusion devices, these plasmas are found in the edge regions, particularly in the divertor, where molecular ions of hydrogen, hydrocarbons, and hydrides of other impurities are the dominant ion species [1]. Hydrocarbon ions have received a fair amount of attention [2, 3, 4] due to the presence of carbon in many plasma-facing components.

Molecular ions are also important constituents of low-temperature plasmas in astrophysics, such as in diffuse interstellar and planetary clouds [5]. Collisions of these ions with electrons greatly influence the chemistry of these clouds [6].

A number of dissociation channels are possible in an electron-molecular ion collision, depending on the energy of the interaction. At the lowest energies, typically below 5 eV, the dominant process is dissociative recombination (DR):

$$e + AB^+ \rightarrow AB^{**} \rightarrow A + B + KER. \tag{1}$$

In this reaction the molecular ion AB^+ recombines with a low-energy electron. The intermediate molecule AB^{**} then dissociates producing neutrals fragments A and B that

CP926, Atomic Processes in Plasmas—15th International Conference on Atomic Processes in Plasmas
edited by J. D. Gillaspy, J. J. Curry, and W. L. Wiese
© 2007 American Institute of Physics 978-0-7354-0436-6/07/$23.00

carry away excess internal energy as kinetic energy of release (KER). The process has no threshold and the DR cross section generally peaks at E=0. At higher energies, channels producing ion fragments are also open and these usually become dominant above 10 eV:

$$e + AB^+ \rightarrow A^+ + B + e + KER \qquad (2)$$

$$e + AB^+ \rightarrow A^+ + B^+ + 2e + KER. \qquad (3)$$

The process in Eq.(2) is known as dissociative excitation (DE) and involves a collision-induced transition from a bound state of AB^+ to a repulsive state that leads to the fragments A^+ and B. Dissociative ionization (DI), given in Eq.(3), results in the release of an electron from AB^+ and yields two ion fragments, but has a higher energy threshold than DE. The fragments produced by DE and DI also take away excess internal energy released in dissociation.

Fundamental quantities measured in dissociation experiments yield information crucial to understanding low-temperature plasmas. Measurements of absolute cross sections and rate coefficients give the probability that certain dissociation reactions will occur. Measured chemical branching fractions tell which fragments and radicals are produced in the reactions and with what relative populations. Imaging studies of dissociation fragments yield excited states populations in the atomic and molecular products. All three lead to better diagnosing, modeling, and controlling of low-temperature plasmas and understanding the chemistry involved. Knowledge of the branching fractions and excited state populations is particularly important when different reaction channels yield products of widely differing reactivity.

DISSOCIATIVE EXCITATION AND DISSOCIATIVE IONIZATION

Experimental investigations of DE and DI, both by merged-beams methods in ion storage rings and by crossed-beams techniques, have been reviewed elsewhere [7, 8], but only the latter method will be discussed here. The ORNL electron-ion crossed-beams experiment has been described in detail previously [9, 10, 11, 12]; a similar apparatus is also used in Louvain-la-Neuve [13] to investigate DE and DI. Molecular ions extracted from an ECR or Colutron ion source at 10 keV are intersected at right-angles by a well-defined electron beam of energy 3-100 eV. Fragment ions produced by DE and DI are magnetically separated from the target ions and detected by an electron multiplier. By measuring the currents and velocities of the two beams and the geometric overlap, one can determine absolute dissociation cross sections.

Absolute cross sections for the production of C^+ fragment ions from CH^+ measured with the crossed-beams method [12] are in very good agreement with DE measurements performed at the ion storage ring TSR [14, 15]. For the dissociation of CH_2^+ producing CH^+ and C^+ fragment ions, however, no storage ring measurements exist for comparison. Our DE+DI results for these channels are shown in Fig. 1 for the energy range 3-100 eV. For comparison, the DR results of Larson *et al.* [16] and the DE+DI results for the $CD_2^+ \rightarrow D^+$ fragment ion channel [17] are also shown. The CH^+ and C^+ channels are dominant in the 5-15 eV range; below that, DR dominates. Cross sections for ionization without dissociation, yielding CH_2^{2+} ions, are an order of magnitude smaller

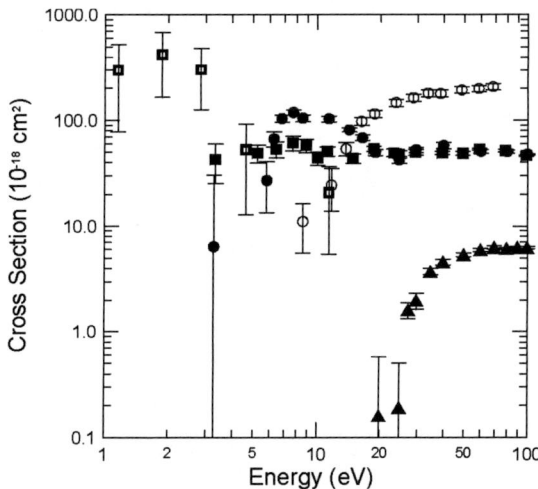

FIGURE 1. Absolute cross sections for the dissociation and ionization of CH_2^+ and CD_2^+ ions by electron-impact as a function of center-of-mass energy. The filled circles and squares are the (DE + DI) measurements for the production of CH^+ and C^+ fragment ions, respectively, shown with one standard deviation relative error bars. The open circles represent the (DE + DI) measurements of Ref. [17] for the production of D^+ fragment ions. The solid triangles are present results for ionization of CH_2^+ and the open squares are the dissociative recombination measurements of Ref. [16].

than those for DE+DI at 100 eV. Fig. 1 represents a nearly-complete picture of electron-impact dissociation of CH_2^+. This sort of "completeness" is a goal of our experimental program.

In a search for systematic trends in DE and DI of molecular ions, a comparison of a number of different sets of experimental data is made, specifically for dissociation resulting from the breaking of a single C-H(D) bond. Crossed-beams measurements have been made for (DE + DI) of CH^+ producing C^+ [12], CH_2^+ producing CH^+ [19], and DCO^+ producing CO^+ [18]. A way to complement these experimental data is using semi-empirical formulas, which try to incorporate the correct behavior of the measured cross section close to threshold and at high impact energies of the electron. The measured cross sections are strongly dependent on initial rovibrational states of the target molecule, which are unknown in our measurement. The initial rovibrational state defines the threshold energy, E_{th}, of the breakup process, which influences the behavior of the cross section both close to the threshold (for example, Wannier laws, E^α) and at high energies (for example, Thompson scaling, E_{th}^{-2}). As a consequence, our scaling reflects the average over the initial internal states of the target molecule and final internal states of the fragments, rather than strict semi-empirical limits. We find that the single-bond breakup cross sections are proportional to E_{th}^{-1}. On the other hand, as found by quantum mechanical considerations of Trevisan and Tennyson [20], the three-body dissociative cross section is proportional to the reduced mass of the impact molecule. We rather find that the breakup cross sections in cases we studied are proportional to the

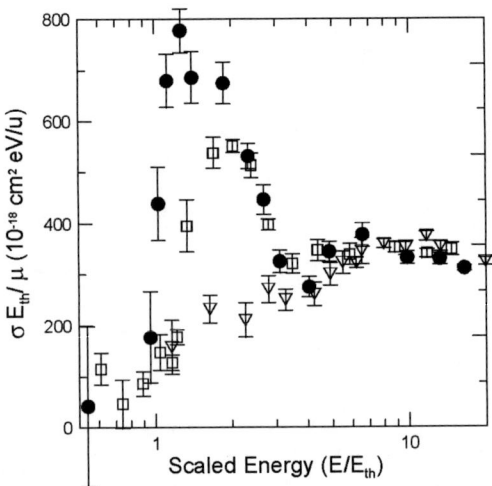

FIGURE 2. Scaled cross sections for dissociative excitation and dissociative ionization of molecular ions by electron-impact as a function of energy in threshold units for systems involving the breaking of a single C-H bond. See the text for a discussion of the scaling. The symbols are the (DE + DI) measurements for specific fragment ion channels shown with one-standard-deviation relative error bars: closed circles, $CH_2^+ \rightarrow CH^+$, Ref. [19]; open squares, $DCO^+ \rightarrow CO^+$, Ref. [18]; open triangles, $CH^+ \rightarrow C^+$, Ref. [12].

reduced mass μ of the breakup fragments. This yields the scaling law in form

$$\sigma = C \frac{\mu}{E_{th}} f \left(\frac{E}{E_{th}} \right), \tag{4}$$

where E is the impact kinetic energy of the electron and f is a function of the scaled energy $(E/E_t h)$. The experimental cross sections for these three systems are shown in Figure 2, with the energy scaled by E_{th} and the cross section scaled by (μ/E_{th}).

DISSOCIATIVE RECOMBINATION

Dissociative recombination experiments using a merged-beams method at ion storage rings have recently been reviewed [21, 22]. Although not discussed in this paper, a flowing afterglow Langmuir probe (FALP) technique is also used to investigate DR [23].

Measurements at CRYRING

The CRYRING heavy ion storage ring [24] at the Manne Siegbahn Laboratory in Stockholm has a circumference of 51.6 m with a magnetic rigidity allowing ions (mass M and charge q) of energy up to $96 \times (q/M)^2$ MeV/amu to be stored. The electron cooler uses an adiabatic expansion factor of 100 to produce a 40-mm diameter beam

TABLE 1. Measured chemical branching fractions for the dissociative recombination of dihydride molecular ions into three- and two-body channels

Ion	X + H + H	XH + H	X + H$_2$	Reference
H$_3^+$	0.77 ± 0.02	–	0.23 ± 0.02	[29]
CH$_2^+$	0.63 ± 0.06	0.25 ± 0.04	0.12 ± 0.02	[16]
NH$_2^+$	0.57 ± 0.01	0.39 ± 0.02	0.04 ± 0.02	[31]
OH$_2^+$	0.71 ± 0.06	0.20 ± 0.05	0.09 ± 0.03	[30]
PD$_2^+$	0.78 ± 0.07	0.14 ± 0.05	0.08 ± 0.02	[28]

with longitudinal and transverse energy spreads as low as 0.05 meV and 1.0 meV, respectively. DR experiments use the electron cooler as a target, with neutral products from the reaction being detected after the first dipole magnet separates them from the stored ions. Similar arrangements are also used at the TSR [25] and ASTRID [26] ion storage rings for studying DR.

A large energy-sensitive surface barrier detector (SBD) is used to measure absolute DR cross sections. These cross sections generally exhibit a (1/E) energy dependence near zero interaction energy [27, 28]. By adding a partial transmission (30%) grid in front of the SBD, it is possible to separate the contributions of different DR channels [29]. For example, for the dihydride ion XH$_2^+$, one can determine how many DR reactions go into each of the XH + H and X + H$_2$ two-body channels and into the X + H + H three-body channel. These are known as the chemical branching fractions. Table 1 shows the chemical branching fractions measured at CRYRING for a number of dihydride ions [16, 28, 29, 30, 31]. The interesting aspect is that the three-body channel is dominant, in contrast to the early theoretical predictions of Bates [32, 33] that the channel with least re-arrangement, i.e., the XH + H channel, would be favored. This dominance of the three-body channel has also been observed in other measurements of DR chemical branching fractions [34, 35].

The dynamics for dissociation of dihydrides into three-body channels are also studied at CRYRING using a fragment imaging technique that has been detailed elsewhere [36, 37, 38]. In short, neutral DR fragments strike a stack of three microchannel plates. The resulting electrons are then accelerated over a very short distance to a phosphor screen where they cause a small light burst. A fast photomultiplier tube viewing the phosphor screen triggers an intensified digital camera that records the event. A thin C or Al foil in front of the center of the imaging detector is selected to allow only the heavy atom from the dihydride to pass while blocking any H atoms. The recorded images are discriminated, choosing only images that show one particle behind the foil (heavy atom) and two outside the foil (H atoms). In this way, only three-body dissociations are studied. By measuring the total displacement (TD) of the three atoms from their center-of-mass, one can determine the KER of dissociation. This in turn yields information about the final state of the heavy atom. Figure 3 shows a measured TD distribution for the DR of CH$_2^+$ at E=0 [31]. The curves represent Monte Carlo simulations of the fragments for the two energetically-allowed final states of C, the ^{3}P ground state and the ^{1}D excited state. By fitting the ratios of these two contributions to the measured distribution, it

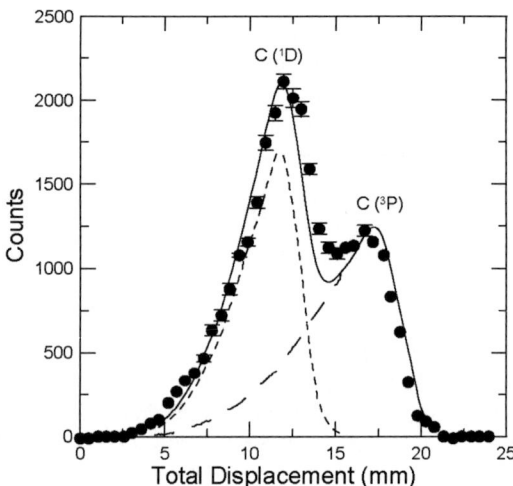

FIGURE 3. Total displacement of the three-body fragments from the DR of CH_2^+. The symbols represent the experimental results. The dashed curves are results of Monte Carlo simulations for the channels leading to the production of $C(^3P)$ (long dashes) and $C(^1D)$ (short dashes) atoms. All data are taken from Ref. [31].

is determined that the $^3P:^1D$ final state ratio is 0.51:0.49. Other information, such as the sharing of KER between the two H atoms and the angle of molecule just prior to dissociation, can be inferred from the imaging measurements [31].

Measurements at MIRF

The merged electron-ion beams energy-loss (MEIBEL) apparatus, originally designed to measure electron-impact excitation of multicharged ions [39], has been adapted to accept ion beams with energies up to 250 kV per charge from the new high-voltage platform at MIRF [40]. This apparatus is shown schematically in Figure 4 and uses trochoidal analyzers to merge and demerge the electrons with the target ion beam. The higher ion energies from the high-voltage platform permit the collection and detection of neutral products of DR reactions, and the merged-beams geometry enables measurements down to zero interaction energy. A similar single-pass, merged-beams experiment was used by Mitchell *et al.* [41] to measure DR rate coefficients, but this research has been discontinued. Presently, the MEIBEL experiment features two neutral detectors: a large discrete dynode multiplier used for particle counting and a 40-mm diameter imaging detector similar to the one used at CRYRING. Energy-sensitive solid state detectors capable of measuring 10 keV H atoms are being developed.

Measurements on the DR of H_2^+ have been made using the MEIBEL apparatus. The measured rate coefficients show the same energy-dependence as the data of Larsson *et al.* [42] measured at CRYRING but are roughly a factor of three higher. The DR

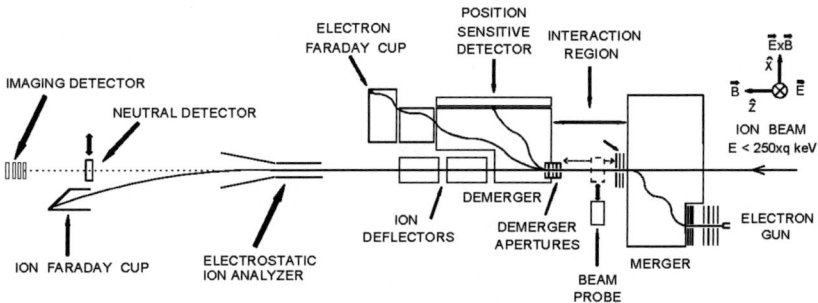

FIGURE 4. Merged electron-ion beams energy-loss (MEIBEL) apparatus with modifications required for dissociative recombination measurements. See text for a description.

results are also more than a factor of two lower than the early measurements of Peart and Dolder [43] and Auerbach *et al.* [44]. Since it has been established that the DR rate coefficients are strongly dependent on the vibrational distribution of H_2^+ [45, 46], our results are not surprising. Larsson *et al.* had only v=0,1 states populated in their H_2^+ ion beam, while the earlier researchers used hotter ion sources and did not employ any vibrational cooling mechanisms. An all-permanent magnet ECR ion source is used to produce the H_2^+ ions for our study of DR on MEIBEL, so one would expect the vibrational distribution to be hotter than for the ions stored and cooled in CRYRING. Preliminary imaging measurements on MEIBEL reinforce this assertion; the measured total displacement distribution is consistent with a vibrational population that peaks around v=4-6.

SUMMARY

The dissociation of molecular ions by electron impact is studied experimentally at ORNL and CRYRING, with crossed-beams measurements of dissociative excitation and dissociative ionization and merged-beams measurements of dissociative recombination, both in an ion storage ring and in a single-pass experiment. In support of low-temperature plasma science, particularly in the applications of fusion edge plasmas and astrophysics, dissociation cross sections, rate coefficients, and chemical branching fractions are measured for systems where no theoretical predictions exist. Further insights about the dynamics of dissociation are gained by studying the kinetic energy of release of the fragments produced. It is our hope that these studies will stimulate theoretical efforts and serve as benchmarks for any future calculations.

ACKNOWLEDGMENTS

This work was supported in part by the Office of Basic Energy Sciences and the Office of Fusion Energy Sciences of the U.S. Department of Energy under Contract No. DE-AC05-00OR22725 with UT-Battelle, LLC, and by the Swedish Research Council. EMB,

MRF, and HA gratefully acknowledge support from the ORNL Postdoctoral Research Associates Program administered jointly by the Oak Ridge Institute for Science and Education and Oak Ridge National Laboratory.

REFERENCES

1. H. Tawara, in *Atomic and Molecular Processes in Fusion Plasmas*, edited by R. K. Janev (Plenum, New York, 1995), pp. 461-496.
2. W. D. Langer, Nucl. Fusion **22**, 751 (1982).
3. R. K. Janev, T. Kato, and J. G. Wang, Phys. Plasmas **7**, 4364 (2000).
4. R. K. Janev and D. Reiter, Phys. Plasmas **9**, 4071 (2002).
5. A. Dalgarno, in *Atomic Processes and Applications*, edited by P. G. Burke and B. L. Moiseiwitsch (North-Holland, Amsterdam, 1976), pp. 109-132.
6. E. Herbst, Annu. Rev. Phys. Chem. **46**, 27 (1995).
7. G. H. Dunn and N. Djurić, "Electron Impact Dissociaive Excitation and Ionization of Molecular Ions," in *Novel Aspects of Electron-Molecule Collisions*, edited by K. H. Becker, World Scientific, Singapore, 1998, pp. 241-281.
8. N. Djurić, "Recent Experimental Studies of Electron-Impact Excitation of Atomic and Molecular Ions," in *Atomic and Molecular Data and Their Applications*, edited by T. Kato et al., AIP Conference Proceedings 771, American Institute of Physics, New York, 2005, pp. 162-171.
9. D. C. Gregory, F. W. Meyer, A. Müller, and P. Defrance, Phys. Rev. A **34**, 3657 (1986).
10. P. A. Schulz, D. C. Gregory, F. W. Meyer, and R. A. Phaneuf, J. Chem. Phys. **85**, 338 (1986).
11. M. E. Bannister, Phys. Rev. A **54**, 1435 (1996).
12. M. E. Bannister, H. F. Krause, C. R. Vane, N. Djurić, D. B. Popović, M. Stepanović, G. H. Dunn, Y.-S. Chung, A. C. H. Smith, and B. Wallbank, Phys. Rev. A **68**, 042714 (2003).
13. J. Lecointre, D. S. Belić, H. Cherkani-Hassani, J. J. Jureta, and P. Defrance, J. Phys. B **39**, 3275 (2006).
14. P. Forck, Ph.D. thesis, Ruprech-Karls-Universität Heidelberg, 1994 (unpublished).
15. Z. Amitay, D. Zajfman, P. Forck, U. Hechtfischer, M. Grieser, D. Habs, D. Schwalm, and A. Wolf, in *Applications of Accelerators in Research and Industry*, edited by J. L. Duggan and I. L. Morgan (AIP Press, New York, 1997), p. 51.
16. A. Larson, A. Le Padellec, J. Semaniak, C. Stromholm, M. Larsson, S Rosén, R. Peverall, H. Danared, N. Djurić, G. H. Dunn, and S. Datz, Astrophys. J. **505**, 459 (1998).
17. N. Djurić, S. Zhou, G. H. Dunn, and M. E. Bannister, Phys. Rev. A **58**, 304-308 (1998).
18. E. M. Bahati, R. D. Thomas, C. R. Vane, and M. E. Bannister, J. Phys. B **38**, 1645 (2005).
19. C. R. Vane, E. M. Bahati, M. E. Bannister, and R. D. Thomas, submitted to Phys. Rev. A (2007).
20. C. S. Trevisan and J. Tennyson, J. Phys. B **34**, 2935 (2001).
21. M. Larsson, "Merged-beams studies of electron-molecular ion interactions in ion storage rings," in *Advances in Gas Phase Ion Chemistry*, Vol. 4, edited by N. G. Adams and L. M. Babcock, (JAI Press Inc./Elsevier Science B. V., USA, 2001), p. 179.
22. M. Larsson, "Ion storage rings," in *The Encyclopedia of Mass Spectrometry. Volume 1 Theory and Ion Chemistry*, edited by P. B. Armentrout, M. L. Gross, and R. Caprioli (Elsevier Ltd, UK, 2003), p. 195.
23. N. G. Adams, "Spectroscopy determination of the products of electron-ion recombination," in *Advances in Gas Phase Ion Chemistry*, Vol 1., edited by N. G. Adams and L. M. Babcock (JAI Press Inc./Elsevier Science B. V., USA, 1992) pp. 217-310.
24. K. Abrahamsson, G. Andler, L. Bagge, P. Carlé, H. Danared, S. Egnell, K. Ehrnstén, M. Engström, C. J. Herrlander, J. Hilke, J. Jeansson, A. Källberg, S. Leontein, L. Liljeby, A. Nilsson, A. Paál, K. G. Rensfelt, U. Rosengård, A. Simonsson, A. Soltan, J. Starker, M. af Ugglas, and A. Filevich, Nucl. Instrum. Methods Phys. Res. B **79**, 269 (1993).
25. P. Forck, M. Grieser, D. Habs, A. Lampert, R. Repnow, D. Schwalm, A. Wolf, and D. Zajfman, Phys. Rev. Lett. **70**, 426 (1993).
26. J. B. A. Mitchell, G. Angelova, C. Rebrion-Rowe, O. Novotny, J. L. LeGarrec, H. Bluhme, K. Seiersen, A. Svendsen, and L. H. Andersen, J. Phys. Conf. Ser. **4**, 198 (2005).

27. F. Hellberg, V. Zhaunerchyk, A. Ehlerding, W. D. Geppert, M. Larsson, R. D. Thomas, M. E. Bannister, E. Bahati, C. R. Vane, F. Österdahl, P. Hlavenka, and M. af Ugglas, J. Chem. Phys. **122**, 224314 (2005).
28. V. Zhaunerchyk, F. Hellberg, A. Ehlerding, W. D. Geppert, M. Larsson, C. R. Vane, M. E. Bannister, E. M. Bahati, F. Österdahl, M. af Ugglas, and R. D. Thomas, Mol. Phys. **103**, 2735-2745 (2005).
29. S. Datz, G. Sundström, Ch. Biedermann, L. Broström, H. Danared, S. Mannervik, J. R. Mowat, and M. Larsson, Phys. Rev. Lett. **74**, 896 (1995).
30. S. Rosén, A. Derkatch, J. Semaniak, A. Neau, A. Al-Khalili, A. Le Padellec, L. Vikor, R. Thomas, H. Danared, M. af Ugglas, and M. Larsson, Faraday Discuss. **115**, 295 (2000).
31. R. D. Thomas, F. Hellberg, A. Neau, S. Rosén, M. Larsson, C. R. Vane, M. E. Bannister, S. Datz, A. Petrignani, and W. J. van der Zande, Phys. Rev. A **71**, 032711 (2005).
32. D. R. Bates, Astrophys. J. **306**, L45 (1986).
33. D. R. Bates, J. Phys. B **24**, 3267 (1991).
34. A. I. Florsecu-Mitchell and J. B. A. Mitchell, Phys. Rep. **430**, 277 (2006).
35. N. G. Adams, V. Poterya, and L. M. Babcock, Mass. Spectrom. Rev. **25**, 798 (2006).
36. Z. Amitay and D. Zajfman, Rev. Sci. Instrum. **68**, 1387 (1997).
37. R. Peverall, S. Rosén, M. Larsson, J. R. Peterson, R. Bobbenkamp, S. L. Guberman, H. Danared, M. af Ugglas, A. Al-Khalili, A. N. Maurellis, and W. J. van der Zande, Geophys. Res. Lett. **27**, 481 (2000).
38. M. Saito, Y. Haruyama, T. Tanabe, I. Katayama, K. Chida, T. Watanabe, Y. Arakaki, I. Nomura, T. Honma, K. Noda, and K. Hosono, Phys. Rev. A **61**, 062707 (2000).
39. E. W. Bell, X. Q. Guo, J. L. Forand, K. Rinn, D. R. Swenson, J. S. Thompson, G. H. Dunn, M. E. Bannister, D. C. Gregory, R. A. Phaneuf, A. C. H. Smith, A. Müller, C. A. Timmer, E. K. Wåhlin, B. D. DePaola, and D. S. Belić, Phys. Rev. A **49**, 4585-4596 (1994).
40. F.W. Meyer, M.E. Bannister, D. Dowling, J.W. Hale, C.C. Havener, J.W. Johnson, R.C. Juras, H.F. Krause, A.J. Mendez, J. Sinclair, A. Tatum, C.R. Vane, E. Bahati Musafiri, M. Fogle, R. Rejoub, L. Vergara, D. Hitz, M. Delaunay, A. Girard, L. Guillemet, and J. Chartier, Nucl. Instrum. Methods Phys. Res. B **242**, 71 (2006).
41. J. B. A. Mitchell, Phys. Rep. **186**, 215 (1990).
42. M. Larsson, L. Brostrom, M. Carlson, H. Danared, S. Datz, S. Mannervik and G. Sundstrom, Phys. Scr. **51**, 354 (1995).
43. B. Peart and K. T. Dolder, J. Phys. B **7**, 236 (1974).
44. D. Auerbach. R. Cacak, R. Caudano, T. D. Gaily, C. J. Keyser, J. W. McGowan, J. B. A. Mitchell, and S. F. J. Wilk, J. Phys. B **10**, 3797 (1977).
45. W. J. van der Zande, J. Semaniak, V. Zengin, G. Sundström, S. Rosén, C. Strömholm, S. Datz, H. Danared, and M. Larsson, Phys. Rev. A **54**, 5010 (1996).
46. L. H. Andersen, P. J. Johnson, D. Kella, H. B. Pedersen, and L. Vejby-Christensen, Phys. Rev. A **55**, 2799 (1997).

A New And Improved Version Of HULLAC

M. Klapisch[a✉], M. Busquet[a,b] and A. Bar-Shalom[a,c]

[a]ARTEP[#],inc. 2922 Excelsior Springs Court, Ellicott City, MD 21042,USA
[b]LERMA, Observatoire de Paris, UPMC, CNRS, Place Jules Janssen, 92190 Meudon, France
[c]NRCN, 9001 Be'er Sheva, Israel.

Abstract: We present a new version of the collisional radiative model (CRM) generator code HULLAC. The main features are: *(i)* input considerably simplified and flexible, *(ii)* capacity of "post-averaging" configurations and superconfigurations in mixed mode, *(iii)* a new fitting formula for cross sections, completely correcting the problems of the classical Sampson and Golden formula, *(iv)* a new algorithm for solving the rate equations of the CRM, more robust and giving more insight in the quality of the model than the biconjugate gradient method, *(v)* thanks to thorough comparisons with the LANL code, some errors were corrected, and very good agreement has been obtained on all types of transitions, *(vi)* finally, most of the code has been re-written according to up-to-date standards. This code is now ready for distribution.

Keywords: Non-LTE, collisional radiative model, collisional cross sections, rate equation solver.
PACS: 33.20Rm, 52.20Fs, 52.25Dg, 52.25Os.

INTRODUCTION

The HULLAC code [1] is an integrated collisional radiative model generator and solver. Its efficiency is due to original methods and algorithms. All relevant processes are computed with the same relativistic wavefunctions [2]. The angular momentum recoupling coefficients are computed using graphical methods [3]. The collision processes use the factorization-linearization method [4],[5]. It has been used by a number of groups, and was advertised at the 2001 APIP [6]. However, because the different modules were developed over the years, most of the coding was obsolescent. This is why it was decided to rewrite[1] most of the code in a modern style, more conform to standards of software engineering, in order for the code source to be understood and compiled easily by potential users. By the same token, many improvements have been made, and errors have been corrected. The rate equation solver, which was missing in previous versions, has been developed, based on a new algorithm. It is the purpose of this paper to describe these improvements and additions. The reader is referred to the above references for the theory.

[✉] to whom correspondence should be addressed: marcel.klapisch@nrl.navy.mil
[#] Contractor to Naval Research Laboratory, Washington, DC, 20375.
[1] mostly done by M.B.

CP926, *Atomic Processes in Plasmas—15th International Conference on Atomic Processes in Plasmas*
edited by J. D. Gillaspy, J. J. Curry, and W. L. Wiese
© 2007 American Institute of Physics 978-0-7354-0436-6/07/$23.00

NEW INPUT POSSIBILITIES

The module `anglar` acts as the master module, in addition to computing all the angular momentum coefficients of the various matrix elements of the energy and of all the transitions, spontaneous and collisional. It reads an input file in which, among other data, the list of configurations to be computed should be listed. This was inefficient and painstaking for complex spectra.

Defining Configurations

It is now possible to use the concept of superconfigurations [7] to generate all the desired configurations in a very compact manner.

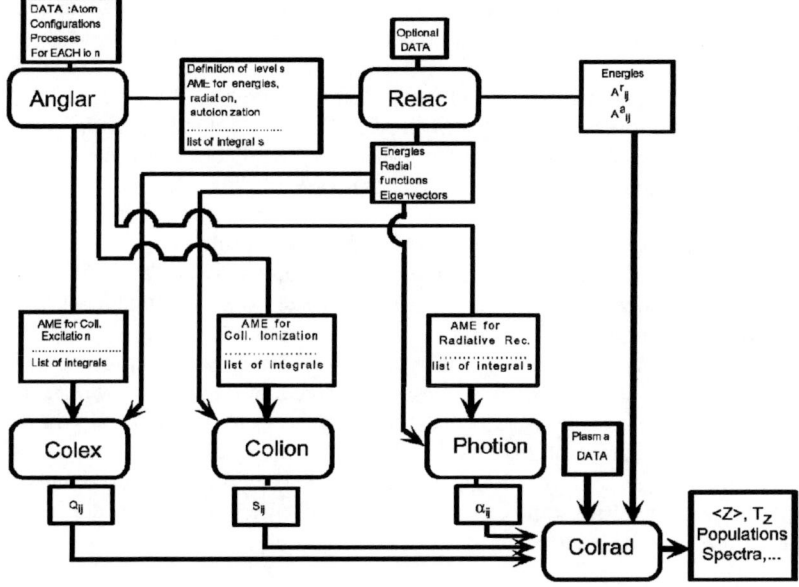

FIGURE 1. Flow chart of the HULLAC package. AME stands for angular matrix elements.

For instance, the following lines:

```
superconf nel=15
abc 1s2 2L [5:8] 3L [0:18] (4L,5L) [0:1]
endsuper
```

generate *all and only those configurations* with a total of 15 electrons distributed between the supershells (2s,2p) – with all combinations from 5 to 8 electrons, (3s,3p,3d) – with zero to 18 electrons, and (4s,4p,4d,4f,5s,5p,5d,5f,5g) supershell with 0 to 1 electron. As a new addition, the symbols "Le" and "Lo" would select only even or odd values of ℓ. The sizes of the arrays have been greatly increased to do very complex computations, but if there is an overflow, it will now be detected and the relevant parameter will be clearly designated in the output file.

207

There was in nearly all versions of HULLAC the possibility of computing different groups of configurations in different potentials. This can greatly increase the accuracy of the energies and wavefunctions. This is now achieved in a very simple way with the keyword: `potential new`. In the same way, the potential to be chosen as the final one – with respect to which the other configurations are shifted to take into account their accurate computations – is simply designated as `potential new final`.

"Direct-" and "Post-" Averaging

In the previous versions, all the input configurations created a list of fine structure levels in *jj* coupling. For complex spectra, this generates too many detailed levels. Therefore, several modes were introduced in the `SETTING` chapter of the input file:

mode = l(evels) – this generates as before fine structure levels for all configurations listed.

mode = c(onfigurations) – this generates directly, by simple sum rules, relativistic configuration averages as basic entities for the whole chain of program.

mode = l+c – makes two sets of output files for all the rates, and two different solutions can be obtained at the level of the CRM solver "`colrad`".

In addition, one can now perform "*post averaging*", that is, `anglar` generates one of the two previous sets, but can then perform additional, "brute force" averaging. Since `anglar` is the less time consuming module of the chain, it is not a significant penalty. Two additional keywords are possible: `nrc` - creates non relativistic configurations averages (NRC) of relativistic configurations by statistical weight, and `sc` - creates superconfigurations as averages of configurations. These two averages can be done starting from mode c or l. In the latter case, the effects of *configuration interaction* will be included in the averages. These averages can be performed on all the levels or configurations in the list, or on *selected* levels/configurations, in effect creating a "mixed model" in which some configurations can be described as fine structure levels, others a configuration averages, and some others as superconfigurations. The reader is referred to the mode of use for more details

A NEW FITTING FORMULA

Collisional excitation cross sections are the most time-consuming part of setting the model. The temperature dependent rates are obtained by integrating them over the collision energy. Therefore, it is has become standard to represent the cross sections by an easily integrable analytical formula, the parameters of which are transmitted from the cross section module to the CRM equation solver module, so the same cross sections can be used to model several temperatures. The classical, nearly ubiquitous, formula is usually called the "Golden and Sampson (GS) formula"[8].

It reads:

$$\Omega(u = \varepsilon / \Delta E) = A + D\ln(u) + \frac{c_1}{(a+u)} + \frac{c_2}{(a+u)^2},$$

where Ω is the collision strength and u is the collision energy divided by the transition energy. A, D, a, c_1, c_2 are the free parameters that are fitted to the computed cross sections for each transition by a least square method. The rates are then obtained as a combination of exponential integrals [9]. In principle, the 4 coefficients, A, D, c_1 and c_2 of the GS fit are positive and one or two are dominant. a has been introduced to further improve the fit and reproduce a possible slope change near threshold. It has to be larger than -1 to prevent singularity. However, in practice the coefficients are obtained with a least square method as suggested in GS, possibly with the Bethe limit constraint, and from a limited range of energies (because a finite number of sampled cross-sections is computed). But the least square method cannot guarantee accuracy of the fit over the whole sampled range, neither does it allow for a correct asymptotic behavior. Different shapes yield fitting inadequacies of different kind as can be seen in two cases displayed in Fig.2. In Fig. 2b, the reconstructed cross section (dashes) differs from the computed values (diamond) by less than 20% of the maximum cross-section but is a very poor approximation. The origin of disagreement is the presence of a maximum of the cross-section at a distance from the threshold, that cannot be reproduced by the GS formula (except using a less than -1 leading to a singularity).

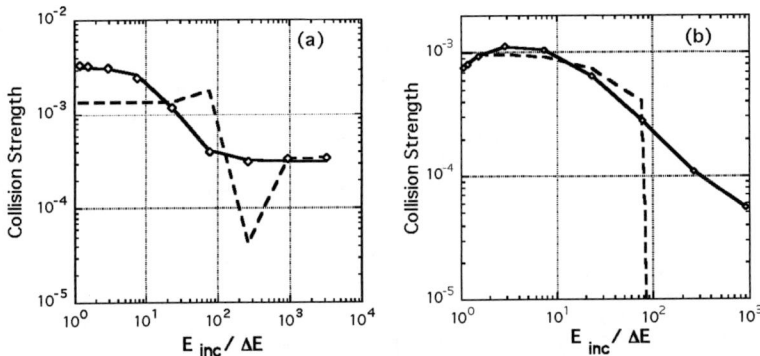

FIGURE 2: Examples of problems with the GS formula. Reconstructed GS fit (dashed lines) and BBTC fit (solid line) compared to initial cross sections (diamonds)

The consequence of using GS fit in the CRM solver is that a number of rates (sometimes up to 20%) turn out negative, and have to be ignored. This leads to errors in the solution or singularities in the matrices.

The BBTC fitting procedure

We developed a new fitting algorithm [10] based on the idea of Burgess, Tully and Chidichimo [11]. The idea is to make a mapping of the incident energy (with range from ΔE to ∞) into the compact support [0,1]. In reference [11], the actual form of the mapping and the position of energy samples are dependent on the transition, on the selection rules which are in force and on the actual values of the cross-sections. In order to make it tractable in our CRM model, we improved their procedure with a more general procedure and a fixed, optimized set of sampling point. We map the function $\Omega(u)$ to a function $z(x)$ with the following transformation:

$$x = 1 - (1 + C) / (u + C),$$

$$z = (\Omega - A \ln u).(u + C)^n.$$

here n is the integer power of the residue (i.e. after the logarithmic term has been taken out) obtained from the asymptotic behavior deduced from the last 2 or 3 computed cross-sections, and C is a free parameter to optimize the fit. The collision strength then reads:

$$\Omega = A \ln u + z / (u + C)^n.$$

Note that our mapping differs from the one proposed by Burgess et al, as we have one transformation only, regardless of the selection rule relevant to the specific transition, and that we allow any integer asymptotic power expansion n. We then sample the cross sections on a fixed set of x values. We found that the set of 9 values $\{x_l\}_{l=0:8} = \{0, 1, 3, 6, 10, 14, 17, 19, 20\} / 20$ is optimum for most cases. The cross sections are now fully described by 13 coefficients, and can be interpolated by splines. An example of the mapping can be seen on figure 3.

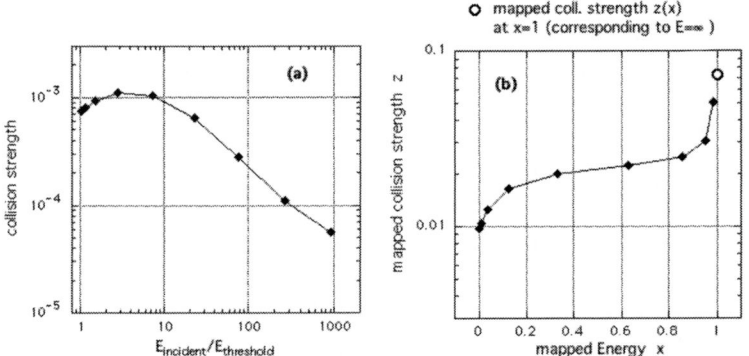

FIGURE 3: Example of mapping of a cross section with the BBTC mapping scheme.

The rates are easily recovered in the CRM solver by a combination of exponential integrals, and spline coefficients depending only on the set $\{x\}$. With this algorithm, we did not find any problematic values of rates in the Xe case with 39000 configurations described below.

COMPARING WITH LANL CODE

In the recent NLTE workshops [12], it became clear that the discrepancy between the various codes for CRM is noticeable. In an effort to understand the source of these differences, we compared HULLAC cross sections for collisional processes with the results of the LANL code[2]. These two codes are based on the same physical assumptions,

---

[2] We are thankful to T. Durmaz and R. Mancini for their help in this comparison.

but use quite different algorithms, so the results should be similar if not identical. We found at first some important discrepancies that forced us to check thoroughly some routines, and eventually we found and corrected previously undetected errors. The DWBA code of LANL was checked at LANL by comparing to another one, based on many body perturbation theory. HULLAC now agrees with both codes, as shown on figures 4 and 5. We are now confident that the cross sections for collisional processes are correct.

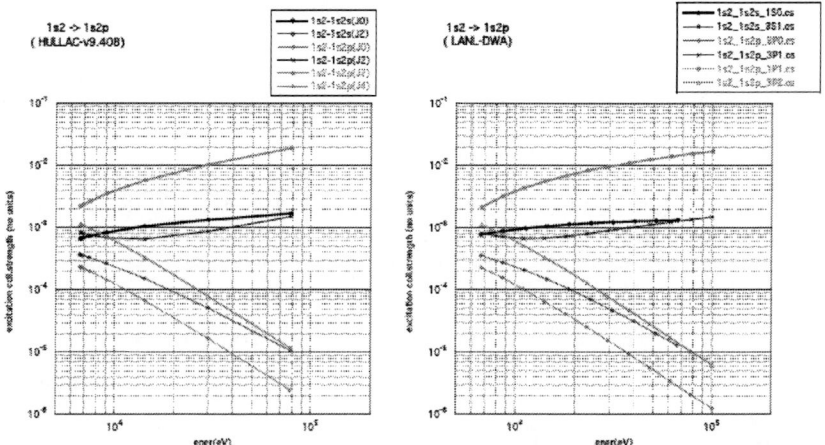

FIGURE 4: Comparison of cross sections for collisional excitation of He-like Fe XXV $1s^2$ -1s2p.

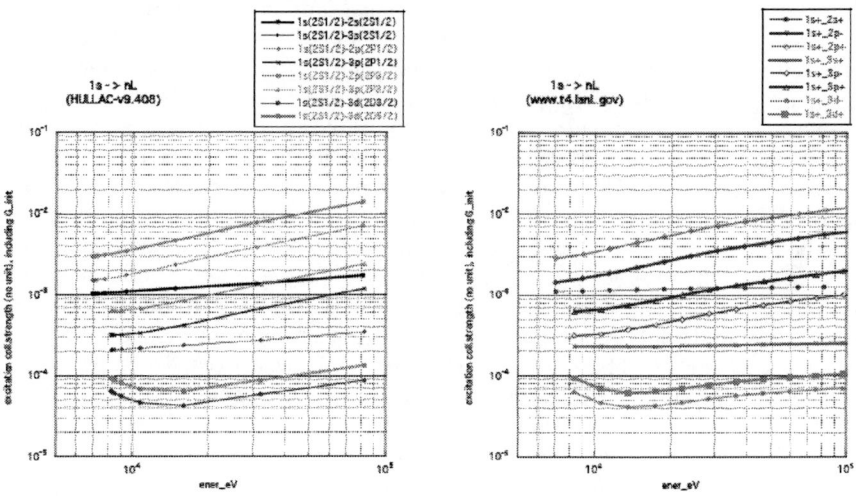

FIGURE 5: Comparison of collisional ionization cross sections on H-like Fe XXVI 1s –*nl*.

Figure 6 shows comparison between the codes on photo-ionization of H-like Fe XXVI. Again the agreement is very good.

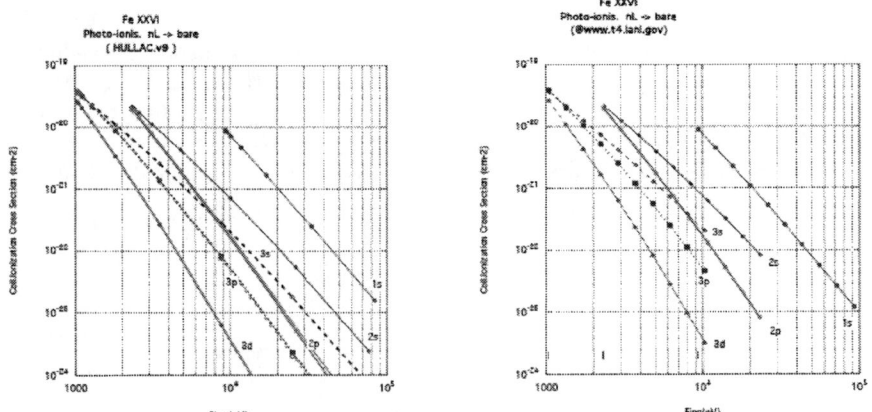

FIGURE 6: Comparison between HULLAC and LANL codes on photo-ionization of H-like FeXXVI.

The figures above show that the agreement between the codes is as good as can be for codes that were written differently by different people with different algorithms.

THE NEW CRM EQUATIONS SOLVER

Generalities

HULLAC was until now distributed without a solver for the rate equations. Our aim is to be able to use HULLAC for building opacities databases for cases where the non-LTE super transition array approach (SCROLL) [13] is not detailed enough. For this purpose, one needs a robust solver for very large linear systems. Indeed, it has been recognized that many ion stages are necessary, as well as many thousands of states for each ion stage. The matrix involved is sparse because of selection rules, and one popular method is the bi-conjugate gradient [14]. The problem with the latter is that when it does not converge, one cannot extract any useful information. This is a real concern, because the lifetimes of metastable and ground states are orders of magnitude longer than those of resonant states. Thus, the condition of the rate matrix can be in excess of 10^{10}. The method adopted here is to solve each ion stage separately, and iterate. Each ion matrix's condition is bound to be smaller than the condition of the whole matrix, and because of the connection to neighboring ions, these ion matrices are not singular. Similar methods [15] involve the separation of ground and metastable states from excited states. This does not converge easily to the LTE limit and ignores transitions from excited states of an ion to excited states of the neighboring ions. Therefore, we factorize the population of individual states as a product of *reduced population* and *total ion population*

$$N_Q = \sum_{p=1}^{M_Q} N_p \; ; \; n_p^Q = N_p / N_Q$$

where N_p, N_Q, n_p are, respectively, individual population of state p, total population of charge Q, and reduced population of level p of charge state Q. The advantage of this factorization is that the global effective rates $W_{Q\rightarrow Q'}$– i.e. temperature and density dependent contribution from an ion to another, depend only on the reduced population:

$$W_{Q\rightarrow Q'} = \sum_{p'\in Q'}\left(\frac{1}{N_Q}\sum_{p\in Q}N_p W_{p\rightarrow p'}\right) = \sum_{p'\in Q'}\sum_{p\in Q}n_p^Q W_{p\rightarrow p'}$$

Therefore, if the reduced populations are found at a given iteration, the global rates can be computed. Then a matrix, the order of which is only the *number of active charge states*, i.e. *about a dozen*, can be easily set up and solved, giving the charge state distribution, i.e. the N_Q of the next iteration. Using a Newton-style development, we find the increments of the reduced populations:

$$\Delta \mathbf{n}^Q = -\frac{1}{N_Q}\left\{\Delta N_{Q'}\mathbf{M}_Q^{-1}\mathbf{R}_Q\tilde{\mathbf{n}}^{Q'} + \Delta N_{Q''}\mathbf{M}_Q^{-1}\mathbf{S}_Q\tilde{\mathbf{n}}^{Q''} + \Delta N_Q\tilde{\mathbf{n}}^Q + \mathbf{M}_Q^{-1}\tilde{\mathbf{G}}_Q\right\}$$

where $Q'=Q+1$, $Q''=Q-1$, $\mathbf{R}_Q$ and $\mathbf{S}_Q$ are, respectively, the rectangular matrices of recombination and ionization *into* ion Q, and $\mathbf{M}_Q$ is the matrix of transitions inside ion Q, including losses to other ions on the diagonal. $\tilde{\mathbf{G}}_Q$ is the vector of the residues for the states *at the previous iteration*. More details can be found in reference [16].

Results

Using this new version of HULLAC, with the new fit and the new solver, we were able to obtain the solution of the CRM for Xe at T_e = 50 eV , for a range of electron densities from 10^{16} to 10^{22} cm^{-3}. The model included 39,683 states (relativistic configurations).

Table 1: Details of number of levels and rates for the full Xe model

Ion	States	Rad. Trans.	Coll. Excit.	Ioniz. & R. R.	Autoioniz.
Xe IX	4 546	35 349	108 159	15 008	31 351
Xe X	1 706	11 727	36 244	6 658	10 340
Xe XI	2 295	16 026	49 505	11 349	7 368
Xe XII	4 669	35 198	106 854	18 341	28 915
Xe XIII	5 867	44 616	135 412	21 936	29 057
Xe XIV	5 984	46 098	139 180	22 176	25 313
Xe XV	5 990	46 155	139 312	21 101	22 199
Xe XVI	4 909	38 163	114 522	16 665	15 833
Xe XVII	3 717	28 799	86 096	---	---
Total	39 683	302 131	915 284	2 x 133 236	170 388

At each iteration, one gets automatically the ion charge distribution, therefore this algorithm allows one to follow the convergence of the process, and it is more meaningful than a "black box" algorithm in which the iteration do not have physical meaning. It is then possible to stop at the required accuracy, and create a compact database of charge distribution. Figure 7 shows the evolution of the charge distribution of a smaller model (with 2386 non-relativistic configuration averages). In this case, the convergence criterion was very stringent, (10^{-4} relative variation of individual populations), to show the stability of the algorithm.

Figure 8 displays the spectra obtained from the full model (39,683 states) at two different densities. There is no statistical approximation besides the configuration

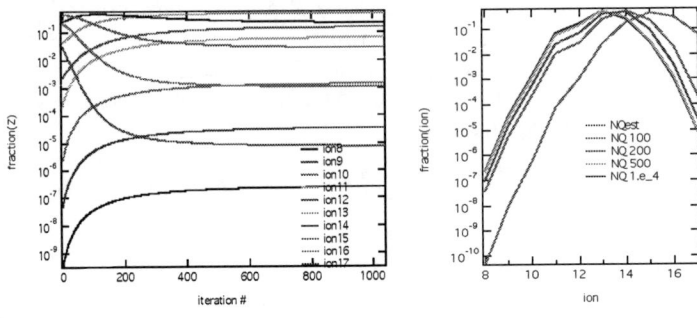

average.

FIGURE 7: Evolution of charge distribution for Xe at $T_e = 50$ eV and $N_e = 10^{20}$ cm^{-3}

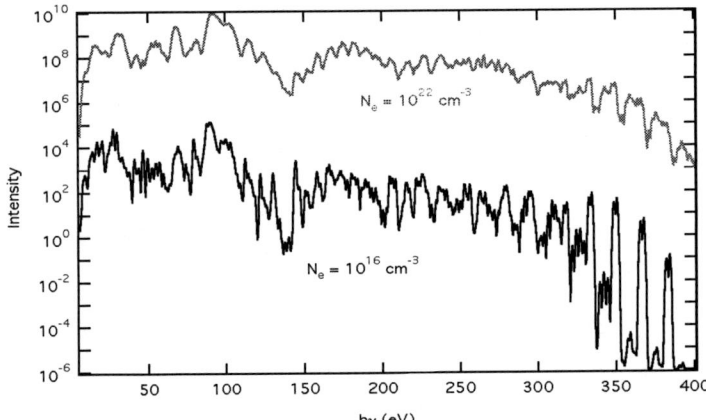

FIGURE 8: Spectra of Xe (39,683 states) at $T_e = 50$ eV and $N_e = 10^{16}$ and 10^{22} cm^{-3}.

SUMMARY AND CONCLUSION

This new version of HULLAC is much more reliable, complete, and easy to use than the previous ones. It was thoroughly checked by comparison to the LANL codes with excellent agreement. The new fitting procedure guarantees a trustworthy evaluation of the rates in the CRM solver. The latter has been shown to be robust and to give insight into the atomic model. All the code is in Fortran – mostly 77 and 95 compatible–, and it was successfully compiled on many platforms. It is now ready for distribution.

ACKNOWLEDGEMENTS

We thank Dr. S. Obenschain, Head of the Laser Plasma Branch (LPB) of the Naval Research Laboratory for his encouragements and support. This work was done under a contract with LPB through a grant from the USDOE. We are grateful to Prof. R Mancini and T. Durmaz for help with the LANL codes.

REFERENCES

1. A. Bar-Shalom, M Klapisch, and J Oreg, *J. Quant. Spectros. Rad. Transf.* **71**, 169 - 88 (2001).
2. M Klapisch, *Comput. Phys. Comm.* **2**, 239 -60 (1971); E. Koenig, *Physica(Utrecht)* **62**, 393 (1972).
3. A. Bar-Shalom and M. Klapisch, *Comput. Phys. Comm.* **50**, 375 - 93 (1988).
4. A. Bar-Shalom, M Klapisch, and J. Oreg, *Phys. Rev. A*. **38**, 1773- 84 (1988).
5. J. Oreg, W. H. Goldstein, M Klapisch et al., *Phys. Rev. A* **44**, 1750 (1991).
6. A. Bar-Shalom, M Klapisch, and J Oreg, presented at the *13th Int'l Conference on Atomic Processes in Plasmas*, Gatlinburg, TN, (2002), D R Schultz, F W Meyer, and F Ownby Eds., AIP Conference Proceedings, **635**, 92 - 100.
7. A. Bar-Shalom, J. Oreg, W. H. Goldstein et al., *Phys. Rev. A*. **40**, 3183 - 93 (1989).
8. L B Golden and D H Sampson, *Astroph. J. Supp.* **38**, 19 (1978); L B Golden, R E H Clark, S J Goett et al., *Astroph. J. Supp.* **45**, 603 - 12 (1981); S J Goett, R E H Clark, and D H Sampson, *Atomic Data Nuclear Data Tables* **25**, 185 (1980).
9. M Abramowitz and I A Stegun, *Handbook of Mathematical Functions.* (Dover Publications, New York, 1965).
10. M Busquet, *High Ener. Dens. Phys.* doi:10.1016/j.hedp.2007.01.007 (2007).
11. A Burgess and J A Tully, *Astronom. & Astrophys.* **254**, 436 - 53 (1992); A Burgess, M. C. Chidichimo, and J A Tully, *Astronom. & Astrophys. Supp.* **131**, 145 - 52 (1998).
12. Y Ralchenko, R W Lee, and C Bowen, presented at the *14th APS Topical conference on Atomic Processes in Plasmas*, Santa Fe, NM, (2004), J. S. Cohen, S. Mazevet, and D. P. Kilcrease Eds., AIP conference proceedings, **730**, 151 -160.
13. A Bar-Shalom, J Oreg, and M Klapisch, *J. Quant. Spectros. Rad. Transf.* **65**, 43 (2000).
14. William H Press, Brian P Flannery, Saul A Teukolsky et al., *Numerical Recipes in Fortran 77 :The art of scientific computing.*, 2nd ed. (Cambridge University Press, Cambridge, UK, 1996).
15. S. D. Loch, C. J. Fontes, J. Colgan et al., *Phys. Rev. E* **69**, 066405 (2004); M Busquet, J P Raucourt, and J C Gauthier, *J. Quant. Spectros. Rad. Transf.* **54**, 81 - 87 (1995).
16. M. Klapisch and M. Busquet, *High Ener. Dens. Phys.* **3**, 143-148 (2007).

Simultaneous COLTRIMS And X-Ray Spectroscopic Studies Relevant To Cometary, Planetary, And Heliospheric X-Ray Emission

Rami Ali

Department of Physics, University of Jordan, Amman 11942, Jordan

Abstract. Recent results from highly differential laboratory measurements of X-rays, scattered projectile, and recoil ions are reported for collisions relevant to cometary, planetary, and heliospheric X-ray emission. The studies employed supersonically cooled targets, position imaging detectors, and time-of-flight coincidence techniques; making it possible to simultaneously perform X-ray and cold-target recoil-ion momentum spectroscopic (COLTRIMS) measurements. The measurements provided unequivocal evidence for the importance of the role played by multiple-electron capture in the case of the many electron targets. Experimental relative n-selective single-electron capture (SEC) cross sections (σ_n^{rel}) are also reported. State-selective He-like X-ray spectra originating in SEC to specific states characterized by the quantum numbers n that represent state-of-the-art testing tools for theories are also presented.

INTRODUCTION

X-ray and extreme ultraviolet (EUV) emission has been observed from over 20 comets since the first observation in 1996 by Lisse et al. [1]. The charge exchange mechanism between highly charged solar wind (SW) minor heavy ions and cometary neutrals suggested by Cravens [2] is now recognized as the primary mechanism responsible for the observed emission lines [3-6]. In the SW charge exchange (SWCX) mechanism, electrons are captured from cometary neutrals by the SW ions into excited states of the resulting ions; which may then decay radiatively and in the process emit X-ray and/or EUV radiation. SWCX has also been suggested as contributing to the soft X-ray background of the heliosphere [7,8]. Very recently, the first definite detection of SWCX induced X-ray emission from the exosphere of planet Mars has been reported by Dennerl et al. [9]. The SWCX mechanism has been invoked with various degrees of sophistication to model and interpret cometary X-ray and EUV emission spectra (see, e.g., references [10,11], and references therein), and has been recently reviewed by Cravens [12] and Krasnopolsky, Greenwood, and Stancil [13].

Accurate modeling, from which reliable information on the cometary and planetary atmospheres and solar wind composition and velocity can be obtained, requires detailed understanding of charge exchange collisions and atomic data on state-selective charge transfer cross sections, as well as relaxation pathways of excited ionic species. This is further necessitated by the very high spectral resolution of the new

CP926, *Atomic Processes in Plasmas—15th International Conference on Atomic Processes in Plasmas*
edited by J. D. Gillaspy, J. J. Curry, and W. L. Wiese
© 2007 American Institute of Physics 978-0-7354-0436-6/07/$23.00

generation of space-based telescopes such as *Chandra* and *XMM-Newton*. The data should cover the pertinent range of solar wind ion velocities, species, and charge states as well as cometary molecular species and all of the individual reaction channels such as single-electron capture (SEC), multiple-electron capture (MEC), Auger and radiative relaxation pathways. Laboratory experiments are needed to benchmark theoretical models that will, in turn, predict this large range of required data.

The need for relevant atomic data for the accurate and reliable modeling of cometary X-ray and EUV emission has prompted several experimental groups to carry out laboratory investigations of relevant collision systems (see, e.g., references [14-21], and references therein). To date, however, all cometary X-ray and EUV emission models invoking SWCX had to rely on the limited relevant atomic data in the literature or on simple charge exchange models. In particular, all models including the most detailed ones (see, e.g., reference [11], and references therein) have assumed that cometary X-ray and EUV emission is the result of SEC only and ignored contributions from MEC. Furthermore, most models have assumed a dominant population of one principal quantum number n in SEC, and did not take into account the variation of n and the angular momentum quantum number l of the captured electron with the collision velocity.

Recent advances in experimental techniques have made it possible to investigate atomic collisions at an unprecedented level of detail. In this paper we describe a combination of experimental techniques that made it possible to simultaneously perform X-ray and cold-target recoil-ion momentum spectroscopic (COLTRIMS) measurements. The COLTRIMS technique has been reviewed by Dörner et al. [22]. We also report recent laboratory simulations of SW–comet interactions. The measurements involved the determination of relative experimental n-selective SEC cross sections (σ_n^{rel}) for a number of collision systems typical of SW-comet interactions. Our highly differential measurements also demonstrate that while the assumption of the dominance of SEC is justifiable to some extent in settings where He, or H, is the predominant target species, it is seriously flawed in the case of the many electron cometary target species such as H_2O, CO, CO_2, OH, and O. Finally, the true power of the technique is demonstrated through the ability to obtain state-selective sub-partial X-ray spectra originating in SEC to specific states characterized by the quantum numbers n.

EXPERIMENTAL SETUP

A schematic of the experimental apparatus is shown in Fig. 1. The experiments were carried out at the University of Nevada, Reno, multicharged ion research facility. The projectile ions used in the measurements were produced by a 14 GHz electron cyclotron resonance (ECR) ion source and extracted by appropriate potentials. The ions were guided to the interaction region where they crossed supersonic gas jets at 90°. The jets were about 2 mm wide with an internal pressure of about 0.1 mTorr. Three detectors were used to detect the collision products. First, an electric field extracted the recoil ions perpendicular to the incident ion beam. They were guided through a time-of-flight (TOF) spectrometer and were detected by a two-dimensional position sensitive microchannel plate detector (2D-PSD). Second, the highly charged projectile ions continued forward after the collision and then entered a region with a transverse electric field. The ions were deflected differently depending on their charge states. They struck another 2D-PSD and were detected. The collision chamber was differentially pumped in order to maintain high ion beam purity both before entering the chamber and after leaving it. Finally, X-rays emitted at 90° relative to the incident ions were detected by a windowless high-purity germanium (HPGe) EG&G IGLET-X detector [23], placed opposite the recoil detector, with a resolution of about 133 eV for the H-like Ne^{9+} Lyα line.

The measurements were performed in a number of modes; the COLTRIMS only mode involving the coincident detection of projectile and recoil ions only, the noncoincident-singles X-ray mode where only X-rays are detected, and the triple-coincidence mode involving the coincident detection of X-rays, projectile, and recoil ions. In the latter, more sophisticated mode, the data acquisition system was triggered by the detection of an X-ray. A fast timing signal derived from the X-ray detector was used to start a time-to-digital converter (TDC) that was stopped by a signal from the projectile detector. Another TDC was started by a fast signal from the projectile detector and stopped by a recoil signal. Analog-to-digital (ADC) converters read the position signals from the projectile and recoil detectors. The TOF of the recoil ions provided their charge states, and together with the impact positions provided the three momentum components of the recoil ions. The impact positions on the projectile 2D-PSD provided the final projectile charge states. The time and position data were stored in event mode for further processing.

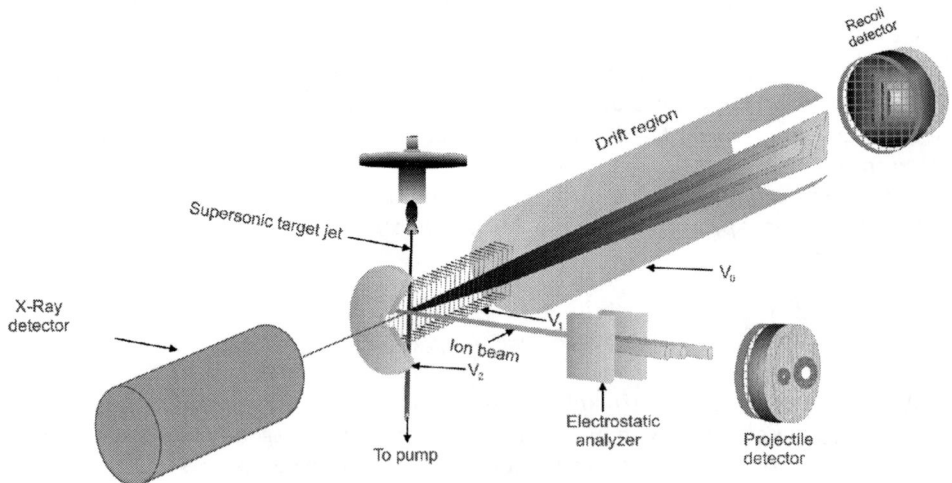

FIGURE 1. A schematic of the experimental setup.

RESULTS AND DISCUSSION

To demonstrate the high differentiation power of the triple-coincidence mode, Fig. 2 shows a multiparameter representation of the measurements for the 4.55 keV/u (933 km/s) $^{22}Ne^{10+}$ on Ar collision system [24]. The scatter plot of Fig. 2a represents coincidences between recoil ions and X-rays. Figure 2b represents coincidences between projectile and recoil ions. Projections onto the appropriate axes provide the recoil ion TOF spectrum (Fig. 2c), the singles X-ray spectrum (Fig. 2d), and the final projectile charge state distribution (Fig. 2e). It should be mentioned that while the targets of interest to cometary X-ray emission are mainly molecular ones, atomic targets were used in some of these studies to better judge the role of MEC since complications would arise due to Coulomb explosions following MEC from molecular targets. Furthermore, the first few ionization potentials of Ar are close to those of the molecular targets, and electron capture processes are expected to be very similar.

Significance of Multiple-Electron Capture

It is evident from Fig. 2a that the singles X-ray spectrum resulted from processes involving the capture of up to six electrons as evidenced by the observation of Ar^{6+}

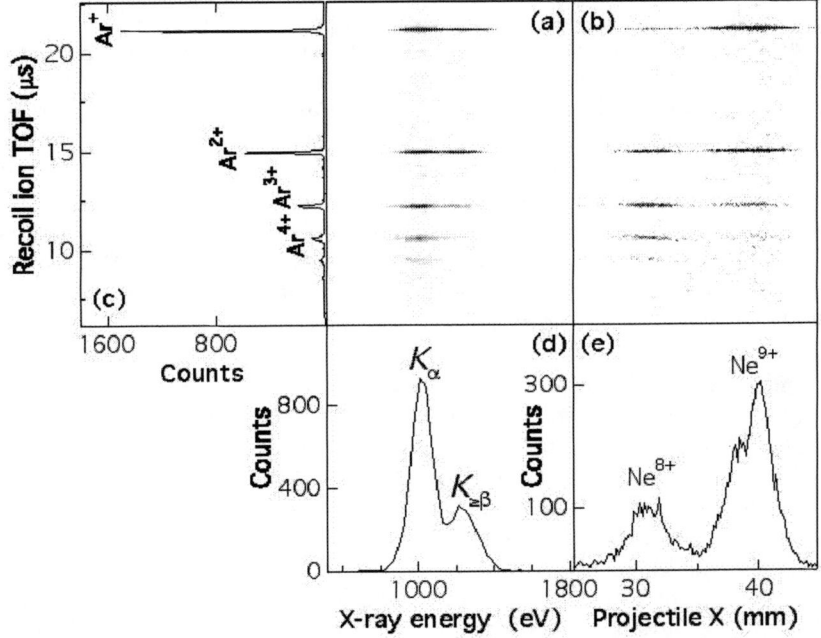

FIGURE 2. Multiparameter representation of the triple-coincidence measurements for the Ne^{10+} on Ar collision system. (a) Coincidences between recoil ions and X-rays. (b) Coincidences between projectile and recoil ions. (c) Recoil ion TOF spectrum. (d) Singles X-ray spectrum. (e) Final projectile charge state distribution. Reproduced by permission of the AAS from Ali et al. [24].

target ions. It is also evident from Fig. 2*b* that the projectile ions keep one or two electrons, resulting in only Ne^{9+} or Ne^{8+} final projectile ions regardless of the initial number of captured electrons. In particular, Fig. 2*b* clearly demonstrates that in collisions leading to the production of Ne^{9+} ions, as many as four electrons may have been initially captured by the Ne^{10+} ions, thus forming up to quadruply excited projectile ions. These multiply excited ions must have then undergone a number of autoionization processes (see, e.g., [25]) ending in singly excited Ne^{9+} ions that have subsequently decayed radiatively. For example, three autoionization processes take place when an Ar^{4+} target ion is produced. Such autoionization processes result in a singly excited state population prior to the radiative transitions that is completely different from that resulting from true SEC where only one electron is captured. Indeed, it is interesting to note that in Fig. 2*a*, the higher recoil ion charge states resulting from MEC are found dominantly in coincidence with $K\alpha$ X-rays. This may be due to a combination of populating lower levels on the projectile in MEC and the role played by the autoionization cascades that also feed lower levels. Both scenarios lead to the dominance of $K\alpha$ emission.

The relative importance of SEC and MEC collisions can be obtained from recoil ion TOF spectra similar to that of Fig. 2*c*, except that the spectra should be obtained from coincidence measurements of recoil ions and scattered projectiles only without regard to whether an X-ray was emitted or not. This is essential in order to account for MEC collisions that may not give rise to X-ray emission. Such TOF spectra have been measured for the He, Ne, and Ar targets and are shown in Fig. 3*a*. By determining the areas under the respective peaks, the fraction of events leading to singly ionized targets (SEC) or multiply ionized targets (MEC) can be found. For the He target, SEC dominates by a large margin, and limiting the models to SEC might be easily justified in environments where He is the prevalent target, such as in the heliosphere. The case is clearly different for the Ne target where the SEC and MEC fractions are close to each other but where SEC events still outnumber MEC events. For the Ar target, however, the scenario has changed, and MEC events outnumber the SEC ones. Clearly, any model ignoring the role of MEC for Ar, or the very similar cometary neutrals, will undoubtedly lead to erroneous conclusions.

A major advantage of the coincidence measurements is that it is possible to obtain partial X-ray spectra corresponding to any recoil charge state. For simplicity, however, we show in Fig. 3*b* two partial X-ray spectra for each atomic target: one corresponding to SEC and the other corresponding to the cumulative MEC. Singles X-ray spectra, which are the sum of SEC and MEC, are also shown. The percentages indicate the fraction of X-rays that resulted from either SEC or MEC collisions. We note that for He, the SEC and singles spectra are almost identical in profile, which supports the earlier argument that ignoring MEC for this target may be justified to first order in models. This is definitely not true for the Ne and Ar targets, where the SEC and the singles profiles are clearly different from each other and from MEC spectra as well. We also note that the SEC profiles for Ne and Ar are different from each other and that the same is true for the MEC profiles. Moreover, we note a shift from high-*n* to low-*n*

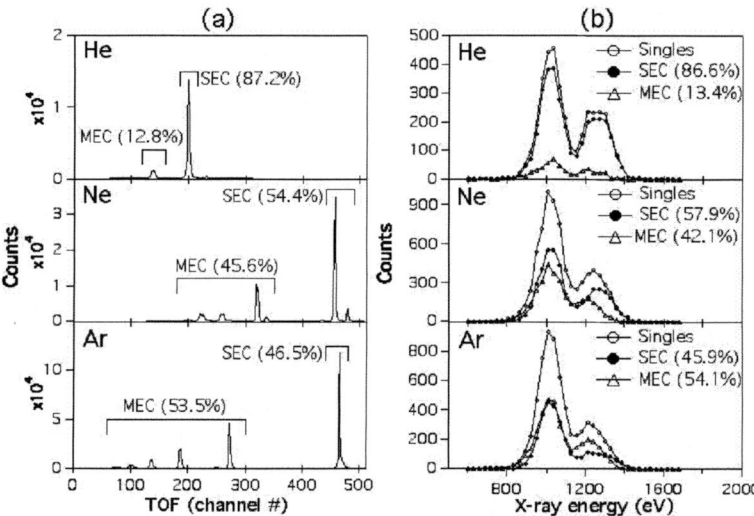

FIGURE 3. (*a*) Recoil ion TOF spectra for the Ne^{10+} on He, Ne, and Ar collision systems. The percentages represent the fraction of SEC and MEC collisions for each target. (*b*) Singles and partial X-ray spectra corresponding to SEC and MEC collisions. The percentages represent the fraction of X-rays resulting from SEC or MEC collisions.. Reproduced by permission of the AAS from Ali et al. [24].

(with $n \geq 3$) emission when comparing the MEC profiles to the SEC profiles, which confirms an earlier suggestion [10] that strong emission from $n = 3, 4$ levels is due to double (or multiple) electron capture.

It order to compare singles X-ray spectra for different targets, we have obtained such spectra for the He, Ne, Ar, CO, and CO$_2$ targets in the noncoincident-singles X-ray mode. Figure 4 compares these X-ray spectra, normalized to the same total number of counts, for all targets. Surprisingly, all targets apart from He give rise to identical spectra. While the first ionization potentials of Ar (15.8 eV), CO (14.0 eV), and CO$_2$ (13.7 eV) are close to each other, and one might expect similar spectra assuming SEC to be dominant, that of Ne (21 eV) is much larger. The similarity of the spectra is due to the complementary roles played by SEC and MEC. In fact, when added together, the SEC and MEC profiles for Ne and Ar in Fig. 3*b* give rise to identical singles profiles. This is unequivocal evidence for the importance of the role played by MEC in the case of the many electron targets. This argument is further supported by the fact that the ionization potential of He (24.5 eV) is much closer to that of Ne, and yet there is a clear difference in their spectra resulting from the dominance of SEC for He. In fact, the He spectrum does not show the low-energy shoulder at 900 eV, a signature of MEC-induced He-like Ne^{8+} X-ray emission, that all other targets show. Therefore, in the case of many electron targets, one cannot simply assume SEC to be dominant and hope to extract accurate information through comparisons of model results with observed spectra. Accurate modeling should take into account MEC and the intermediate autoionization processes that alter the radiative state population.

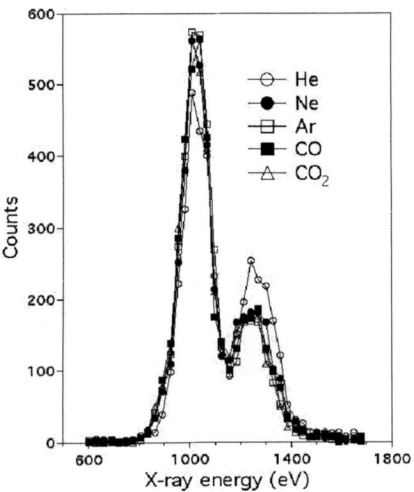

FIGURE 4. Singles X-ray spectra obtained in collisions of Ne^{10+} with He, Ne, Ar, CO, and CO_2. The spectra are normalized to the same total number of counts. Reproduced by permission of the AAS from Ali et al. [24].

State-Selective Single-Electron Capture

The multiparameter representation of Fig. 2 does not display all the information obtained in the triple-coincidence measurements. For example, the recoil ion impact position image is not displayed. Such an image is shown in Fig. 5, although for different collision systems, namely; the 4.67 keV/u (946 km/s) $^{15}N^{7+}$ on He and H_2O collision systems [26]. Supersonic jets of H_2O were produced using He as a carrier gas. From the recoil impact position and TOF, the three components of the momentum transfer to the recoil ion can be calculated.

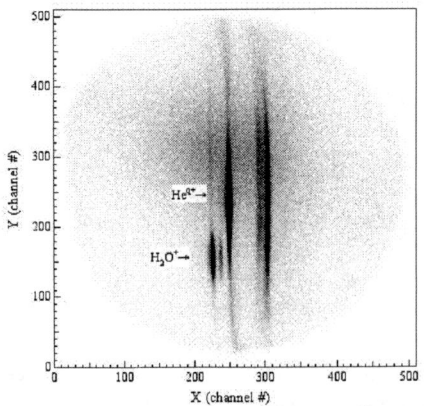

FIGURE 5. A two-dimensional impact position image for the He^{q+} and H_2O^+ recoil ions produced in the 4.67 keV/u $^{15}N^{7+}$ + He and H_2O collision systems.

In pure charge exchange collisions, typical of slow collisions, where no electrons are directly ejected to the continuum, the change in electronic energy of the collision system, or the Q-value, is a direct measure of the projectile state population. Q-value spectra, therefore, provide the experimental n-state-selective relative cross sections σ_n^{rel}. The Q-value is given by $Q \approx - (P_{long} v + iv^2/2)$ [27], where P_{long} is the longitudinal (i.e., parallel to the incident projectile direction) momentum transfer to the recoil, v is the projectile velocity, and i is the number of electrons captured by the projectile. For SEC processes not accompanied by dissociation of the residual molecular ions, COLTRIMS can be applied to obtain the Q-value when molecular targets are used.

Figure 6 shows state-selective results for SEC from H_2O and other cometary neutrals, as well as from He, in collisions with N^{7+} and O^{7+} ions, which are typical SW constituent ions. Two different collision velocities were used for some collision systems. From these spectra, Hasan et al. extracted σ_n^{rel} and compared them to predictions of a number of models. In particular, the classical trajectory Monte Carlo model agreed reasonably well with the measurements. Figure 6 clearly demonstrates the strong dependence of σ_n^{rel} on the target ionization potential. It also demonstrates that with decreasing collision energy, the capture becomes increasingly selective and seems to converge to a dominant n. In addition, the close similarity of the Q-value distributions for the N^{7+} and O^{7+} projectile ions suggests that the n-state–selective electron capture process is essentially independent of the internal projectile structure

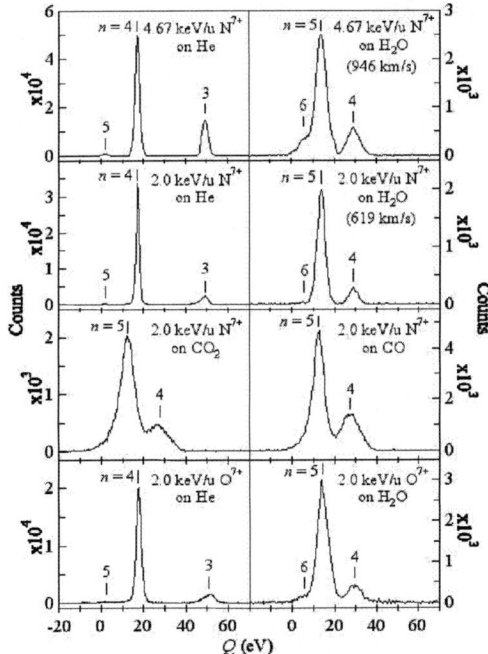

FIGURE 6. Experimental Q-value spectra for SEC for a number of collision systems; n is the energy level to which the electron is captured. The spectra for the molecular targets are for the nondissociating channels. Reproduced by permission of the AAS from Hasan et al. [26].

for these high-charge states. However, the internal structure of the projectile can have an effect on the product l-distribution (and spin statistics), which in turn strongly influences the radiative relaxation pathways [16,17,28].

State-Selective X-Ray Spectra

A prominent feature in cometary X-ray and EUV spectra is that some of the most intense emission lines are those from He-like ions following electron capture by H-like ions. In particular, the X-ray spectra exhibit strong contributions from the forbidden transitions $1s2s\ ^3S_1 \rightarrow 1s^2\ ^1S_0$, as well as contributions from the intercombination transitions $1s2p\ ^3P_1 \rightarrow 1s^2\ ^1S_0$. It is therefore essential for the accurate modeling of cometary X-ray emission to be able to account for the spin multiplicity of the product He-like states following charge exchange, in addition to the distributions of the principal and angular momentum quantum numbers (n,l).

A major advantage of the triple-coincidence measurements is that it is possible to obtain state-selective X-ray spectra corresponding to each populated n-level in the case of SEC. This is accomplished by considering only those X-rays resulting from SEC collisions whose Q-values are consistent with a particular n. Such He-like spectra are shown in Fig. 7 for SEC in collisions of Ne^{9+} with He at a velocity of 800 km/s [29]. Clearly, SEC to different n-levels results in different X-ray profiles. Such state-selective spectra provide the ultimate testing tools for the ability of any theoretical model to account for the relative population of triplet to singlet states as well as the (n,l) distributions in He-like product ions. The analysis of these and other similar spectra is still in progress and final results will be reported in a future publication [29].

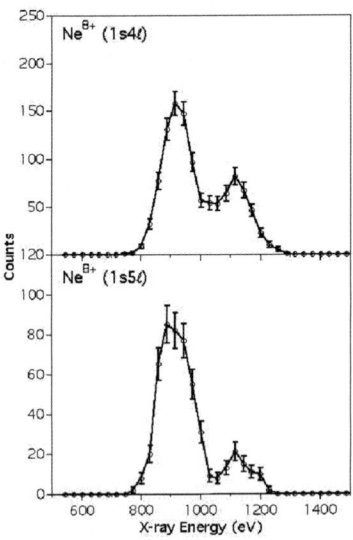

FIGURE 7. Experimental state-selective He-like neon X-ray spectra following SEC in Ne^{9+} with He collisions at a velocity of 800 km/s.

224

CONCLUSIONS

We have presented recent results from highly differential laboratory measurements of X-rays, scattered projectile, and recoil ions produced in collisions typical of SW interactions with cometary, planetary atmospheres, and heliospheric neutrals. Unequivocal evidence for the importance of the role played by MEC in the case of the many electron targets has been presented. Relative experimental n-selective SEC cross sections (σ_n^{rel}) for a number of collision systems have also been presented. State-selective He-like X-ray spectra originating in SEC to specific states characterized by the quantum numbers n have also been presented and will serve as stringent testing tools for future theoretical calculations.

ACKNOWLEDGMENTS

The author gratefully acknowledges the collaboration of P.A. Neill, P. Beiersdorfer, C.L. Harris, M.J. Rakovic, J.G. Wang, D.R. Schultz, and P.C. Stancil, A.A. Hasan, F. Eissa, and R.A. Phaneuf.

REFERENCES

1. Lisse, C. M., et al., *Science* **274**, 205 (1996).
2. Cravens, T. E., *Geophys. Res. Lett.* **24**, 105 (1997).
3. Lisse, C. M., et al., *Science* **292**, 1343 (2001).
4. Krasnopolsky, V. A., and Mumma, M. J., *ApJ* **549**, 629 (2001).
5. Krasnopolsky, V. A., et al., *Icarus* **160**, 437 (2002).
6. Sasseen, T. P., et al., *ApJ* **650**, 461 (2006).
7. Cravens, T. E., *ApJ* **532**, L153 (2000).
8. Pepino, R., Kharchenko, V., Dalgarno, A., and Lallement, R., *ApJ* **617**, 1347 (2004).
9. Dennerl, K., et al., *A&A* **451**, 709 (2006).
10. Beiersdorfer, P., et al., *Science* **300**, 1558 (2003).
11. Kharchenko, V., Rigazio, M., Dalgarno, A., and Krasnopolsky, V. A., *ApJ* **585**, L73 (2003).
12. Cravens, T. E., *Science* **296** 1042 (2002).
13. Krasnopolsky, V. A., Greenwood, J. B., and Stancil, P. C., *Space Sci. Rev.* **113**, 271 (2004).
14. Greenwood, J. B., Williams, I. D., Smith, S. J., and Chutjian, A., *ApJ* **533**, L175 (2000).
15. Greenwood, J. B., Williams, I. D., Smith, S. J., and Chutjian, A., *Phys. Rev. A* **63**, 062707 (2001).
16. Beiersdorfer, P., et al. 2000, *Phys. Rev. Lett.* **85**, 5090 ().
17. Beiersdorfer, P., Lisse, C. M., Olson, R. E., Brown, G. V., and Chen, H., *ApJ* **549**, L147 (2001).
18. Gao, H., and Kwong, V. H. S., *ApJ* **567**, 1272 (2004).
19. Gao, H., and Kwong, V. H. S., *Phys. Rev. A* **69**, 052715 (2004).
20. Bodewits, D., Juh_asz, Z., Hoekstra, R., and Tielens, A. G. G. M., *ApJ* **606**, L81 (2004).
21. Bodewits, D., et al., *ApJ* **426**, 593 (2006).
22. Dörner, R., et al., *Phys. Reports* **330**, 95 (2000).
23. Smith, A. A., et al., *Rev. Sci. Instrum.* **66**, 2333 (1995).
24. Ali, R., et al., *ApJ* **629**, L125 (2005).
25. Emmons, E.D., Hasan, A.A., and Ali, R., *Phys. Rev. A* **60**, 4616 (1999).
26. Hasan, A. A., Eissa, F., Ali, R., Schultz, D. R., and Stancil, P. C., *ApJ* **560**, L205 (2001).
27. Ali, R., et al., *Phys. Rev. Lett.* **69**, 2491 (1992)
28. Kharchenko, V., and Dalgarno, A., *ApJ* **554**, L99 (2001).
29. Eissa, F., et al., (to be submitted).

PLASMA DIAGNOSTICS

Monochromatic Soft X-Ray Self-Emission Imaging in Dense Z Pinches

B. Jones*, C. Deeney[†], C. J. Meyer**, C. A. Coverdale*, P. D. LePell**,
J. P. Apruzese[‡], R. W. Clark[‡], J. Davis[‡], and K. J. Peterson*

*Sandia National Laboratories, PO Box 5800, Albuquerque, NM 87185 USA
[†]National Nuclear Security Administration, Washington, DC 20585 USA
** Ktech Corp., Albuquerque, NM 87123 USA
[‡]Naval Research Laboratory, Washington, DC 20375 USA

Abstract. The Z machine at Sandia National Laboratories drives 20 MA in 100 ns through a cylindrical array of fine wires which implodes due to the strong $\mathbf{j} \times \mathbf{B}$ force, generating up to 250 TW of soft x-ray radiation when the z-pinch plasma stagnates on axis. The copious broadband self-emission makes the dynamics of the implosion well suited to diagnosis with soft x-ray imaging and spectroscopy. A monochromatic self-emission imaging instrument has recently been developed on Z which reflects pinhole images from a multilayer mirror onto a 1 ns gated microchannel plate detector. The multilayer can be designed to provide narrowband (~10 eV) reflection in the 100-700 eV photon energy range, allowing observation of the soft emission from accreting mass as it assembles into a hot, dense plasma column on the array axis. In the present instrument configuration, data at 277 eV photon energy have been obtained for plasmas ranging from Al to W, and the z-pinch implosion and stagnation will be discussed along with > 1 keV self-emission imaging and spectroscopy. Collisional-radiative simulations are currently being pursued in order to link the imaged emissivity to plasma temperature and density profiles and address the role of opacity in interpreting the data.

Keywords: X-ray imaging, plasma diagnostics, wire array z pinch, multilayer mirror.
PACS: 52.58.Lq, 52.59.Qy, 52.70.-m, 52.70.La, 73.21.Ac

INTRODUCTION

The Z machine [1] at Sandia National Laboratories is a pulsed power generator capable of driving a range of powerful, efficient z-pinch radiation sources for high-energy density physics experiments [2]. Compact tungsten wire arrays produce soft x-ray photons in the < 1 keV range at powers > 200 TW [3] for inertial confinement fusion research [4], while Al, Ti, stainless steel, and Cu wire arrays and Ar gas puff z pinches have been studied for generation of 1-8 keV K-shell radiation [5].

All classes of pinches studied on Z generate copious broadband radiation in the 100-1000 eV photon energy range, and time-resolved self-emission imaging through the use of a filtered pinhole camera is a standard diagnostic technique [6]. Filtered x-ray imaging can reject lower energy photons, providing a broadband higher photon energy image with a cutoff in the > 1 keV range. However, filtration alone cannot pass lower energy photons while rejecting higher energy x-rays in order to achieve narrowband monochromatic self-emission imaging, which is desirable for imaging the

CP926, *Atomic Processes in Plasmas—15th International Conference on Atomic Processes in Plasmas*
edited by J. D. Gillaspy, J. J. Curry, and W. L. Wiese
© 2007 American Institute of Physics 978-0-7354-0436-6/07/$23.00

cooler trailing mass present during the implosion. This can be accomplished by combining filtration with reflection from an x-ray mirror, and has been studied using grazing incidence reflection from a Si mirror [7, 8] or from a multilayer mirror (MLM) [9] in z-pinch or laser-driven plasma experiments.

In this paper, we briefly describe an instrument employing MLM reflection for monochromatic imaging in the 100-700 eV range on the Z machine. This diagnostic has been described in greater detail in References [10-12]. In the following section, we describe the instrument and show an example of data from a Z wire array implosion. Then, future directions for quantitative analysis of the imaging data are discussed.

DESCRIPTION OF THE X-RAY IMAGING DIAGNOSTIC

The Z MLM imager diagnostic combines a standard, filtered pinhole camera [Fig. 1(a)] with two pinhole cameras whose images are reflected from a MLM [one is shown in Fig. 1(b)]. Each MLM provides narrowband (~5 eV) reflection and serves as a monochromator. MLM reflectivity (R) and filter transmission (T) for a 277 eV photon imaging configuration that has been successfully fielded on Z are shown in Fig. 2. The calculation of this reflectivity curve and those for other untested configurations (Table 1) are discussed in Ref. [12]. For each configuration, the designed MLM must be matched with a filter that can pass the MLM reflection at a photon energy slightly less than an absorption edge energy of the filter material, but which rejects UV/visible specular reflection from the MLM and suppresses second order reflection. The spatial resolution (Table 1) is limited by diffraction for these low photon energies given the 3.56 m distance from source to pinholes [10] that is desired for ease of fielding outside the Z vacuum chamber and mitigation of debris damage of the device. As will be illustrated, however, the resolution is adequate for resolving the dynamics of the several-millimeter-scale imploding z pinches. Each camera employs eight pinholes and an eight-frame microchannel-plate detector for multi-point time-resolved (< 1 ns) imaging [10].

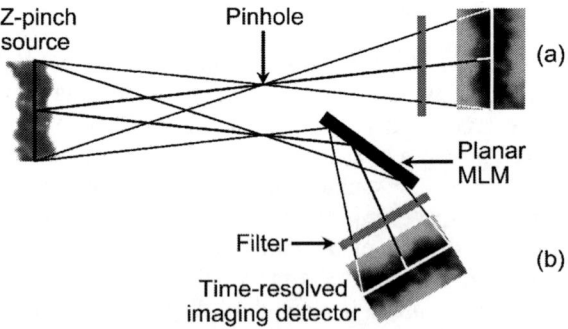

FIGURE 1. Instrument combines (a) a standard pinhole camera with (b) pinhole imagers reflecting from a multilayer mirror (MLM) monochromator. Reprinted with permission from B. Jones *et al.*, "Monochromatic X-Ray Self-Emission Imaging of Imploding Wire Array Z-Pinches on the Z Accelerator," IEEE T. Plasma Sci. **34**, 213 (2006), Fig. 1. ©2006, IEEE.

Figure 3 shows imaging data collected over two nominally identical Z shots in which 60 mm on 30 mm diameter nested copper wire arrays were employed to generate ~8 keV Cu K-shell emission [13]. These false-color, intensity-rescaled images display K-shell self-emission (standard pinhole configuration with 127 μm Kapton filter) in green, overlayed with 277 eV monochromatic emission in red; yellow indicates the superposition of both photon energies. The harder K-shell emission is seen to originate only from a narrow column on the wire array axis from early in time through the main x-ray power peak. In contrast, the 277 eV MLM-reflected images allow the observation of the cooler plasma that is accreting (and depositing energy) on axis over the rise of the main x-ray pulse. The radial distribution of the imploding mass is impacted by magnetic Rayleigh-Taylor instabilities growing from the start of the implosion [14], and the distribution of mass and its arrival time on axis is believed to determine the rise time of the x-ray pulse [4, 11]. Detailed study of the implosion

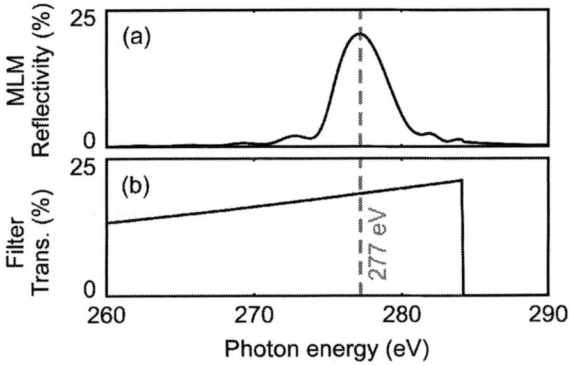

FIGURE 2. (a) Calculated reflectivity of the Cr/C MLM employed in the instrument (~40 Å period, ~34° grazing angle). (b) 4 μm parylene-N + 1000 Å Al filter passes the mirror reflection but suppresses visible light and second order reflection. Reprinted with permission from B. Jones *et al.*, "Monochromatic X-Ray Self-Emission Imaging of Imploding Wire Array Z-Pinches on the Z Accelerator," IEEE T. Plasma Sci. **34**, 213 (2006), Fig. 2. ©2006, IEEE.

TABLE 1. Proposed MLM/filter configurations for monochromatic imaging at various photon energies. These involve a simple swap of the Config. 2 filter and 34° grazing incidence MLM that is presently fielded on the Z machine. Reused with permission from B. Jones, C. Deeney, C. A. Coverdale, C. J. Meyer, and P. D. LePell, Review of Scientific Instruments 77, 10E316 (2006). Copyright 2006, American Institute of Physics.

Config.	Photon Energy (eV)	FWHM (eV)	Peak R (%)	Multilayer Materials	Bi-layer Period (Å)	Filter	Spatial Resolution (μm)
1	96.2	8.3	59.9	Mo/Si	125	2.5 μm Be	835
2	277.6	4.2	24.1	Cr/C	40.42	4 μm Parylene-N + 1000 Å Al	482
3	442.1	5.4	4.1	W/Si	25.35	1 μm Ti	382
4	500.0	5.0	4.9	W/Si	22.375	1 μm V	359
5	527.6	3.9	5.1	W/Si	21.18	1 μm Cr	350
6	700.0	4.3	4.2	W/Si	15.935	1 μm Fe	304
7	768.7	3.0	4.4	W/Si	14.5	1 μm Ni	290

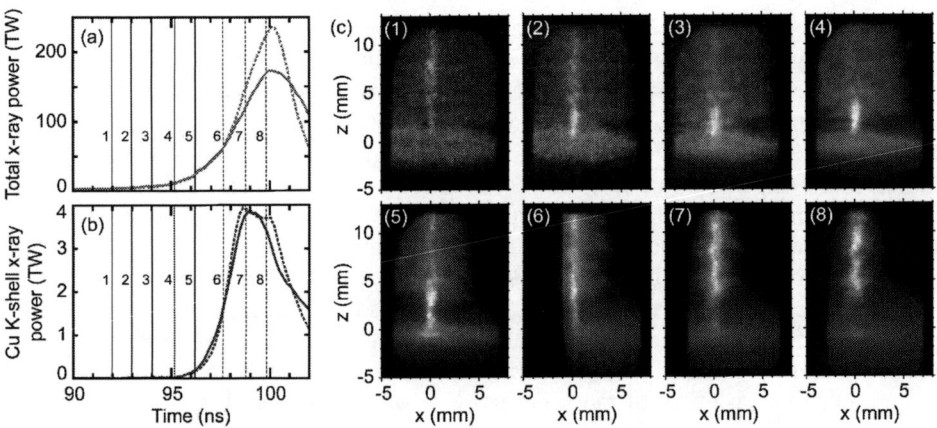

FIGURE 3. (color online) (a) Total radiated x-ray power and (b) ~8 keV Cu K-shell x-ray power with associated imager frame timing indicated for Z shots 1616 (dashed lines) and 1617 (solid lines). (c) False-color overlay of 277 eV (red) and Cu K-shell (green) self-emission images. Yellow indicates both 277 eV and K-shell emission superimposed. MLM-reflected images track the implosion of cooler trailing mass, which accretes on axis where K-shell emission is excited.

kinetic energy, plasma energy deposition, and x-ray generation is ongoing.

The images in Fig. 3 also show a premature pinching of the plasma within 4 mm of the cathode, which is the glowing surface viewed at 12° from the horizontal at the bottom of each image. Strategies to mitigate this electrode effect are presently being studied [15].

In addition to allowing observation of the cooler trailing mass, the 34° grazing angle MLM reflection also allows the detector to be shielded from direct line of sight from the x-ray source [10]. Background exposure of the detector caused by bremsstrahlung emission has been problematic with standard filtered pinhole cameras, but these hard-energy photons pass through and are not reflected from the x-ray mirror and thus do not reach the detector, resulting in significant improvements in signal-to-noise in imaging configurations featuring mirror reflection [8, 11].

COLLISIONAL-RADIATIVE SIMULATION AND FUTURE DIRECTIONS FOR DATA ANALYSIS

Images as shown in Fig. 3 paint a valuable qualitative picture of z-pinch dynamics, but we also wish to process the data to extract quantitative information. The diameter of the radiating, stagnated column can be measured in a straightforward manner, and is a necessary quantity for inferring plasma properties and relating them to radiated power [16]. In addition, Abel inversion of the images has been pursued in order to track the evolution of the radial profile of plasma emissivity at 277 eV photon energy [11]. Emissivity profiles can be useful in benchmarking magnetohydrodynamic (MHD) simulations of the z-pinch implosion, but the monochromatic nature of the imaging data lends itself to further analysis in order to link emissivity with plasma density and temperature. We must also address whether opacity plays a role at the photon energy being imaged in order to justify Abel inversion, which implicitly assumes optically thin emission in most algorithms.

In this section, we discuss an example of analysis that begins to address these issues for a wire array z-pinch plasma of alloy Al 5056 (95 % Al, 5 % Mg). This material was chosen for this discussion as it is the lowest atomic number (and thus has the fewest atomic levels) of interest as a K-shell wire array radiator. Non-LTE collisional-radiative simulation was performed for such a plasma with the PrismSPECT code [17] which included all ground state levels, and all available Al and Mg levels for B-like through H-like charge states (4194 levels). This model is expected to be adequate for the range of plasma conditions of interest in imploding fast z pinches, except for electron temperature (T_e) less than 100 eV where C-like states are significantly populated in the code (and only the ground state is included here). This model could be further refined for accuracy in this region, and to speed calculation by judicious elimination or bundling of atomic levels. We do not address here the accuracy of atomic rates and completeness of level structure in the ATBASE atomic database [18] used in the simulation. The model incorporates radiation transport across a one-dimensional slab (1 mm path length is used, as relevant length scales in z pinches are of this order) in order to estimate opacity versus photon energy.

The strategy pursued here is to use modeled emissivity $\varepsilon(h\nu)$ and opacity $\kappa(h\nu)$ along with the bandpass of a given imaging configuration in order to relate measured signal and optical depth to T_e and plasma ion density n_i. Figure 4(a) shows MLM reflectivity (R) and filter transmission (T) for configurations 2 and 6 of Table 1; the product RT gives the instrument response versus photon energy. An example of modeled emissivity [Fig. 4(b)] and optical depth [Fig. 4(c)] over 1 mm path length are shown for plasma conditions of $T_e = 316$ eV, $n_i = 10^{19}$ cm^{-3}. The effective optical depth, indicated for the two imager configurations, is calculated using an effective opacity $\kappa_{eff} = \int_0^\infty RT\varepsilon dE \Big/ \int_0^\infty RT\varepsilon/\kappa dE$, integrated over photon energy $E = h\nu$

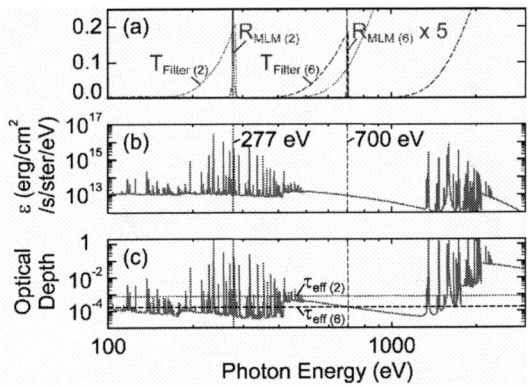

FIGURE 4. (a) MLM reflectivity (R) and filter transmission (T) versus photon energy for the imaging configurations 2 and 6 from Table 1. (b) PrismSPECT modeled emissivity for $T_e = 316$ eV, $n_i = 10^{19}$ cm^{-3}, with config. 2 and 6 imaging energies indicated. (c) Modeled optical depth through 1 mm path length, with the effective optical depth (τ_{eff}) of each imager configuration indicated by dashed horizontal lines.

similarly to a Rosseland mean opacity but weighted also by instrument response. The quantity $\int_0^\infty RT\varepsilon dE$ is proportional to the quasi-monochromatic emissivity that is recovered from imaging data after Abel inversion.

Collisional-radiative simulations were carried out for a range of T_e, n_i with the quantity $\int_0^\infty RT\varepsilon dE$ representative of the measured signal shown in Fig. 5 (a) and (b) for imager configuration 2 and 6 respectively. The effective optical depth is shown versus T_e, n_i in Fig. 5(c,d). As noted earlier, these models should be reasonable for $T_e \geq 100$ eV, which is likely the case in the final implosion stage of Al 5056 wire arrays on Z based on the observation of Al and Mg K-shell line emission from large radius at that time (detailed analysis is in progress) and on MHD simulation [15].

The first observation from Fig. 5(c,d) is that the emission for both photon energies considered is robustly optically thin for $T_e \geq 100$ eV, justifying the Abel inversion of the measured images. The simulation includes only two points per decade in T_e, n_i space, but this conclusion seems convincing nonetheless based on the extremely low values of τ_{eff} for all points with $T_e \geq 100$ eV.

The second goal in discussing Fig. 5 is to address the relationship between T_e, n_i and ε in order to identify how to extract density or temperature profile information from Abel-inverted emissivity profiles. If the contours in Fig. 5(a,b) were vertical lines with $\varepsilon \propto n_i^2$, for example, then we could simply reconstruct the density profile by taking the square-root of the reconstructed $\varepsilon(r)$ profile (in arbitrary units in the present diagnostic implementation) and normalizing to the total initial wire array mass. Figure 5(a,b) exhibits temperature dependence as well, however, with a more complicated dependence at 277 eV (the present configuration implemented on Z) due to the line structure at that photon energy as seen in Fig. 4(b). By implementing configuration 6 at 700 eV in the future, it might be possible to image in the Al L-shell

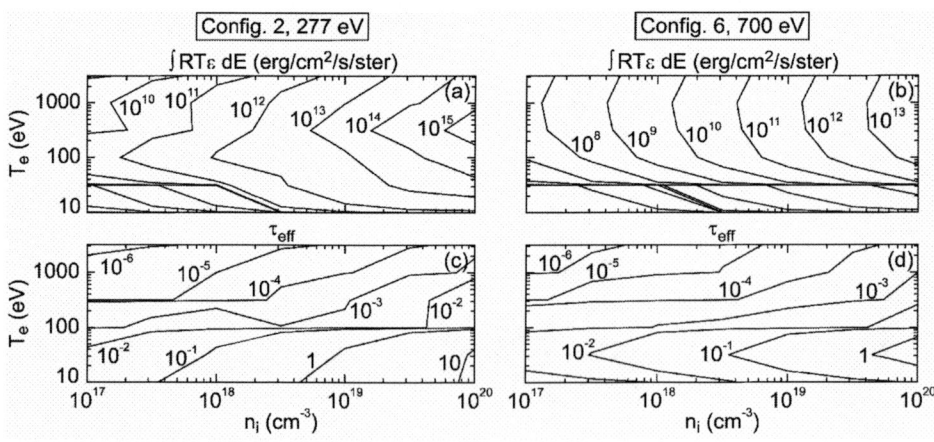

FIGURE 5. Representative measured signal (a, b) and effective optical depth (c, d) versus electron temperature and ion density for imager configurations 2 (a, c) and 6 (b, d), calculated for a 95 % Al, 5 % Mg plasma using the PrismSPECT non-LTE collisional-radiative code. Calculations may not be valid for $T_e < 100$ eV, where the atomic model provides inadequate detail in the populated C-like states.

continuum, where a simple $\varepsilon \propto n_i^2$ dependence is seen in Fig. 5(b) along with a weak temperature dependence. Temperature dependence might be corrected for using $T_e(r)$ inferred through K-shell spectroscopy using a time- and space-resolved crystal spectrometer [19] (as mentioned, Al is overdriven on Z and is seen to emit K-lines at large radius from early in time).

To address these same issues of opacity and the relation of emissivity to plasma conditions in higher-atomic-number-plasmas, the same techniques could be applied, though additional challenges would be presented by the increased atomic complexity. The number of levels included in the atomic physics models would likely need to be higher to capture the behavior of the emission spectrum with fidelity. Preliminary modeling for Cu indicates that the plasma is optically thin at 277 eV throughout the implosion [10], but adequacy of the L-shell model needs to be carefully considered. Another concern is that K-shell emission from Cu loads, for example, is seen only from on axis as discussed for Fig. 3. L-shell or even UV spectroscopy (which would also be more challenging due to the number of lines) might have to be explored for mid- to high-atomic-number elements in order to observe line emission from large radius early in time and have a hope of inferring the temperature profile.

Another approach might be profitable both for Al and higher-atomic-number plasmas—if monochromatic imaging could be obtained with absolute calibration at two nearby photon energies in the 100-1000 eV range, it might be possible to infer temperature profiles based on the ratio of the Abel-inverted emissivities. This has been previously suggested for monochromatic imaging of free-free continuum emission at higher x-ray photon energies ($E \gg Z^2 \times 13.6$ eV) [20], and an analogous technique could be possible using either non-LTE collisional-radiative modeling or analytical theory in the Al L-shell continuum region, for example. The required absolute calibration could be obtained by normalizing each image to the power measured by an absolutely calibrated XUV diode viewing the z-pinch self-emission reflected from the same MLM (and passed by the same filter) that is used in the imaging instrument.

SUMMARY

An x-ray self-emission imager has been fielded on Sandia's Z machine (a broadband source) that reflects pinhole images from a multilayer mirror in order to obtain monochromatic images with < 10 eV bandwidth in the 100-1000 eV range. Data have been successfully obtained at 277 eV photon energy from Al, Ar, stainless steel, Cu, and W z pinches. A similar instrument is being developed for the 8 MA Saturn generator; in addition to filtered K-shell and 277 eV monochromatic imaging, an imaging configuration at 528 eV corresponding to a prominent Ar L-shell line [12] will also be evaluated for imaging Ar gas puff z pinches. Time-resolved (1 ns), eight-frame microchannel plate detectors provide imaging at three separate energies (two monochromatic MLM-reflected, one filtered for broadband response) for both of these diagnostic instruments.

Monochromatic imaging through x-ray reflection provides several key advantages. The background exposure of the detector can be significantly reduced by shielding from hard x-rays, which do not reflect from the mirror. The use of a multilayer reflecting in the 100-1000 eV photon energy range allows the cooler trailing mass to be observed during the z-pinch implosion, while > 1 keV filtered imaging preferentially sees the hot, dense mass accumulating on the z-pinch axis. Abel inversion of the monochromatic self-emission images allows dynamics of the implosion phase to be explored and compared to MHD code simulations.

Future work will pursue quantitative measurement of temperature and density profiles by relating measured emissivity to plasma conditions via non-LTE collisional-radiative simulation. Preliminary analysis indicates that the density profile can be inferred through the combined use of monochromatic self-emission imaging, Abel inversion to obtain $\varepsilon(r)$, $T_e(r)$ measured through K-shell spectroscopy (or two-color emissivity ratios), and collisional-radiative modeling to relate emissivity and plasma conditions. Additional imager configurations as in Table 1 will be tested on Z and Saturn, and wire array and gas puff z-pinch dynamics will be further investigated through this x-ray imaging technique.

ACKNOWLEDGMENTS

The authors would like to thank G. Dunham (Ktech), L. P. Mix, G. A. Rochau, J. E. Bailey (Sandia), R. Mancini (University of Nevada, Reno) and J. J. MacFarlane (Prism Computational Sciences, Inc.) for helpful discussions and assistance with data analysis; D. Petmecky, P. Gard, and C. Ball (TMI, Inc.) for valuable contributions to the design and manufacture of the instrument; and T. C. Moore (Ktech), P. W. Lake, and N. R. Joseph (Sandia) for assistance in fielding the instrument on the Z machine. This work was supported by Sandia National Laboratories, a multiprogram laboratory operated by Sandia Corporation, a Lockheed Martin Company, for the United States Department of Energy's National Nuclear Security Administration under contract DE-AC04-94AL85000.

REFERENCES

1. Spielman, R. B., *et al.*, *Phys. Plasmas* **5**, 2105-2111 (1998).
2. Matzen, M. K., *Phys. Plasmas* **12**, 055503-1—055503-16 (2005).
3. Deeney, C., *et al..*, *Phys. Rev. Lett.* **81**, 4883-4886 (1998).
4. Cuneo, M. E., *et al.*, *Phys. Plasmas* **13**, 056318-1—056318-18 (2006).
5. Jones, B., *et al.*, *J. Quant. Spectrosc. Radiat. Transfer* **99**, 341-348 (2006).
6. Nash, T. J., *et al.*, *Rev. Sci. Instrum.* **72**, 1167-1172 (2001).
7. Wenger, D. F., *et al.*, *Rev. Sci. Instrum.* **75**, 3983-3985 (2004).
8. Nash, T. J., *et al.*, *Rev. Sci. Instrum.* **77**, 10E319-1—10E319-4 (2006).
9. Koch, J. A., *et al.*, *Rev. Sci. Instrum.* **76**, 073708-1—073708-4 (2005).
10. Jones, B., *et al.*, *Rev. Sci. Instrum.* **75**, 4029-4032 (2004).
11. Jones, B., *et al.*, *IEEE T. Plasma Sci.* **34**, 213-222 (2006).
12. Jones, B., *et al.*, *Rev. Sci. Instrum.* **77**, 10E316-1—10E316-4 (2006).
13. Coverdale, C. A., *et al.*, *IEEE T. Plasma Sci.*, accepted for publication (2007).
14. Sinars, D. B., *et al.*, *Phys. Plasmas* **12**, 056303-1—056303-8 (2005).
15. Jennings, C. A., private communication (2007).
16. Apruzese, J. P., *et al.*, *J. Quant. Spectrosc. Radiat. Transfer* **57**, 41-61 (1997).

17. MacFarlane, J. J., *et al.*, "Simulation of the ionization dynamics of aluminum irradiated by intense short-pulse lasers" in *International Conference on Inertial Fusion Sciences and Applications-2003*, edited by B. A. Hammel *et al.*, La Grange Park, IL: American Nuclear Soc., 2004, pp. 457-460.
18. Wang, P., *et al.*, *Phys. Rev. E* **48**, 3934-3942 (1993).
19. Bailey, J. E., *et al.*, *Phys. Rev. Lett.* **92**, 085002-1—085002-4 (2004).
20. Koch, J.A., *et al.*, *J. Quant. Spectrosc. Radiat. Transfer* **88**, 433-445 (2004).

Inference of Mix from Experimental Data and Theoretical Mix Models

L. Welser-Sherrill[1], D.A. Haynes[1], R.C. Mancini[2], J.H. Cooley[1], S.W. Haan[3], I.E. Golovkin[4]

[1] Los Alamos National Laboratory, P.O. Box 1663, Los Alamos, NM 87545
[2] University of Nevada Reno Physics Department, 1664 Virginia St., Reno, NV 89557
[3] Lawrence Livermore National Laboratory, 7000 East Ave., Livermore, CA 94550
[4] Prism Computational Sciences, 455 Science Drive, Suite 140, Madison, WI, 53711

Abstract. The mixing between fuel and shell materials in Inertial Confinement Fusion implosion cores is a topic of great interest. Mixing due to hydrodynamic instabilities can affect implosion dynamics and could also go so far as to prevent ignition. We have demonstrated that it is possible to extract information on mixing directly from experimental data using spectroscopic arguments. In order to compare this data-driven analysis to a theoretical framework, two independent mix models, Youngs' phenomenological model and the Haan saturation model, have been implemented in conjunction with a series of clean hydrodynamic simulations that model the experiments. The first tests of these methods were carried out based on a set of indirect drive implosions at the OMEGA laser. We now focus on direct drive experiments, and endeavor to approach the problem from another perspective. In the current work, we use Youngs' and Haan's mix models in conjunction with hydrodynamic simulations in order to design experimental platforms that exhibit measurably different levels of mix. Once the experiments are completed based on these designs, the results of a data-driven mix analysis will be compared to the levels of mix predicted by the simulations. In this way, we aim to increase our confidence in the methods used to extract mixing information from the experimental data, as well as to study sensitivities and the range of validity of the mix models.

Keywords: inertial confinement fusion, x-ray spectroscopy, mix models
PACS: 52.57.-z, 47.20.Ma, 47.51.+a

INTRODUCTION

In Inertial Confinement Fusion (ICF) implosions, the mixing between fuel and shell materials is an important characteristic to quantify [1]. Mixing occurs as a result of hydrodynamic instabilities which tend to smear the interface between regions with different densities. In previous work [2-6], we have demonstrated a spectroscopic technique which allows the extraction of quantitative information on mixing directly from experimental data gathered at the Laboratory for Laser Energetics' OMEGA facility. Using this method, the quantity calculated is a spatial mixing profile that can be further reduced to a "mix width", which is defined here as the physical extent of the layer that contains material from both the shell and the fuel. For comparison to theory, we previously developed post-processors [2-6] for hydrodynamic simulations which were based on the principles of two commonly used mix models, Youngs'

CP926, *Atomic Processes in Plasmas—15th International Conference on Atomic Processes in Plasmas*
edited by J. D. Gillaspy, J. J. Curry, and W. L. Wiese
© 2007 American Institute of Physics 978-0-7354-0436-6/07/$23.00

phenomenological model [7] and the Haan saturation model [8]. In this way, we estimated the amount of mixing expected from the experimental design of interest. The theoretical mix width is generally smaller than the mix width derived from the experimental data analysis, which is consistent with the way the theoretical models measure mix.

Since the capability of the theoretical mix models has been established, our new goal is to use the models to design direct drive ICF implosions which exhibit measurably different levels of mixing, while simultaneously maintaining similar temperature and density conditions. The work presented here is compared to a so-called "nominal" direct drive implosion which was designed to achieve electron temperatures and densities that would be consistent with the conditions necessary to employ Ar K-shell x-ray line emission spectroscopy. By keeping the temperatures and densities approximately constant between shots, we hope to focus on the experimental signal differences caused purely by mix amounts.

An important outcome of this work is an assessment of our ability to predict the extent of mix in ICF experiments by using a combination of hydrodynamic simulation post-processing and off-line mix models. The idea is to use the theoretical mix models *a priori* to design experiments with measurably different levels of mix. First we post-process a number of hydro simulations with the mix models to study the parameter space, in order to decide which physical characteristics of the lasers and targets are most sensitive to mix. Next, we perform detailed hydro simulations of the appealing cases to settle on a series of experimental conditions. Once the experiments are conducted based on these assessments, the mix width will be directly extracted from the experimental data and compared against the theoretical results. In essence, the mix models are used as a guide to design the experiments, which will in turn be used to check the validity of the mix models.

LINKS BETWEEN EXPERIMENT AND THEORY

For comparison, we begin by defining a nominal case, in which a spherical plastic microballoon filled with D_2 and doped with Ar is driven by a 1 ns square 23 kJ laser pulse. A cross-section of the nominal target is shown in Figure 1. For the experimental design project reported here, we confine ourselves to direct drive cases similar to that of the nominal direct drive implosion, i.e., we use the same gas and shell materials and the same laser pulse shape.

Since the mix widths are measured in two distinct ways, experimentally and theoretically, and are ultimately compared, it is important to maintain consistency between the two perspectives. In the case of the relevant experiments, the data are recorded for a 50 ps time interval starting at the time of peak emission from the Ar dopant, which is used to spectroscopically diagnose the plasma conditions and ultimately the mix. For consistency, Haan's and Youngs' theoretical mix models, which will be discussed in more detail below, were set up to calculate the mix width 50 ps after the time of peak Ar emission. This time was chosen because the experimental data were integrated over a 50 ps duration, and therefore this connects to the amount of mix calculated by the models at the end of the 50 ps time interval.

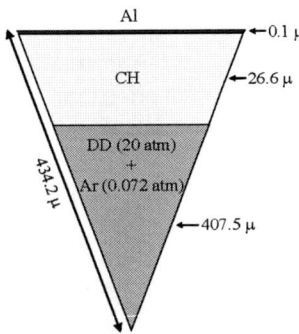

FIGURE 1. Target cross-section for the nominal direct drive implosion that is used for comparison throughout this work.

We emphasize that the theoretical mix models are used as a guide in designing a set of experiments with different amounts of mix. While the value of the mix width may not be accurately predicted by either of the models, both Haan's and Youngs' models incorporate physics which allows us to design implosions with measurably different mix widths. Since we are using the same models and the same implementations across all of the simulations, we are therefore able to distinguish differences in the amount of mix when one experimental design is compared to others.

HYDRODYNAMIC SIMULATIONS USED FOR DESIGN

In designing these experiments, a number of experimental parameters were identified which could potentially affect mix. The laser pulse shape, pulse duration, and total energy are possible candidates. In addition, the mix could be sensitive to target characteristics such as fill gas material and amount, dopant material and concentration, and the shell material and thickness. Only a few of these avenues were pursued, since it can be difficult to change pulse shapes and dopant specifications from shot to shot. It is important to be realistic in modifying the experimental parameters, because too many changes can affect the quality of the campaign. In addition, although changing characteristics such as the dopant material might dramatically affect the mix, the experimental analysis will be accomplished using Ar K-shell x-ray spectroscopy, making it necessary to use Ar.

The approach taken in this experimental design work involved several steps. First, a number of HELIOS [9] hydrodynamic simulations were performed and post-processed with Youngs' and Haan's models to test the sensitivity to mix of three main parameters: laser energy, fill gas amount, and shell thickness. Next, the parameter space was narrowed down to do more detailed simulations of the parameters that appeared to be most sensitive to mix. Finally, in-depth modeling of the best pool of candidates for the design was used to make a final determination for the experimental specifications. For the first two phases described above, HELIOS modeling was used to efficiently study the parameter space. The in-depth modeling was accomplished using the full collisional-radiative atomic kinetic code HELIOS-CR.

THEORETICAL MIX MODELS USED FOR DESIGN

Each hydro simulation was post-processed using two independent mix models in order to extract mix widths. Since a number of simulations had to be post-processed, IDL codes with graphical user interfaces were built for both Youngs' and Haan's models to systematically calculate the mix widths.

Youngs' model [7] is the simpler to employ. It is based on a phenomenological description of the growth of the mixing region as an effect of the Rayleigh-Taylor instability. The mix width h is a simple function of an adjustable fluid dynamics parameter α, the Atwood number A (density contrast between fluids), the acceleration g (assumed constant), and the time t,

$$h = \alpha A g t^2 .$$

Youngs' model is implemented here for a case with variable acceleration, in which the incremental width added in each time step of the simulation is defined as

$$Dh = \frac{dh}{dt} \cdot Dt = Dt \cdot \alpha \cdot 2 \cdot g(t) \cdot A(t) \cdot t .$$

Input for Youngs' model therefore consists of an acceleration profile $g(t)$ and an Atwood number profile $A(t)$ from the hydro simulation. The parameter α was set as 0.04 to be consistent with the current general usage. The mix widths grow when the pressure and density gradients at the interface are opposed, which occurs during the deceleration phase of the implosion. The result of the post-processing procedure is a mix width profile as a function of time, $h(t)$. The mix at 50 ps after the time of peak Ar emission is extracted from this profile.

The Haan saturation model [8] is based on the idea that instabilities are seeded by a number of experimental effects. If there are very minor initial seeds, then Youngs' model accounts for the most important contributions. On the other hand, in an experimental situation such as laser-induced fusion, the initial perturbations can be significant and can arise from several factors. Haan's model calculates a mix width by estimating the growth of multi-mode initial perturbations on the fuel-shell interface. For the direct drive case discussed here, the initial target surface roughness, the laser beam power imbalance, and the laser imprint spectra as a function of mode number were used. The modes grow exponentially until saturation occurs, at which time the mode growth becomes linear in time. Each initial perturbation spectrum $|R_{lm}(0)|$ is subjected to the growth formula

$$|R_{lm}| = e^{\gamma t} |R_{lm}(0)| ,$$

where

$$\gamma = \sqrt{gkA}$$

is the growth factor, which is a function of the acceleration g, the wavenumber $k=2\pi/\lambda$, and the Atwood number A. These terms are based on output from the hydro simulations, and are calculated as an average over the period of time from the

241

beginning of the deceleration phase to the time of interest (in this case, 50 ps after the time of peak Ar emission). After the modal growth is accomplished, the perturbations are added in quadrature, producing a single R_{lm} which is subjected to saturation modeling. If individual modes have saturated, the amplitudes are relaxed using the formula

$$R_{lm}(l) = S_l \left[1 + \ln \frac{R_{lm}}{S_l} \right],$$

where

$$S_l = \frac{2R_0}{l^2}$$

is the saturation level as a function of mode number l, and R_0 is the capsule radius at the time of interest (again, 50 ps after the time of peak Ar emission). The root mean square perturbation σ is extracted from

$$\sigma = \sqrt{\frac{1}{4\pi} \sum_l (2l+1)R^2_{lm}} \ .$$

The total mix width is then calculated to be

$$h = \sqrt{2}\sigma(2 + A).$$

The description just given is called the 'nominal' Haan model [2, 10]. This refers to the idea that if a multiplicative factor (referred to here as the Haan multiplier) is applied to the initial perturbation spectra, the mix width can change dramatically. The nominal Haan model result is calculated with a multiplier of 1.0.

A good way to use Haan's model is to think of the nominal result (with a multiplier of 1.0) as a baseline, and to recognize that the mix width could increase or decrease based on additional experimental effects. It would not be unreasonable to consider that one or more of the initial perturbations are in fact enhanced by experimental imperfections. A value for the multiplier other than 1.0 would fold in any additional uncertainties in the initial perturbation amplitudes used by Haan's model.

In general, the mix width predicted by Haan's model shows more sensitivity to variation in the experimental design. This may be due to the fact that the model is based on experimentally derived perturbations and thus includes more effects than Youngs' phenomenological model. Haan's model requires not only the initial perturbation spectra (the same are used for every simulation), but also a number of specific pieces of hydro output from each simulation.

PARAMETER SPACE STUDY USING HELIOS

We began by conducting a study of the parameter space in order to determine which characteristics of the experimental design are particularly sensitive to mix. All simulations were ultimately compared to the nominal case, as described above in Figure 1. Simulations were grouped into clusters that shared common inquiries into certain parameters. The core sizes, times of peak neutron production and compression, electron temperatures and densities, mass densities, and a number of additional quantities were extracted from each simulation and examined in detail. A very important consideration was the temperatures and densities at peak x-ray emission. The main goal was to search for experimental designs that yielded mix widths that were considerably different from that of the nominal case, but that retained approximately the same temperatures and densities. This constrained the design significantly by only allowing small deviations in the experimental parameters.

The results of post-processing the hydro simulations demonstrated that changing the shell thickness (and keeping all other aspects of the nominal capsule identical) led to the most noticeable sensitivity to mix, particularly when using the Haan model. Therefore, in-depth studies of the mix as a function of shell thickness were performed. Eleven simulations crossing the parameter space were investigated first, after which another thirty simulations filled in the parameter space. In totality, the mix width was studied for shell thicknesses between 10 and 40 microns, with a step size of 1 micron between simulations.

We now focus in detail on the mixing post-processor results from the simulations based on changing shell thickness. Youngs' model, as implemented in this work, gives the growth of the mix layer as a function of time. Figure 2 is an amalgamation of thirty simulations, and shows the mix width accumulation as a function of time for all studied shell thicknesses. Notice the trend of a smooth accumulation of mix for thinner shells. However, for shells thicker than the nominal case (>26 microns), the mix accumulation is more sluggish. This suggests that the nominal case is at a kind of threshold, above and below which the mix accumulation trends are noticeably different. This point will be returned to in the discussion of Haan's model results.

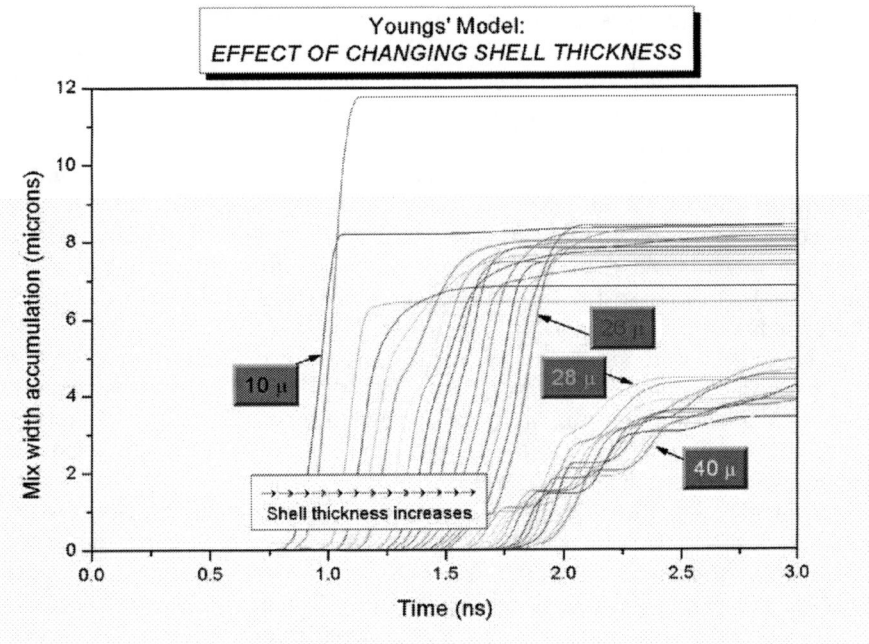

FIGURE 2. Youngs' model mix accumulation as a function of time for a range of shell thicknesses. The numbers indicate shell thickness, which increases to the right.

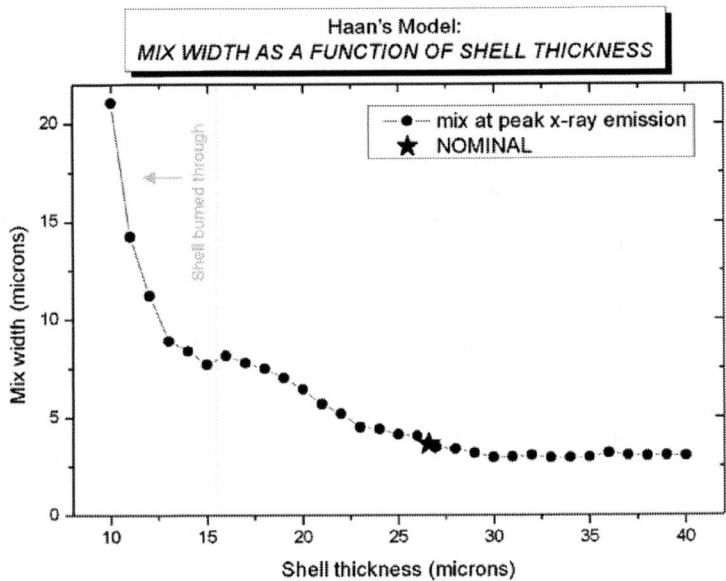

FIGURE 3. Haan's model mix width as a function of shell thickness. The star represents the nominal case described in Figure 1.

Figure 3 shows the mix width as a function of shell thickness resulting from the Haan model. For all cases, the Haan model predicts mix widths that are larger than the Youngs' model results. This is not surprising, since Haan's model includes the effects of perturbations from the initial target surface roughness, laser beam power imbalance, and laser imprint. Therefore it is intrinsically higher than the mix given by Youngs' model, which is considered to be a lower bound [10].

It is important to ensure that the thinner shells are not burned through. An indication of this is found when the bang time and peak compression time are simultaneous. If the shell is burned through, there is no material left to continue the momentum inward, and therefore the core cannot continue to compress after peak burn. This was the case for the shells with a thickness of 15 microns or less, as indicated in Figure 3.

It is not sufficient to simply search for the largest mix width, in order to contrast with the relatively small mix width of the nominal case. In fact, it is necessary to determine the amount of mix as a percentage of the core size. A mix width approaching the size of the core itself would be very difficult if not impossible to detect using the experimental data analysis.

Figure 4 shows a direct comparison between the results of post-processing the hydro simulations with Youngs' and Haan's models. This plot is shown in the form of the percentage of the core radius that forms the mixed layer, giving a more consistent picture for comparing the mix in all simulations. It demonstrates that according to Haan's model, the shell thickness provides a significant lever arm for varying the amount of mix. Two cases of interest are circled, the 19 micron shell, which according to Haan's model exhibits 55% more mix than nominal, and the 33 micron shell, which displays 28% less mix than nominal. Youngs' model results are shown for 50 ps after the time of peak x-ray emission, in order to correspond to the Haan model analysis and the data analysis. It is noticeable, as in Figure 2, that according to Youngs' model the nominal case appears to be just at the edge of a region of parameter space in which the mix widths drop significantly. In fact, the nominal case has one of the smallest mix widths of any of the simulations performed.

The two shell thicknesses of interest, 19 and 33 microns, have been examined for their temperatures and densities. All conditions lie within the sensitivity range of Ar spectroscopy, so experiments designed around these parameters would be viable for the experimental data analysis.

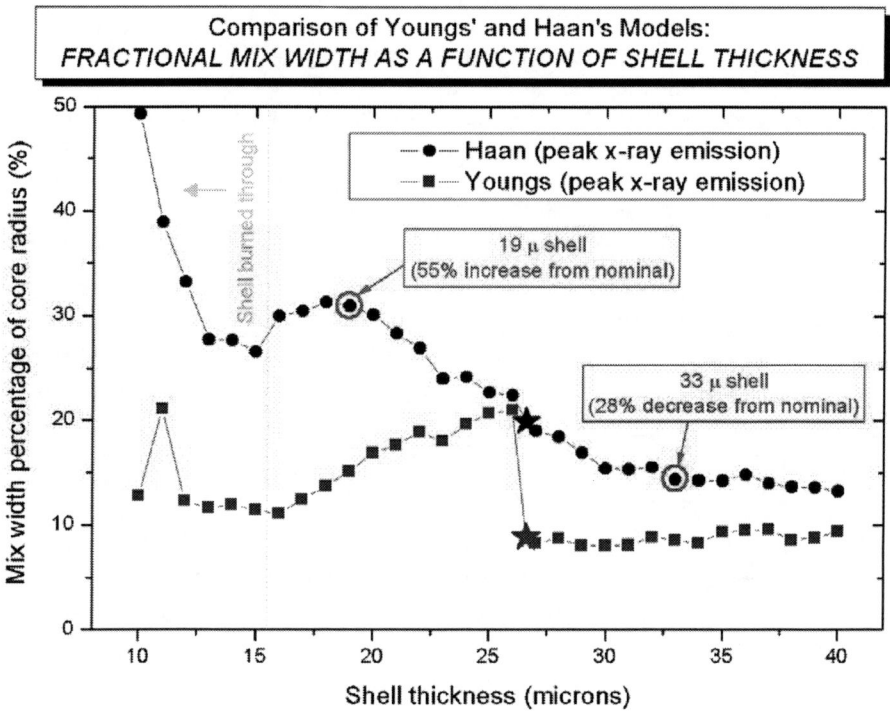

FIGURE 4. A comparison of the fractional mix widths resulting from Youngs' and Haan's models. The cases circled in red are of interest for varying the level of mix. The star represents the nominal case described in Figure 1.

245

DETAILED STUDY OF MIX-SENSITIVE EFFECTS

The cases mentioned in the previous section, with shell thicknesses of 19 and 33 microns, represent two situations that demonstrate, according to Haan's model, significantly different amounts of mix from nominal. They also satisfy the condition that the temperatures and densities be similar. As a check on these preliminary findings, HELIOS-CR calculations using inline collisional-radiative atomic kinetic modeling were performed. HELIOS-CR is generally expected to be more accurate than the baseline HELIOS approach, because the atomic physics and the radiation cooling effects of the Ar dopant can be more appropriately dealt with [9]. These simulations take approximately 30 hours to run. It should be emphasized that we use HELIOS-CR only for an additional check: the fill conditions of the experiments were carefully conceived so that the Ar would have as little effect as possible on the implosion dynamics. Therefore, it was not surprising that the HELIOS-CR results were very similar to the HELIOS results of the previous section. Figure 5 presents the attributes of the main three HELIOS-CR simulations. The Haan mix width for the nominal case exhibited a 12% difference between using HELIOS and HELIOS-CR, while the 19 micron case had an 8% difference and the 33 micron case had a 6% difference. The simulations in Figure 5 also correspond to the three experimental designs chosen from this study. In these cases, the outer capsule radius was set as constant in order to conform to GA target fabrication issues. Two shots of each flavor will be performed for reproducibility, and the results of the experimental data analysis will be compared to the mix width values predicted by Haan's and Youngs' models.

Simulation	Laser	D_2 fill pressure	Ar fill pressure	Inner radius	Shell thickness	Core size	Haan mix
1CR - NOMINAL	1 ns square 23 kJ	20 atm	0.2%	407.5 μm	26.6 μm CH	20.7 μm	4.1 μm (20% of core)
2CR	1 ns square 23 kJ	20 atm	0.2%	415.1 μm	19 μm CH	23.2 μm	6.9 μm (30% of core) (50% above nominal)
3CR	1 ns square 23 kJ	20 atm	0.2%	401.1 μm	33 μm CH	19.4 μm	3.0 μm (15% of core) (30% below nominal)

FIGURE 5. Characteristics of three HELIOS-CR simulations and chosen designs. Resulting mix predicted by Haan's model is also shown.

CONCLUSIONS

The ultimate goal of this work is to improve the predictive capability for mix. Previous work [2-6] demonstrated our capability to extract information on mixing directly from a spectroscopic data analysis and to calculate a corresponding mix width from theoretical mix models. The current work uses the predictive capability of Youngs' and Haan's models to design direct drive implosions which exhibit measurably different levels of mixing, while simultaneously maintaining the plasma temperature and density conditions. The goal is not to predict absolute mix widths, but rather to examine trends relating to larger and small mixing regions. In addition, interpreting these experiments could help to validate and/or determine applicability of Youngs' and Haan's models.

It is also important to demonstrate experimental reproducibility for the spectroscopic analyses. Only one indirect drive experiment has been exhaustively studied in order to demonstrate the spectroscopic techniques discussed in [2-6]. Analyzing a number of good-quality direct drive experiments would broaden the scope of the technique, and would provide an important test-bed for the spectroscopic analysis.

We have demonstrated the feasibility of designing ICF experiments based on the phenomenological approaches of Youngs' and Haan's mix models. This series of experiments will contribute to the field by continuing to validate our techniques, and could go even further by investigating some of the deep details of the mix models.

ACKNOWLEDGEMENTS

This work was supported by DOE NLUF grants DE-FG52-05NA-26012 and DE-FG52-07NA-28062, and by NNSA Campaign 10.

REFERENCES

1. J. Lindl, Phys. Plasmas **2**, 3933 (1995).
2. L.A. Welser, Ph.D. thesis, University of Nevada, Reno (2006).
3. L. Welser-Sherrill, R.C. Mancini, et al, Phys. Rev. E, submitted for publication (2007).
4. L. Welser-Sherrill, R.C. Mancini, et al, Phys. Rev. Lett., submitted for publication (2007).
5. L. Welser-Sherrill, R.C. Mancini, et al, High Ener. Dens. Phys., in publication (2007).
6. L. Welser-Sherrill, R.C. Mancini, et al, Phys. Plasmas, submitted for publication (2007).
7. D.L. Youngs, Physica D **12**, 32 (1984).
8. S.W. Haan, Physical Review A, **39**, 5812 (1989).
9. J.J. MacFarlane, et al, Phys. Plasmas **12**, 032702 (2005).
10. S.W. Haan, private communication.

Development of Compton radiography using high-Z backlighters produced by ultra-intense lasers

Riccardo Tommasini, Hye-Sook Park, Prav Patel, Brian Maddox, Sebastien Le Pape, Stephen P. Hatchett, Bruce A. Remington, Michael H. Key, Nobuhiko Izumi, Max Tabak, Jeffrey A. Koch, Otto L. Landen, Dan Hey, and Andy MacKinnon

Lawrence Livermore National Laboratory, Livermore CA

John Seely and Glenn Holland

Naval Research Laboratory, Washington DC

Larry Hudson, Csilla Szabo

National Institute of Standards and Technology, Gaithersburg MD

Abstract. High-energy x-ray backlighters will be valuable for radiography experiments at the National Ignition Facility (NIF), and for radiography of imploded inertial confinement fusion cores using Compton scattering to observe cold, dense plasma. Key considerations are the available backlight brightness, and the backlight size. To quantify these parameters we have characterized the emission from low- and high-Z planar foils irradiated by intense picosecond and femtosecond laser pulses from the TITAN laser facility at Lawrence Livermore National Laboratory. Spectra generated by a sequence of elements from Mo to Pb, spanning the x-ray energy range from 17 keV to 75 keV, have been recorded using a Charged Coupled Device (CCD) in single hit regime and a Dual Crystal Spectrometer (DCS). High-resolution point-projection 2D radiographs have also been recorded on Fuji BaFBr:Eu2 image plates using calibrated resolution grids. We discuss the results in light of the requirements for applications at NIF.

Keywords: laser-plasma interaction, hard x-rays, x-ray spectroscopy, radiography.
PACS: 52.50.Jm; 87.59.Bh; 52.70.La

CP926, *Atomic Processes in Plasmas—15th International Conference on Atomic Processes in Plasmas*
edited by J. D. Gillaspy, J. J. Curry, and W. L. Wiese

INTRODUCTION

The National Ignition Facility (NIF) [1] is being built to achieve fusion ignition [2] by imploding capsules containing deuterium and tritium atoms. In the indirect drive configuration, the capsule will be at the center of a high-Z hohlraum, and about 1MJ of energy will be delivered on target, using 192 UV laser beams. The plasma resulting from such implosions will have extreme conditions: electron temperatures from 10 keV to few tens of keV, electron densities of the order of 2e26 cm^{-3}, areal densities of about 3 g/cm^2. The implosion can be spoiled by a number of reasons, for instance asymmetries in the laser drives or hydrodynamic instabilities. An important tool to understand the reasons for possible failures will be time-resolved-image diagnostics using photons with energies high enough to penetrate the hohlraum walls and the imploding capsule. The goal is imaging the dense cold fuel surrounding the hot spot, by recording radiographs from x-ray backlighters.

The usual approach to obtaining backlit radiographs of an imploding object at different times is to use multiple shots having different time delays of the backlighting source or to use streak or framing camera methods to show changes during the emission time of a long pulse backlighter. The major drawbacks of using several different shots to obtain a time sequence are: an increase in number of shots required and a possible change of the experimental conditions from shot to shot. The first point is a major concern in large-scale facilities like NIF. The second has a large scientific impact since the exact conditions in the sample and in the drive may have changed and are difficult to trace.

X-RAY COMPTON RADIOGRAPHY

Transmission radiography of NIF ignition capsules requires high photon energies because the imploded cores are opaque below about 15 keV and because the anticipated self-emission would overwhelm the backlighter brightness below some threshold >70 keV. Standard framing and streaking techniques cannot be used at high photon energies (70 keV to 200 keV) because no practical source of sufficiently long pulse x-rays in this energy range is known to exist. In this photon energy range the main source of opacity is Compton scattering, hence the designation ''Compton radiography''.

NIF is being integrated with high-energy short pulse laser beams. This advanced radiography capability (ARC) [3,4] will have up to 8 beams with energies between 0.5 and 1.6 kJ depending on the pulse duration, between 0.7 and 50 ps. ARC laser beams can be used to irradiate solid metal foils and create multiple backlighting x-ray sources having about the same duration of the laser pulses. Therefore we can produce a series of radiographs of the dense fuel as the capsule evolves through compression, with picosecond temporal resolution. As the x-rays from the backlighters traverse the dense fuel, Compton scattering will reduce the net flux of photons towards the detector, according to the density of the traversed regions, resulting in radiographs which are a map of the fuel density. The Compton scattering cross section [5] can be approximated as $\sigma \sim \sigma_T[1 - 2 (h\nu/mc^2) + (26/5) (h\nu/mc^2)^2 + ...]$, where h$\nu$ is the x-ray photon energy, σ_T is the Thomson scattering cross section and m is the scatterer mass.

Therefore in the energy range of interest, the process is largely independent of probing energy so that we can choose the energy of the x-ray photons merely according to signal-to-background considerations. Moreover, the optical depth, at the expected areal density, is nearly ideal: 0.5.

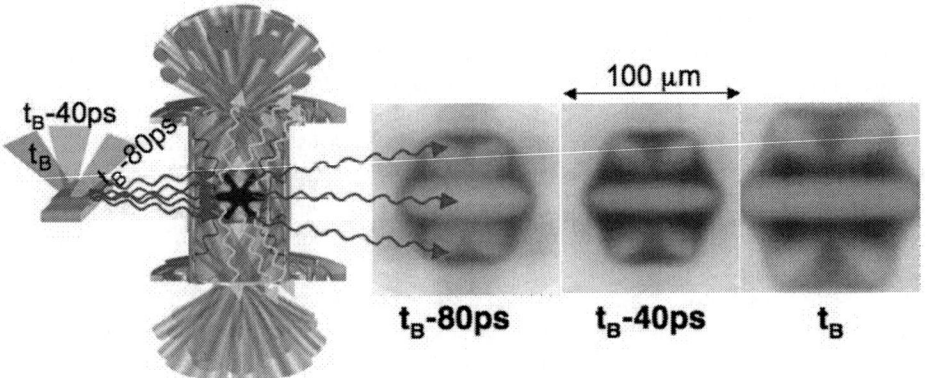

FIGURE 1. The main concept of Compton radiography applied to record the time history of imploding capsules using multiple radiographs from ARC backlights on NIF. In this particular case, three radiographs are recorded, with a time delay of 40 ps, according to the firing sequence of ARC pulses. For simplicity, the picture shows only one backlighter. Radiographs are simulated using LASNEX. The temporal resolution of each radiograph is about the same as the ARC pulse duration, i.e. ~10 ps.

Figure 1 shows a typical setup for the proposed multi-pulse backlighter on NIF-ARC, in the form of a diagnostic for ICF implosions. Micro-wire backlighters [6], oriented parallel to the line of sight (edge-on), are sequentially hit by ARC pulses, providing the sources for point-projection radiography. The backlighter size is properly chosen in order to provide the required resolution at the capsule plane. An energy band-pass device (not shown) provides a means to increase the signal over background (S/B) at the detector. The radiographs give a map of the density of the imploding core, corresponding in this particular case to an ignition attempt that fails from too much drive asymmetry (in the Y60 spherical-harmonic component), with a temporal resolution of about 10 ps. The radiograph images shown are calculated using LASNEX radiation-hydrodynamics simulations [7].

The immediate advantages of multi-pulsed radiographs are evident: constant conditions for the driving laser and a reduced number of shots to gain the actual time history of the sample under the same experimental conditions. We believe that this novel diagnostic will enable better tuning of the hohlraum during the National Ignition Campaign (NIC) and will be a powerful tool for high energy density physics experiments, such as those studying material strength. Its realization, however, is challenging due to the high levels of background expected during the implosion. Radiographing the hydrogen fuel, through peak compression, means dealing with the background from hard x-rays emitted by the hot core. This will be an extremely bright background, even from non-igniting implosions with neutron yields of ~100 kJ. Another source of background are the hard x-rays from the hot electrons traversing the hohlraum walls. Both are strong and with a spectrum extending from a few keV to a few hundred keV, therefore overlapping with the photon energy of any x-ray backlighting source. A collimator can be used to reduce the background from hard x-rays produced by the hot electrons in the hohlraum walls, but nothing can be done to reduce the signal from the core self-emission, since this is entirely emitted within the field of view of the radiographs. Therefore it is important to estimate the signal that can be produced from high-energy x-ray backlighters.

EXPERIMENTAL RESULTS

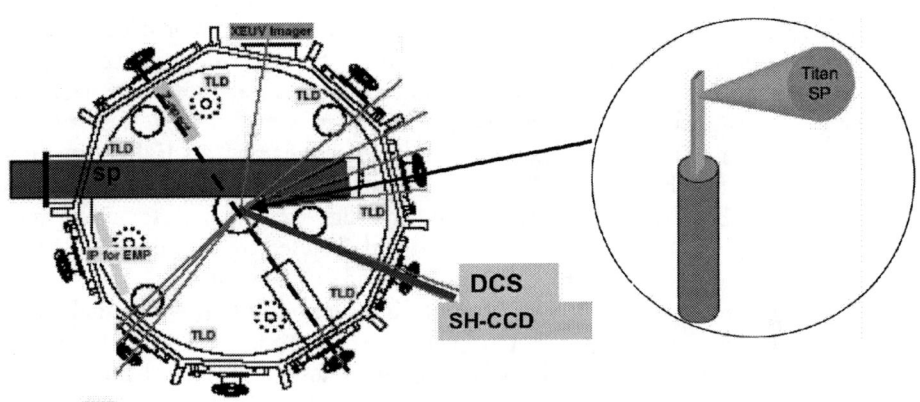

Figure 2. Experimental setup on TITAN for the characterization of 25 μm thick slab backlighters.

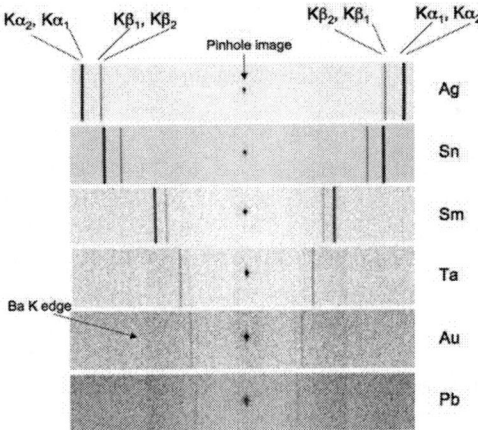

FIGURE 3. The raw spectra recorded with the DCS for the sequence of elements Ag, Sn, Sm, Ta, Au and Pb. Kα and Kβ (up to Au) are clearly visible. The approximate photon energies for the Kα₁ lines are: 22.2, 25.3, 40.1, 57.5, 68.8, 75.0 keV for Ag, Sn, Sm, Ta, Au and Pb, respectively.

In order to characterize the emission and source size of high-energy x-ray backlights, we performed experiments on the TITAN laser facility at the Lawrence Livermore National Laboratory. The experimental set-up is shown in Figure 2. The short pulse, 1054 nm wavelength, beam of TITAN was used to irradiate the 25 μm thick targets, about 0.5 mm × 4 mm in size. The laser parameters were maintained constant during the experiment, with a spot size of ~ 50 um, a pulse duration of 40 ps and an intensity on target of ~ 2e17 W/cm^2. The K-shell lines and the continuum Bremsstrahlung, emitted by a sequence of elements (Ag, Sn, Sm, Ta, Au, Pb), have been recorded using the Dual Crystal Spectrometer (DCS) [8] and a Charged Coupled Device (CCD) in the single hit regime. The DCS implements two quartz crystals (10-11) in transmission (Laue) geometry. One crystal is bent to a radius of 119 mm and covers 11.5 keV to 45 keV, and the other crystal has a radius of 254 mm and covers 18 keV to 120 keV. A pinhole is on the axis of the spectrometer and provides an image of the source. The measurements reported here were obtained using this second crystal. The DCS was positioned outside the Titan target chamber and viewed the front side of the targets through a port with a Lexan vacuum window, with a source-to-crystal distance of 1.2 m. The spectral images were recorded using Fuji BaFBr:Eu$_2$ imaging plates near the Rowland circle. The delay between the shot and the imaging plate scanning procedure was kept constant during the experiment, to avoid read-out variations due to the temporal decay of the signal detected by the imaging plates. The raw spectra recorded by the DCS, for the sequence of elements Ag, Sn, Sm, Ta, Au and Pb, are shown in Figure 3. The Kα_1 and Kα_2 lines (from 2p$_{3/2}$-1s$_{1/2}$ and 2p$_{1/2}$-1s$_{1/2}$ transitions, respectively), with photon energies from 22 keV to 75 keV, are clearly visible for all elements. The Kβ lines are also well defined, for all elements up to Au. Notice that the DCS provides two spectra that are symmetric about the central pinhole image. Also visible is the K absorption edge of Ba, an element of the imaging plates.

The spectral axis of the DCS was calibrated according to the procedure outlined in Ref.[7]. The energy calibrated spectra for the six elements, limited to regions including the respective Kα energies, are shown in Figure 4, and were obtained by column summation of the imaging-plate scans reported in Figure 3. All the lines appear bright and distinct.

We also performed experiments to test the resolution capabilities of the backlights, or, equivalently, the achievable source size. For this we used micro-wire backlights [6] to radiograph calibrated resolution grids in a point projection geometry. In this case the backlighters consisted of a 300 μm long Ag wire on a 300 μm × 300 μm square plastic substrate. The wires were 10 μm thick and had square cross section, while the substrates had a thickness of 5 um. The 2D radiographs were recorded on imaging plates looking at the wire, edge-on, through the resolution grids. Figure 5 shows the experimental setup and resulting radiograph using the Kα line from Ag (22 keV). The 10 μm pitch lines from the resolution grid are clearly resolved. A detailed analysis of the radiograph results in a source point spread function slightly less than 10 μm (FWHM).

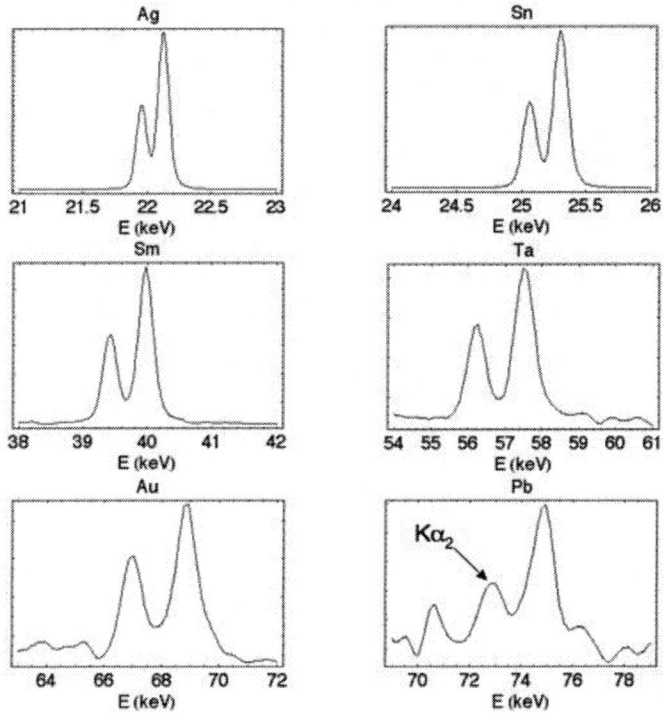

FIGURE 4. Energy calibrated spectra for the six elements as obtained by column summation of the imaging-plate scans reported in Figure 3.

DISCUSSION AND CONCLUSIONS

The exact determination of the conversion efficiency of the backlighters would require the measurement of the efficiency for the imaging plates and the DCS crystal. Experiments to accomplish these tasks are in progress. To get an estimate of the achieved values we calculated the efficiencies (relative to the Ag Kα lines) of the imaging plate and of the DCS crystal [9]. To obtain calibration points for the calculations, we used the single-hit spectra from a CCD, fielded on the same line of sight as the spectrometer, measuring the strength of the Mo, Ag and Sn K-shell lines. The result of this procedure is shown in Fig. 5, showing values of a few 1e-5 for the conversion efficiencies to Kα emission lines from Ta, Au and Pb, i.e. in the 60 keV to 75 keV region. In the same way reported above, we also estimated the conversion efficiencies to broadband Bremsstrahlung, between 75 keV and 100 keV, from the same series of targets and the same laser parameters. The analysis shows, also in this case, nearly constant conversion efficiencies, reaching values of about 2e-4 for Ta and Au. Note that in the case of line emission, the conversion efficiency has been calculated as the ratio of the energy of x-ray photons contained in the Kα₁ and Kα₂ lines and the energy of the laser pulse; in the case of Bremsstrahlung emission the conversion efficiency has been calculated as the ratio of the energy of the x-ray photons contained in the 75 keV-100 keV band and the energy of the laser pulse. In both cases the x-rays are emitted over the full solid angle. It is important to realize that, in view of our intended application and due to the independency of Compton radiographs from probing energy, a Bremsstrahlung backlighter in the 75 keV - 100 keV range is an extremely interesting option. Indeed the estimated CE value would result in a backlighter signal more than a factor of ten higher than the background from the simulated core self-emission, for sub-ignition failures (100 kJ). Moreover, a broadband backlighter, in contrast to a line-emission one, has the advantage of requiring only a moderate band-pass for the detector to be employed in the Compton radiography on NIF implosions. Therefore the presented results are very promising towards the development of Compton radiographs for laser-fusion implosions with good signal to background ratio and capable of 10 μm spatial resolution.

ACKNOWLEDGMENTS

This work was performed under the auspices of the U.S. Department of Energy by University of California, Lawrence Livermore National Laboratory under Contract W-7405-Eng-48.

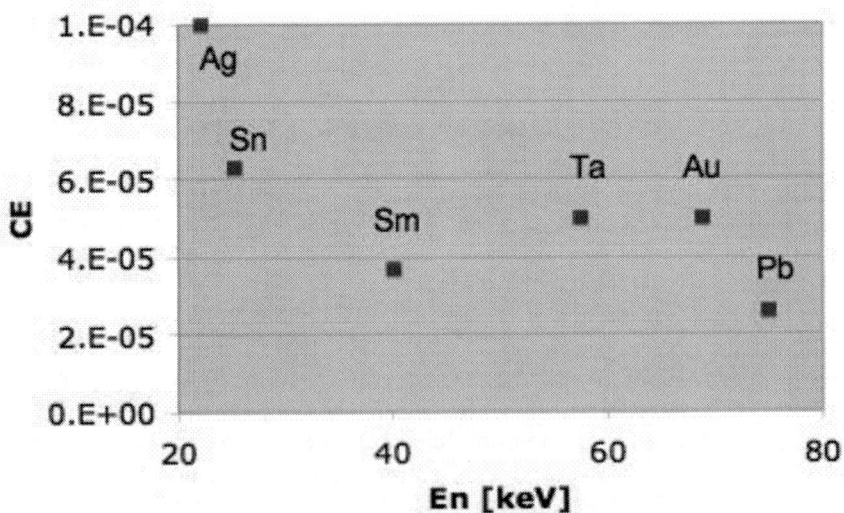

FIGURE 5. Conversion efficiency to Kα line emission between 22 and 75 keV, for the sequence of elements Ag, Sn, Sm, Ta, Au and Pb. The values are estimated from calculated detection efficiency of the DCS crystal spectrometer and the Fuji BaFBr:Eu2 imaging plate.

REFERENCES

1. G. H. Miller, E. I. Moses, and C. R. Wuest, Nucl. Fusion **44**, S228 (2004).
2. J. D. Lindl et al., Phys. Plasmas **11**, 339 (2004).
3. Barty, C P J, UCRL-JC-152483, 2003
4. Pennington, D M et al., UCRL-JC-153260-ABS, 2003.
5. Jackson, John D. , Classical Electrodynamics (3rd ed.). Wiley. ISBN 047130932X, 1998
6. Park HS, et al., Phys. Plasmas **13**, 056309 (2006)
7. G.B Zimmerman and W.L.Kruer, Comments Plasma Phys. Control. Fusion **2**, 51 (1975).
8. Seely J., et al, to appear in High Energy Density Physics (2007).
9. M. Sánchez del Río, et al., SPIE proceedings vol. 3152, 1997.

INDUSTRIAL PLASMAS

Discharge plasmas as EUV Sources for Future Micro Lithography

Thomas Kruecken

Philips Research Laboratories, Weisshausstr. 2, Aachen, Germany

Abstract. Future extreme ultraviolet (EUV) lithography will require very high radiation intensities in a narrow wavelength range around 13.5 nm, which is most efficiently emitted as line radiation by highly ionized heavy particles. Currently the most intense EUV sources are based on xenon or tin gas discharges. After having investigated the limits of a hollow cathode triggered xenon pinch discharge Philips Extreme UV favors a laser triggered tin vacuum spark discharge. Plasma and radiation properties of these highly transient discharges will be compared. Besides simple MHD-models[1] the ADAS software package [2] has been used to generate important atomic and spectral data of the relevant ion stages. To compute excitation and radiation properties, collisional radiative equilibria of individual ion stages are computed. For many lines opacity effects cannot be neglected. In the xenon discharges the optical depths allow for a treatment based on escape factors. Due to the rapid change of plasma parameters the abundancies of the different ionization stages must be computed dynamically. This requires effective ionization and recombination rates, which can also be supplied by ADAS.

Due to very steep gradients (up to a couple orders of magnitude per mm) the plasma of tin vacuum spark discharges is very complicated. Therefore we shall describe here only some technological aspects of our tin EUV lamp: The electrode system consists of two rotating which are pulled through baths of molten tin such that a tin film remains on their surfaces. With a laser pulse some tin is ablated from one of the wheels and travels rapidly through vacuum towards the other rotating wheel. When the tin plasma reaches the other electrodes it ignites and the high current phase starts, i.e. the capacitor bank is unloaded, the plasma is pinched and EUV is radiated. Besides the good spectral properties of tin this concept has some other advantages: Erosion of electrodes is no severe problem as the tin film is regenerative and protects the electrode material. The electrical connections to the rotating electrodes are easily made via the tin baths. As the liquid tin cools very effectively the rotating wheels, the concept is scalable to higher powers. The disadvantage, however, is the rather large amount of tin debris that must be kept away from the optics. Even very thin layers of condensed tin would severely reduce the reflectivity of the mirrors.

Keywords: EUV, micro lithography, xenon, tin, pinch discharge.

1. INTRODUCTION

The search for the optimum radiation source at 13.5 nm wavelength is one of the main challenges in EUV lithography today. To be profitable, the semiconductor industry asks for very high radiation intensities [1]. As EUV radiation is almost everywhere absorbed, the entire wafer stepper operates in vacuum and the optics consists of reflecting mirrors. Most of them are normal incidence refractive multilayer mirrors of Silicon and Molybdenum stacks. These mirrors, however, reflect the EUV radiation

CP926, *Atomic Processes in Plasmas—15th International Conference on Atomic Processes in Plasmas*
edited by J. D. Gillaspy, J. J. Curry, and W. L. Wiese
© 2007 American Institute of Physics 978-0-7354-0436-6/07/$23.00

only in a very narrow spectral range of 13.5 nm ± 1 %. The 13.5 nm was chosen as it corresponds to the 2p-1s line of Li^{2+}. Hence the EUV radiation must not only origin from a very small (for optical reasons) and intense source but its spectrum should also contain as much radiation as possible in the so called inband spectral range of 13.5 nm ± 1 % wavelength. As heat management and erosion limit the input power, the spectral efficacy of the source must be optimized. These requirements force the developers of EUV sources to obtain a better understanding of the underlying physics. On the other hand the very tight schedule does not allow thorough investigations but asks rather for quick, more qualitative, sometimes empirical approaches. In the next chapter we will discuss the EUV radiation properties of highly ionized atoms. xenon, which was the first candidate as EUV radiator, will serve as example.

Besides spectral properties also material properties needed by plasma models can be derived from basic atomic physics. As laser produced plasmas suffer from enormous power inputs and costs, pulsed discharge plasmas are favored as EUV sources. Among others Philips Extreme UV first pursued the development of a xenon pinch discharge. A rather simple but illustrative model of such a discharge will be discussed in chapter 3. The rather poor spectral radiation properties of xenon led to the development of a EUV source with tin as radiator. The radiation properties of various tin ions and the basic features of the Philips Extreme UV source based on a laser triggered tin vacuum spark discharge will be discussed in chapter 4.

2. ATOMIC PHYSICS OF EUV RADIATORS

The 13.5 nm wavelength corresponds to a photon energy of 91.9 eV. As the excited states' energy levels of neutral atoms are much smaller, only highly ionized particles can emit line radiation in this spectral region. Li^{2+} has its 2p-1s resonance line exactly at 13.5 nm. Other ionization stages of Lithium do not have lines in the inband region. It is, however, rather difficult to build a source on only this one line. It will also be subject to strong self-absorption. Moreover Lithium is very difficult to handle. But also ions with nuclear charge around 50 and ionization stages near Z=10 are good candidates: They have ground state configuration with (half filled) 4d-shells and with increasing ionization stage the 4f and 5p excited states' configurations have the required energy levels. Hyperfine structure splits these high angular momentum configurations into hundreds of energy levels with hundreds of thousands of transitions leading to "grass-like" spectra. To calculate the radiation properties one has to first compute the excited states' energy levels, the electron excitation cross sections and the photonic transition probabilities. Hartree Fock atomic structure codes as described in Cowan's text[2] can do this with sufficient accuracy. Because the electron densities and temperatures in typical EUV producing discharges allow neither for a Corona nor LTE description of the excited states populations, for each ionization stage full Collisional Radiative (CR) equilibria are computed which balance the various electronic processes between excited states. These equilibria and the resulting optically thin radiation spectra can be calculated by the atomic physics package ADAS [3].

Xenon As EUV Radiator

As an example we show the results for various EUV emitting xenon ions. To limit the number of energy levels and transitions we limit the number of basic electron configurations has been limited. Note that, even if the energy levels corresponding to the omitted electron configurations are too high to be populated, the resulting reduced basis sets of configurations might shift the locations of lower energy levels corresponding to included configurations. The configurations included for example in the case of X^{+10} are:

- $4s^2\ 4p^6\ 4d^8$ (ground state configuration)
- $4s^2\ 4p^6\ 4d^7\ 4f^1$
- $4s^2\ 4p^6\ 4d^7\ 5s^1$
- $4s^2\ 4p^6\ 4d^7\ 5p^1$
- $4s^2\ 4p^6\ 4d^7\ 5d^1$
- $4s^2\ 4p^5\ 4d^9$

The configurations for the other ionization stages are analogous. The assumed electron temperature and density are $T_e = 32$ eV and $n_e = 10^{17} \text{cm}^{-3}$. Fig. 1 shows the computed EUV spectra of the important xenon ionization stages $Xe^{+8} - Xe^{+13}$.

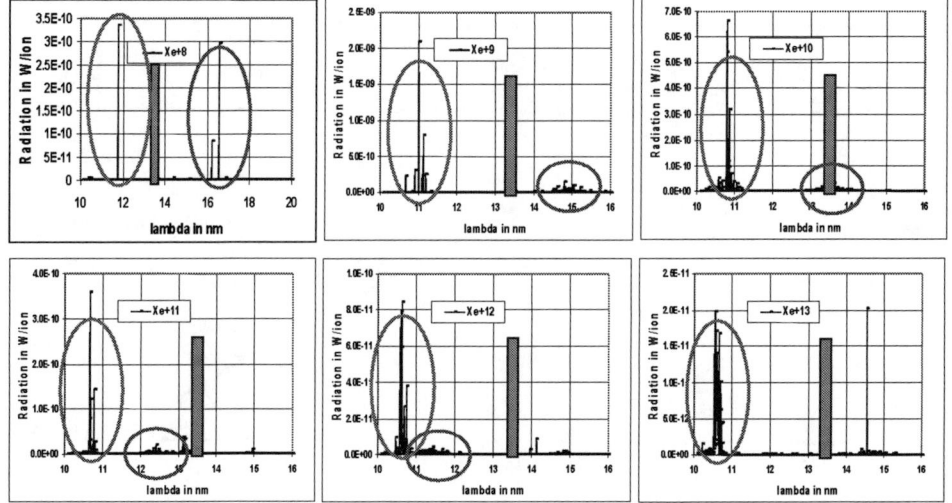

FIGURE 1. Optically thin Spectra of Xe+8 - Xe+13 ions

The inband EUV region near 13.5 nm (orange bar) and the two main features, 5p-4d (blue ellipses), and (4d-4f,4p-4d)-transitions (red ellipses) are marked. Xe^{+8} has a closed shell ground state configuration. Hence there is no Hyperfine splitting of the excited states and its spectrum shows only a few lines compared to the many lines of the other ionization stages. The 4d-5p transitions move with increasing ionization stages to shorter wavelengths. This effect is much less pronounced for the stronger (4d-4f,4p-4d)-transitions. Note that only the 4d-5p transitions of Xe^{+10} contribute to the needed inband EUV radiation.

In a pinch discharge with rather high densities parts of the spectrum might be self absorbed. The optical depth s of line radiation can be estimated using the Ladenburg relation:

$$\frac{1}{s} = \frac{\lambda^4}{8\pi c} A_{21} \frac{g_2}{g_1} \frac{n}{\Delta\lambda} \tag{1}$$

where g_2 and g_1 denote the statistical weight of upper and lower level, A_{21} the transition probability in 1/sec, n the (ion ground state) density, and $\Delta\lambda$ the line width. With an assumed ion temperature $T_i=1000$ eV ($T_e \ll T_i$ is typical for fast pinch discharges, see also next chapter) the Doppler broadening for xenon amounts to $\Delta\lambda \sim$ 3pm, which agrees with measurements.

Fig. 2 shows the resulting optical depth at $\lambda=13.5$ nm as function of A_{21} for $g_2=g_1$ assuming an ion density of $n_i=10^{17}cm^{-3}$ together with the range of A-values for 5p-4d and (4f-4d,4d-4p) transitions as computed by ADAS. As the dimensions of EUV discharge plasmas are of order 0.5 mm one might expect that the (4d-4f,4p-4d) lines are strongly suppressed while the 5p-4d lines are only little affected allowing for a description using escape factors[4].

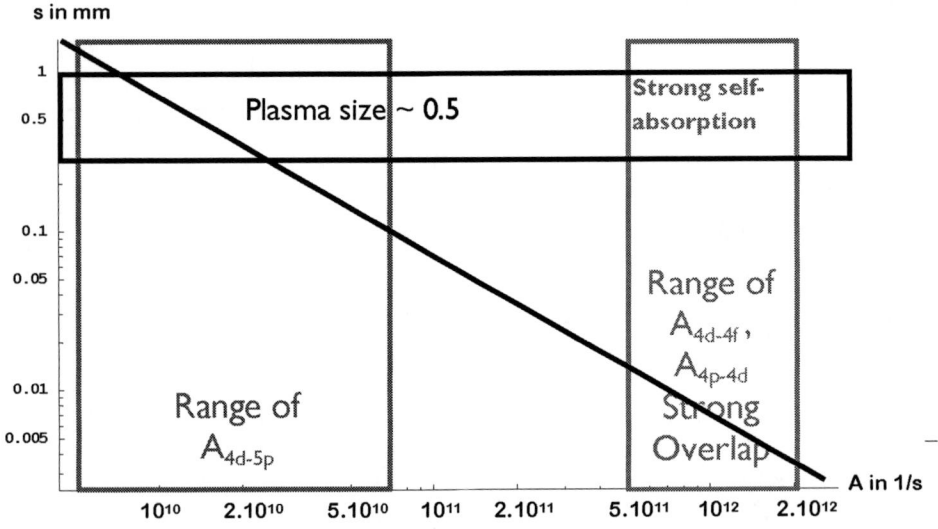

FIGURE 2. Optical Depth of EUV Radiation

This explains why in the measured spectrum shown in Fig. 3 the features near 11 nm are not as pronounced as one might expect from the optically thin spectra of Fig. 1. The self absorption of the (4d-4f,4p-4d) lines of various xenon ions is rather favorable for the inband efficiency.

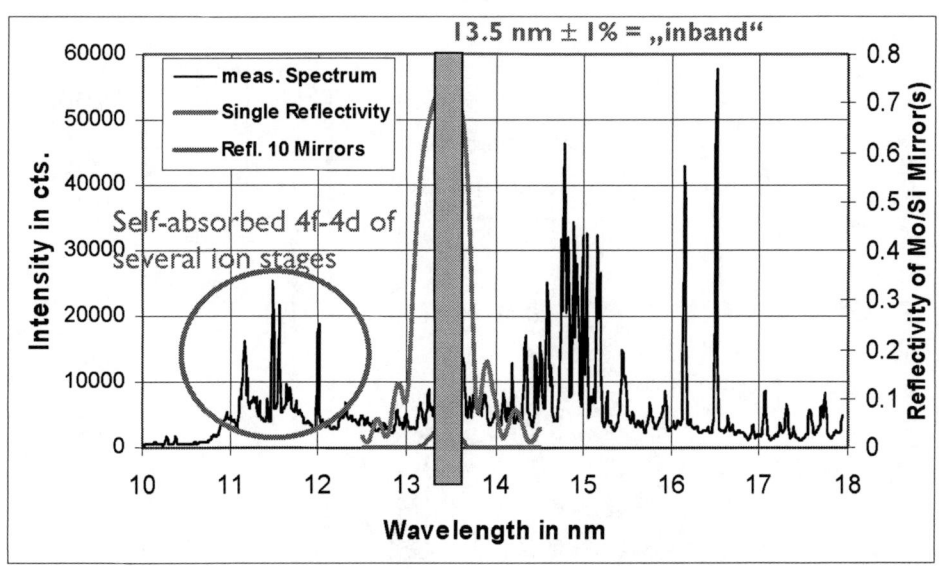

FIGURE 3. Measured EUV Spectrum of xenon together with Mirror Reflectivity

On the other hand Fig. 4 shows a calculated spectrum from a mix consisting of 17% Xe^{+8}, 20% Xe^{+9}, 25% Xe^{+10}, 20% Xe^{+11} 13% Xe^{+12} at $T_e = 32$ eV and $n_e = 10^{17} cm^{-3}$ with the measured spectrum. For the 4d-5p lines with little self absorption the agreement is rather good. For the calculated 4d-4f line of Xe^{+8} the wavelength is shifted considerably: This is a consequence of the incomplete set of basic configurations which has a stronger impact on the 4f levels than on the 5p levels. The corresponding A-value calculated by ADAS is $2.23 \cdot 10^{12}$ 1/sec, indicating that line shape and maximum intensity are changed due to self absorption. As Xe^{+8} has only 2 4d-5p transitions the corresponding A values are also in the range of self absorption. Therefore their actual relative strength in the plasma cannot be derived from optically thin radiation modeling.

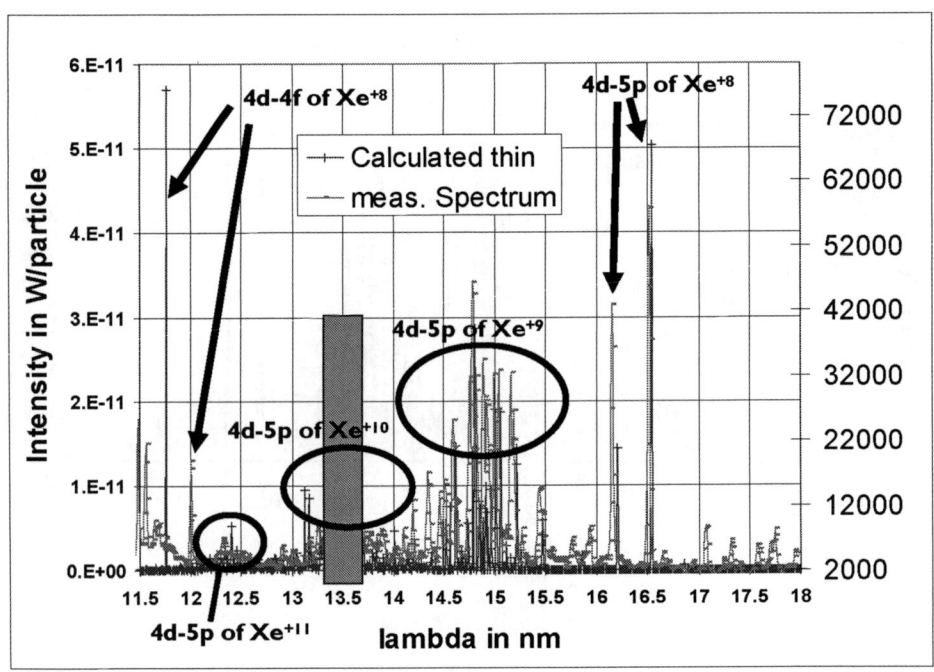

FIGURE 4. Measured spectrum and calculated thin Spectrum of Xe-ion Mixture near 13.5 nm

Ionization and Recombination rates

Dielectronic processes have, especially at rather high energies of the free electron, a strong contribution to effective ionization and recombination rates. The current version of ADAS uses projection techniques to take these effects into account. Also populations of metastable excited states might modify these effective rates significantly, which is not included in the ADAS rates shown as function of the plasma's electron temperature in Fig. 5. Considering this very difficult situation, which is also subject to current research, the rates shown can be used for estimates and simple plasma models as decribed in the following chapter but are probably not accurate enough for detailed plasma simulations.

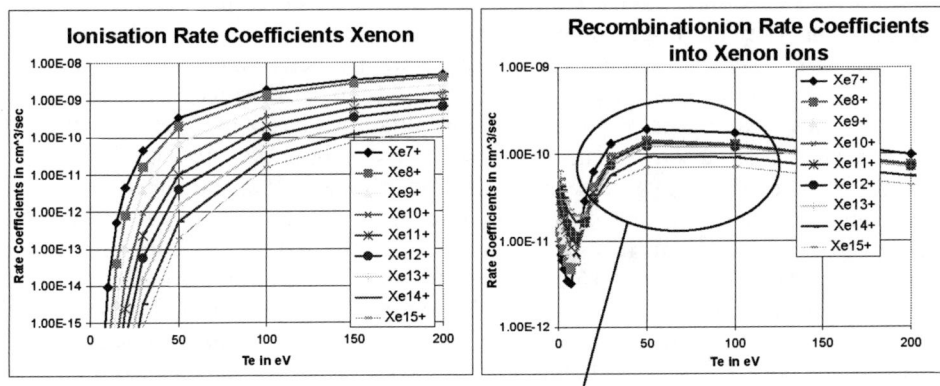

Dielectronic contributions

FIGURE 5. Effective Ionization and Recombination rates for several xenon ions

3. A SIMPLE 2D MODEL OF THE XENON PINCH DISCHARGE

As one of the first candidates for a EUV discharge, Philips Extreme UV [5] and others investigated a pulsed, hollow cathode triggered xenon Z-pinch discharge: In such a device the high pressures and temperatures required to obtain the high ionization stages and EUV radiation are achieved by magnetic compression. Upon ignition a capacitor bank which is coupled with low inductance to the electrode system is unloaded. The magnetic fields induced by the discharge currents of order 10000 A, rapidly compress the plasma which then emits the EUV radiation. A complete simulation of the pinch phase, during which the EUV radiation is emitted, requires the solution of the MHD equations in at least 2-d axi-symmetric geometry. Moreover the electron energy equation and the ionization dynamics must be coupled to the radiation transport equations to model the energy transfer by photons and the impact of photoionization in such highly transient, inhomogeneous plasmas. Finally the plasma simulations must be coupled to a (rather simple) simulation of the external electrical circuit. It is, however, possible to draw some important conclusions from rather simple zero dimensional transient plasma models. We will summarize here only the most important features and results of a model described in more detail in Ref.[6].

Main Features Of A Simple Zero-dimensional Transient Discharge Model

As the coupling of plasma and external circuit time scales is of crucial importance, a simulation of the pinch phase of an EUV source must be coupled to the solution of the electrical circuit equations. The pinch plasma is modeled in a snow plough description as a radially imploding cylinder without axial or radial profiles. The implosion time must be matched with the current phase such that maximum compression occurs at the current maximum. This issue can be modeled very well with such a simple approach. Within the compressed plasma region we solve energy equations for ions and electrons. The electrons provide via inelastic collisions also the power for ionization and radiation. They are heated ohmically and by the compression. The total radiative power loss term in the electron energy equation has been tabulated from ADAS calculations based on the assumption of a CR equilibrium as described above. Opacity effects are modeled by means of escape factors although this approach is questionable for the strongly self absorbed part of the radiated spectrum, cfr. previous chapter. A typical feature of these pinch discharges are rather high ion temperatures exceeding the electron temperatures significantly. In our case $T_i \sim 1$ keV while $T_e \sim 50$ eV. The ionization is treated dynamically by a separate set of differential equations using rates tabulated by ADAS, cf. Fig.5. This is necessary since the plasma parameters change on timescales of order 5 ns which is much faster than the characteristic time scales for ionization and recombination which are in the range of 100 ns. All these simulations were carried out in close collaboration with experiments. To match measured and simulated current pulse shapes, for example, it was necessary to introduce bypass currents which flow not through the pinched plasma, in the simulation.

4. THE LASER TRIGGERED VACUUM SPARK DISCHARGE AS EUV SOURCE

Besides xenon's poor spectral match which limited the inband conversion efficacy near 13.5 nm to <0.5 %, the hollow cathode triggered Xe pinch discharge had other severe problems: attempting to increase the power resulted in erosion the cathode hole accompanied by an unacceptable growth of the EUV emitting pinch region..

Tin As EUV Radiator

The first problem, the relatively small amount of inband radiation from xenon, led to the choice of tin as radiating element. As first pointed out in Ref. [7], the energy levels of 4f excited states of highly ionized atoms between tin and xenon scale with the nuclear charge such that the strong (4d-4f,4p-4d) lines of various tin ions sit just at 13.5 nnm. This has been confirmed by ADAS calculations: Fig. 6 shows the spectra of tin ions Sn+8 - Sn+13 with many lines in the inband region. This is verified in measured spectra, Fig. 7, which show a concentration of many lines near 13.5 nm, emitted from different ionization stages. This feature, which is sometimes called unresolved transition array, makes it very difficult to assign individual lines. Moreover the inaccuracy of the calculated 4f energy levels make the spectra shown in Fig. 6 inappropriate to identify lines.

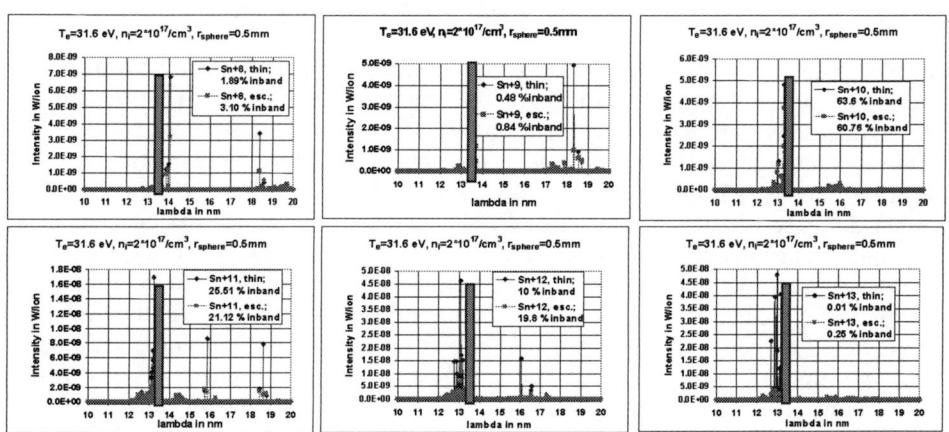

FIGURE 6. Optically thin Spectra of Sn^{+8} - Sn^{+13} ions and Modification due to Escape Factors

cts/pix

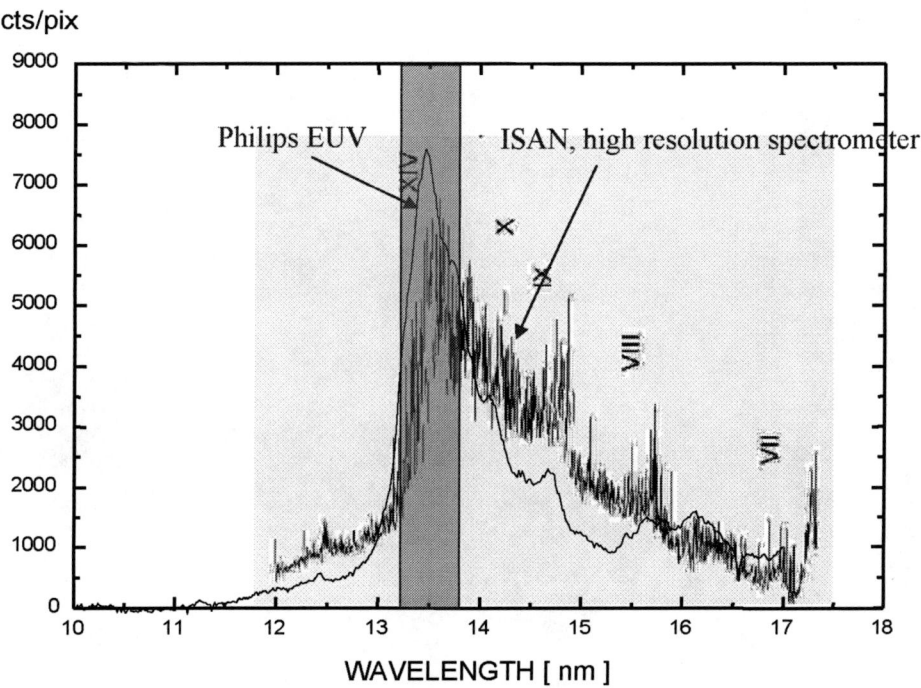

FIGURE 7. Measured tin EUV spectra.

New High Power EUV Source Concept

The electrode system consists of two rotating wheels. To generate a tin plasma these wheels are rotated through baths of molten tin such that a thin film of tin remains on their surfaces. With a laser pulse tin is ablated from one of the wheels and explodes rapidly through vacuum towards the other rotating wheel. When the tin plasma reaches the other electrodes it ignites and the high current phase starts, i.e. the capacitor bank is unloaded, the plasma is pinched and EUV is radiated. Besides the good spectral properties of tin this concept has some other advantages: Erosion of electrodes is no problem anymore as the tin film is regenerative and protects the electrode material. The electrical connections to the rotating electrodes are easily made via the tin baths. As the liquid tin cools very effectively the rotating wheels, the concept is scalable to higher powers. Fig. 8 shows the result of a thermal and flow simulation for a wheel electrode of a 40 kW input power EUV source. Even for such extremely high electrode loads temperatures of the wheel remain almost everywhere below 500°C.

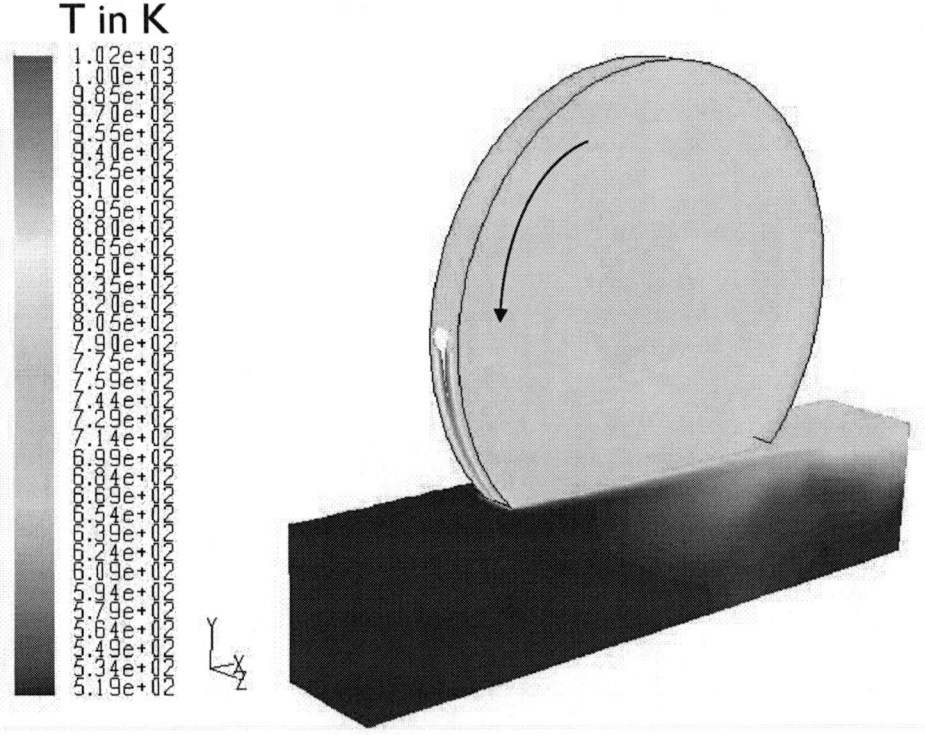

FIGURE 8. Temperature distribution in wheel electrode and liquid metal bath.

The disadvantage, however, is the rather large amount of tin debris that must be kept away from the optics. Even very thin layers of condensed tin would severely reduce the reflectivity of the mirrors. But recently very efficient debris mitigation have been demonstrated.[8]

In a laser ablated vapor plasma density gradients are as large as a couple orders of magnitudes per millimeter and expand with velocities of a few 10 km per second. As this causes severe problems to even very simplified plasma modeling approaches, we cannot show any results from simulations beyond the more general radiation properties discussed above.

ACKNOWLEDGEMENTS

This work was done in close collaboration with Philips Extreme UV GmbH. Part of this work was supported by European Union and Deutsches Ministerium fuer Forschung and Technology (BMBF).

REFERENCES

1. V. Banine, R. Moers, J. Phys. D, Vol. 37, p. 3207-3212
2. R.D. Cowan, The Theory of Atomic Spectra, University of California Press, Berkeley, California (1981)
3. H. Summers et al., Atomic Data and its Utilization at the Jet experiment, Plenum Press Series Physics of Photons and Molecules, Photon and electron collisions with atoms and molecules, edited by. P.G. Burke, C.J. Joachain (1997), p. 265, see also http://adas/phys/strath.ac.
4. F.E: Irons, The Escape Factor in Plasma Spectroscopy – 1. The Escape Factor defined and evaluated, JQSRT 22, p. 1 (1979).
5. J. Pankert, K. Bergmann, J. Klein, W. Neff, O. Rosier, S. Seiwert, S. Probst, D. Vaudrevange, G. Siemons, E. Bosch, P. Zink, Th. Krücken,C. Smith, R. Apetz, J. Jonkers, M. Loeken, G. Derra, " Physical Properties of the HCT source", SPIE Microlithography 2003, March 25-27, 2003, Santa Clara, Proc. SPIE 5037, (2003) p. 15
6. T. Kruecken, K. Bergmann, L. Juschkin, R. Lebert, J. Phys. Vol. 37 (2004), p. 3213-3224.
7. G. O'Sullivan, P.K. Carroll; Opt. Soc. Am 71, (1980) p. 227
8. M. Corthout, M. Loeken, P. Zink, The Philips Extreme UV Sn source: on the way to an integrated system, SPIE Advanced Lithography Conference, San Jose, 2007

Theoretical and Experimental Databases for High Average Power EUV Light Source by Laser Produced Plasma

H. Nishimura, K. Nishihara, S. Fujioka, T. Aota, T. Ando,
M. Shimomura, K. Sakaguchi, Y.Simada[1], M. Yamaura[1], K. Nagai,
T. Norimatsu, A. Sunahara[1], M. Murakami, A. Sasaki[2], H. Tanuma[3],
F. Koike[4], K. Fuijma[5], C.Suzuki[6], S. Morita[6], T. Kato[6], T. Kagawa[7],
T. Nishikawa[8], N. Miyanaga, Y. Izawa, and K. Mima

Institute of Laser Engineering, Osaka U., [1]Institute for Laser Technology, [2]Advanced Photon Research Center, JAEA Kansai, [3]Tokyo Metropolitan U., [4]Kitazato U., [5]Yamanashi U., [6]National Institute for Fusion Science, [7]Nara Women's U., [8]Okayama U.

Abstract. Extreme ultraviolet (EUV) radiation from laser-produced plasma has been thoroughly studied for application in mass-production of the next generation semiconductor devices. Comprehensive experimental databases are provided for a wide range of parameters of lasers and targets. The atomic models are benchmarked with spectroscopic measurements not only for laser-produced plasma (LPP) but also for EUV emissions from magnetic-confinement plasmas or the charge exchange for uniquely ionized ions colliding with rare-gas targets. These experimental data are utilized in the industry as well as used to benchmark the radiation hydrodynamic code, including equation-of-state solvers and advanced atomic kinetic models, dedicated for EUV plasma predictions. Present status of the LPP EUV source studies is presented.

Keywords: Extreme Ultraviolet (EUV), laser produced tin plasma, tin opacity, charge-exchange spectroscopy
PACS: 32.30.Rj, 32.80.Fb, 39.30.+w, 52.38.Ph, 52.50.Dg, 52.65.-y, 52.77.-j

INTRODUCTION

EUV lithography (EUVL) has been situated as one of key technologies to extend further integration of electric circuit manufacturing along the technical path, commonly referred to as the roadmap of the semiconductor industry. In order to print node sizes less than 40 nm, EUVL systems will require intimate technical transfer between industry and research institutes. In Japan, EUVL Association (EUVA) was established in June 2002 under the auspices of the Ministry of Economy, Trade and Industry (METI) to accelerate the development of EUV lithography systems. In the

CP926, *Atomic Processes in Plasmas—15th International Conference on Atomic Processes in Plasmas*
edited by J. D. Gillaspy, J. J. Curry, and W. L. Wiese
© 2007 American Institute of Physics 978-0-7354-0436-6/07/$23.00

fiscal year 2003, following to the start of a new leading project for development of EUV light source for advanced lithography technology under auspices of MEXT, the Institute of Laser Engineering, Osaka University has started intensive research on the laser-produced plasma (LPP) source. [1-3].

LITHOGRAPHY SYSTEM WITH LPP EUV SOURCE

The principle of the LPP EUVL system is shown in Fig, 1. The source plasma is generated with laser pulses synchronized at a high repetition-rate with successively supplied targets of small mass. EUV light from the plasma is collected with a multilayer-coated EUV mirror to transfer the light to the illumination system via intermediate focus (IF). Circuit-patterns recorded on a reflection reticle are finally imaged on an EUV resist layer coated on the silicon wafer. In this system, nearly ten EUV mirrors are unavoidably used from the source plasma to the wafer so that overall transfer efficiency of EUV light from the source to the wafer becomes less than 1% since the highest reflectivity is 65% at 13.5 nm for a Mo/Si mirror. Specifications of EUV source are definitely fixed through discussions among industries [4]. These are listed in Table 1. Since the transfer efficiency from the first mirror to IF is around 30% at highest, over 300 W EUV source power is required at plasma.

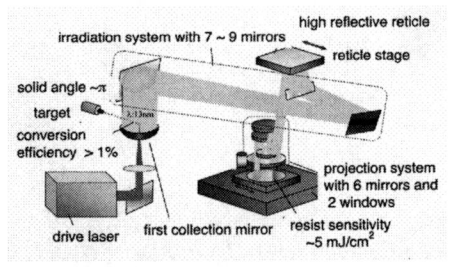

FIGURE 1. EUV light from the source plasma is collected by the first mirror, and transferred to the intermediate focus, at which the source power is defined.

TABEL 1. Specifications of EUV light source

wavelength	13.5 nm @2%BW
EUV power	> 180 W (@IF)
repetition rate	7-10 kHz or more
etendue	1-3.3 mm^{2}sr
stability	+/- 0.3% 3σ over 50 pulses
source cleanliness	>30,000 hours at IF
spectral purity	
130-400 nm	<1% of EUV at wafer
> 400 nm	< 10-100% at wafer

*IF : intermediate focus, a point transferring EUV light to the illumination system.

Prediction of Plasma Parameters

Among several candidates, Sn, Xe, and Li are considered to be the most promising materials due to the matching of their plasma spectra with reflection by Mo/Si multilayer mirror. Suppose an etendue (defined by the product of the source area size and the light extraction solid angle) of 1 mm^{2}sr and an extraction solid angle of π; the area corresponds to that of a laser spot of 600 μm in diameter. It is requested to extract EUV energy E_{EUV} of 30 to 40 mJ per pulse for 10 kHz repetition rate. Take EUV pulse duration τ_{EUV} of 1 to 10 ns, EUV flux I_{EUV} will be 10^9 to 10^{10} W/cm^2. If the conversion efficiency (CE) from laser to 13.5 nm EUV light in a 2% bandwidth (BW), laser irradiance I_{LASER} will be 10^{11-12} W/cm^2. This flux is high enough to attain the plasma temperature of 60-70 eV.

The optimum ion density can be estimated from EUV photon flux Ψ_{EUV} (the number of photons emitted per unit area and unit time). Namely, Ψ_{EUV} is equal to I_{EUV} /hν=7x10^{25}-7x10^{26} photons/scm^2. This must balance with the value $(n_{EUV} \cdot l_{EUV})_{OD=1} \cdot A$, where n_{EUV} is the density of ions effectively contributing to EUV emission (define the efficiency as f), l_{EUV} is EUV plasma depth. Their product must be unity represented as optical density of 1, and A is Einstein A-coefficient. Since A becomes around several 10 ps, the density-depth product $n_{EUV}l_{EUV}$ will be 10^{15-16} cm^{-2}. For f=0.1 and l_{EUV}=100 μm, the optimum ion density n_i will be around 10^{18-19} cm^{-3}.

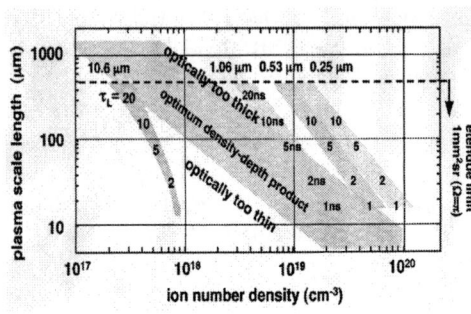

FIGURE 2. Optimum plasma conditions predicted by a simple model. For shorter wavelength lasers, due to higher critical densities, shorter scale length is preferable for efficient EUV generation, and so.

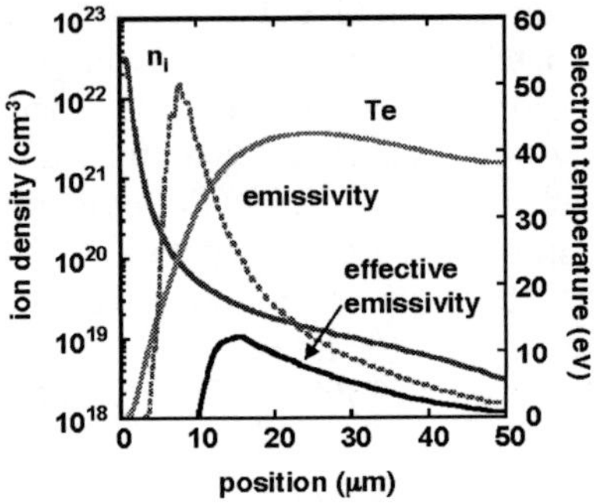

FIGURE 3. Density, temperature, emissivities for the optically thin and practical cases are produced for Sn plasma with SATR [5].

Since laser intensity is at most 10^{11-12} W/cm^2 to generate EUV plasmas, the dominant absorption mechanism is the classical. The ideal condition is that the absorption region corresponds to the EUV dominant region. l_{EUV} is simply given by the product of plasma sound velocity and laser pulse duration. In this way, one can obtain the optimum region as shown in Fig. 2 assuming that the product of l_{EUV} and the absorption coefficient is 1 to 2. From this picture, the optimum laser wavelength will be 2-3 μm although there are no candidates available for practical use. If laser wavelength is shorter than 1 μm, the shorter pulse duration is preferable. For long wavelength laser such as CO_2 laser, energy is deposited in a lower density region than that of the EUV dominant region, then absorbed energy is transported mostly by thermal electrons. For shorter wavelength laser such as excimer laser, energy is deposited in a higher density region, then EUV radiation generated there must pass through the EUV optimum region. This makes the EUV conversion efficiency (CE) lower due to the self-absorption (i.e., opacity effect) in the EUV dominant region.

A Power Balance Model

In order to provide optimum plasma parameters, radiation hydrodynamic code STAR [5] was developed. In parallel, a simple theoretical analysis has been made [6]. In the case of Sn plasma, EUV emission in 13-14 nm is arising from $\Delta n=0$ atomic transitions (n is the principal quantum number). Figure 3 shows typical density and temperature profiles of laser-produced Sn plasma. Also shown are emissivity profiles for the optically thin case and for the practical plasma case. Note that large difference is seen between them, particularly at higher density region due to the opacity effect.

A power balance model was made to deal with the loss fluxes due to ionization, plasma expansion, and radiation in EUV plasma. Physical parameters for the model calculation were derived from advanced atomic models such as HULLAC [7]. As is discussed below the model predictions were confirmed with simulations done with STAR. CEs at 13.5 nm within 2% bandwidth have been provided theoretically as a function of ion densities and electron temperatures for variety of laser conditions for Sn, Xe, and Li. The results are shown in Fig. 4.

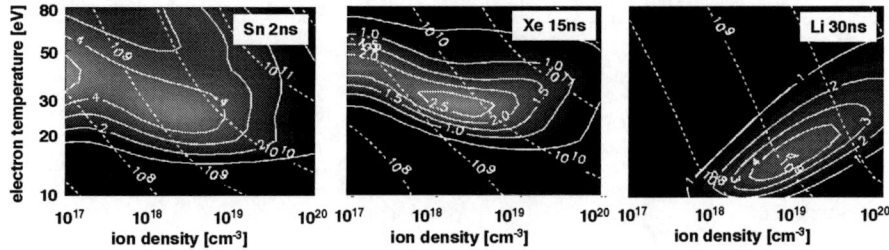

FIGURE. 4 EUV conversion efficiencies at 13.5 nm in a 2% BW are predicted with a simple power balance model as a function of ion density and electron temperature. The numbers on the solid circular lines represent the conversion efficiency in % and the dotted lines absorbed laser intensities needed to create plasma densities and temperatures in these maps.

LASER PLASMA EXPERIMENTS

Spherical Plasma Generation

In previous EUV plasma experiments, a Joule-class, single laser beam was used. Therefore experimental results might be substantially influenced by energy flow along the target surface via plasma expansion and/or thermal conduction (so-called 2-D effect). In order to eliminate the problem for comparison with theoretical interpretations, spherical targets were irradiated uniformly with GEKKO XII, a multiple-laser dedicated to laser fusion research [8]. Spherical plastic targets coated with a 1-μm-thick Sn layer, thick enough to generate EUV light, were used in the experiment. Figure 5 shows the CE dependence as a function of laser intensity. The CEs obtained with the two instruments agree quite well, showing the validity of mutual calibration processes. The highest CE of 3% was attained at an intensity of 0.5-1×10^{11} W/cm^2.

The CE dependence on the laser intensity was analyzed using the power balance model above. The result is also shown in Fig. 5 with a solid curve. A good agreement between the experiment and the prediction is obtained. Comparing the results of this study with Spitzer's investigation [9], we infer that even with a large laser spot size, the 2-D effect is not completely excluded, and thus a higher irradiance was needed. These results conclusively show that the formation of spherical plasma enables us to provide a good benchmark for theoretical models included in simulation codes.

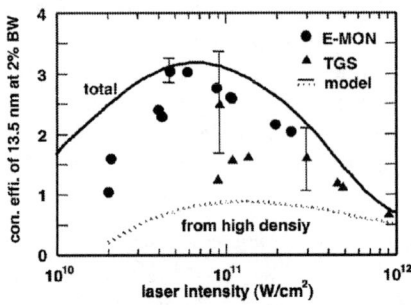

FIGURE 5. Dependence of CE on laser intensity. The maximum CE of 3% was archived at the laser intensity of 0.5×10^{11} W/cm^2 [8].

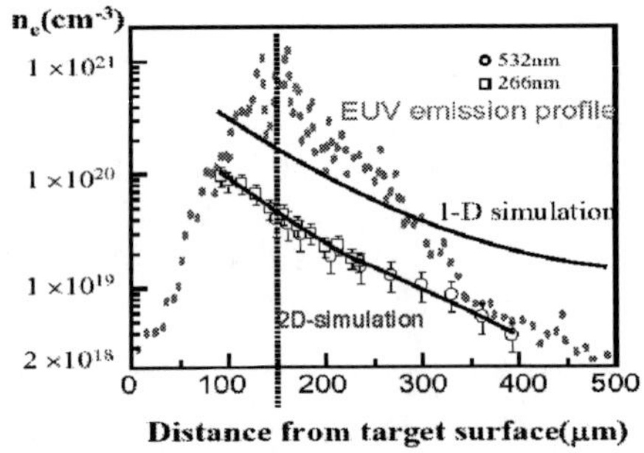

FIGURE 6. Electron density profiles from the 532 nm (circle), 266 nm (square) inteferometers, 1-D and 2-D radiation hydrodynamic simulation STAR (solid lines), and EUV emission profile (dots). [10, 11].

Density Profile Measurement

Electron density profile of laser-produced Sn plasma was investigated using two interferometers to cover a wide range of density profile [10]. Density profiles along the line of laser incidence extracted are plotted in Fig.6. Computer simulations were made using STAR code for 1-D and 2-D geometry (details are described below). A good agreement is obtained for the 2-D case. Discrepancy in 1-D simulation arises from the multi-dimension plasma expansion in experiment.

EUV emission profile at the peak of laser pulse, measured with an EUV imager attached with an x-ray streak camera [11], is plotted in Fig.6 with dots. It is found that most of the EUV light comes from well under-dense plasma region with electron densities from 1×10^{19} cm^{-3} to 1×10^{20} cm^{-3}. The possible reason is the opacity effect of the Sn plasma. The heavy re-absorption induced by the out-layer plasma will reduce CE significantly. EUV emission near 13.5 nm can be attributed to $4d$-$4f$, $4p$-$4d$, and $4d$-$5p$ transitions in Sn plasma with optimum temperature around 30 eV.

CE Dependence on Laser Wavelength and Pulse Duration

Dependence of EUV CE on laser wavelengths was studied using ω, 2ω, and 4ω light of Nd:YAG laser [12] and CO$_2$ laser [13]. EUV spectra generated with 532 nm and 266 nm lasers showed spectral dips around 13.5 nm and these features are well replicated in computer simulations. This infers existence of opaque plasma surrounding the EUV emission region. CO_2 laser produced plasma has shown even

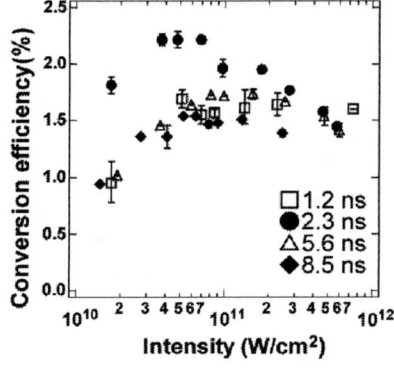

FIGURE 7. CE dependence on laser pulse duration [14].

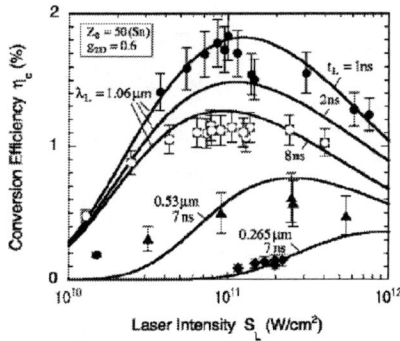

FIGURE 8. Comparison of CE dependences on laser wavelengths and pulse duration between the model and experiments [15].

higher CE than those generated with YAG ω light so that EUV source developers have chosen this laser as the most practical driver for EUV source generation.

EUV CEs were also measured by changing laser pulse durations for Sn plasma. Experimental results shown in Fig. 7 indicate that the optimum pulse duration is determined by two parameters: one is the optical depth of tin plasma for 13.5 nm light and the other is laser absorption rate in the EUV dominant region. The maximum CE of 2.2% is obtained with pulse duration of 2.3 ns [14].

A simple analytical model was presented for the conversion of laser to EUV radiation [15]. The model is based on consideration of the Planck mean-free-path for optically thick materials. Comparison of the model prediction with experimental results (see Fig. 8) shows a good agreement under different conditions of plasma generation such as laser wavelength, intensity, and pulse duration, and target atomic numbers. It is found that relatively high CEs are obtained when the Planck optical thickness of plasma is 0.3-0.5.

CE Dependence on Target Density

Influence of initial density of Sn targets was quantitatively investigated [16]. The low-density targets were fabricated using mono-dispersed polystyrene nano-particles and liquid Sn chloride (see Fig. 9) [17, 18]. As shown in Fig. 10, with decrease in the initial density of the target, CE from incident laser energy to output 13.5 nm light energy in a 2% bandwidth increases; the peak CE of 2.2% was attained with use of 7% low-density SnO_2 targets of 0.49 g/cm^3 irradiated with a Nd:YAG laser of 10 ns at 5×10^{10} W/cm^2. The peak CE is 1.7 times higher than that obtained with the use of solid density Sn targets. Experimental results may be attributed to the influence of the initial density and/or microstructure of the targets on expansion dynamics of the plasmas. This dependence is also interpreted by optimization of the Sn plasma optical depth at 13.5 nm by the plasma density.

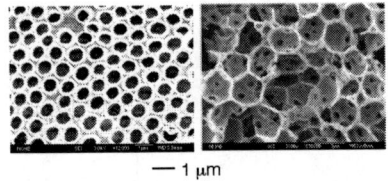

— 1 μm

FIGURE 9. Low density SnO2 target [17, 18].

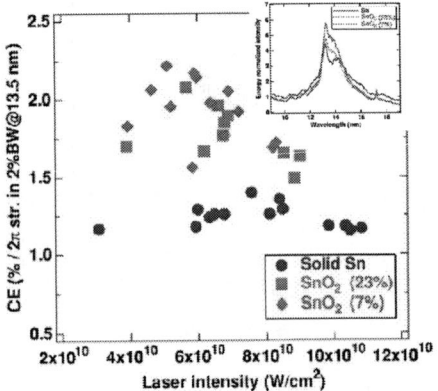

FIGURE 10. CE dependence on the target initial density [16].

SPECTROSCOPIC DATABASE MEASUREMENT

Tin Opacity Measurement

The Sn opacity measurements were performed on Gekko-XII [19]. A radiation confining gold cavity (shown in the inset of Fig. 11) was used to uniformly heat an opacity sample, consisted of a thin Sn layer sandwiched between two 1000 nm-thick parylene (C_8H_8) tampers. The area density of the Sn layer in the sample was $2.04+/-0.18 \times 10^5$ g/cm^2. Radiation temperature on the sample was 50 eV. The radiatively heated sample was backlit with broadband EUV light from another Sn plasma generated with a laser pulse. Radiation-hydrodynamic simulation (ILESTA-1D) [20] predicts an averaged temperature and density of the heated Sn sample of 30 eV and 0.01 g/cm^3 at 1 ns after the peak of the heating x-ray pulse. The dots in Fig. 11 represent the raw measured spectrum of the transmission, while the solid lines are synthesized spectra calculated by HULLAC with electron temperatures of 24, 28, 32, 36 eV, respectively. The configuration interaction between the $4d^n$, $4d^{n-1}4f$, $4d^{n-1}5p$, $4d^{n-1}5f$, and $4p^54d^n$ configurations were taken into account, since the interaction changes the wavelength and strength of emission lines considerably [21, 22]. The population of the ionization state was calculated under the collisional radiative equilibrium condition. Strong absorption is seen around 13.5 nm, arising mainly due to the $4p$–$4d$ and $4d$–$4f$ transitions of Sn^{8+} to Sn^{13+}. Several dips seen between 15 and 18 nm are due to the $4d$–$4f$ and $4d$–$5f$ transitions of Sn^{4+} to Sn^{6+} ions. The transmission at 13.5 nm, the most interesting wavelength, shows quite good agreement between the measurement and calculation, demonstrating that the HULLAC calculation is useful for optimizing 13.5 nm light generation although some discrepancies are still remaining. The measured mass absorption coefficient of the Sn plasma at 13.5 nm is $0.96+/-0.18 \times 10^5$ cm^2/g. Disagreements may be rectified by taking more configuration interactions into account in the calculation, and this work is now in progress.

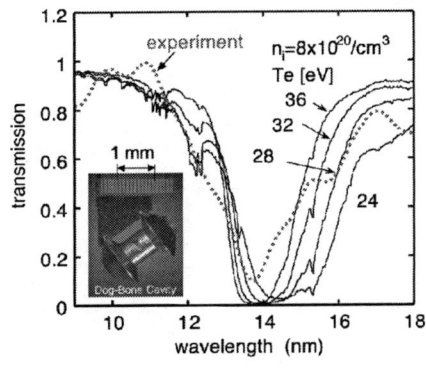

FIGURE 11. Sn opacity measurement [19].

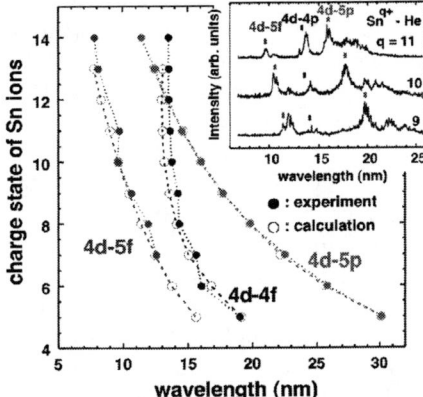

FIGURE 12. Charge exchange spectroscopy of Sn^{q+} ions colliding with He gas [25].

Charge Exchange Spectroscopy

In addition to emission spectroscopy for low density plasma confined in a magnetic confinement machine [23, 24], EUV spectra of multiply charged tin ions were measured in the wavelength range of 5–38 nm following the electron capture into excited states of slow Sn^{q+} (q = 6– 15) ions passing through He gas target [25]. The Sn ions were produced in a 14.25 GHz ECR (electron cyclotron resonance) ion source at Tokyo Metropolitan University [26]. The charge-state dependence of $4d$-nl (nl = $4f$, $5p$ and $5f$) transitions was obtained assuming the single-electron capture. Identification of the transitions has been carried out by comparison with the results of the theoretical calculation with HULLAC. In the inset in Fig. 12, short bars indicate the averaged wavelengths of the $4d$-$4f$, $4d$-$5p$ and $4d$-$5f$ transitions for Sn$^{(q-1)+}$ ions calculated with the HULLAC. Qualitatively good agreement in the comparison of the observation with theoretical wavelengths, indicating that the prominent emission peaks are attributed to the single electron capture and/or the transfer ionization which provides the charge states of $(q-1)+$ in collisions of Sn^{q+} ions with He. In contrast to the excellent agreements in the $4d$-$5p$ transitions, the $4d$-$4f$ transitions have approximately constant differences about 0.5 nm in higher charge states. This discrepancy in the $4d$-$4f$ transitions had also observed in the cases of Xe ions [27]. As the reason of this finding, we consider the strong interactions between $4p^6 4d^{k-1} 4f^1$ and $4p^5 4d^{k+1} 4f^0$ configurations, which have been pointed out by O'Sullivan and Faukner [28]. The theoretical results with the HULLAC code in this work are similar to those with the Cowan code [29].

2D RADIATION-HYDRODYNAMIC SIMULATIONS

EUV emissions from laser-produced Sn plasma were reproduced using 2-D radiation hydrodynamic code STAR. In this code, heat transport is solved with a flux-limit model with Spitzer's conductivity. Radiation transport is solved with a multi-group diffusion approximation adopting 1500 groups in 0-1.5 keV. But the energy interval for EUV region of interest is chosen to be 0.2 eV. Tabulated emissivity and opacity of Sn ions are generated with HULLAC. The difference in wavelength found between the calculation and the experiment is taken into consideration by shifting the calculated wavelength to the red-side by 0.5 nm. Population kinetics is solved including photo-excitation process (PE) by the scheme proposed by F. A. Novikov [30], where an intermediate mixed state of the collisional-radiative equilibrium (CRE) and local-thermodynamic equilibrium (LTE) is assumed.

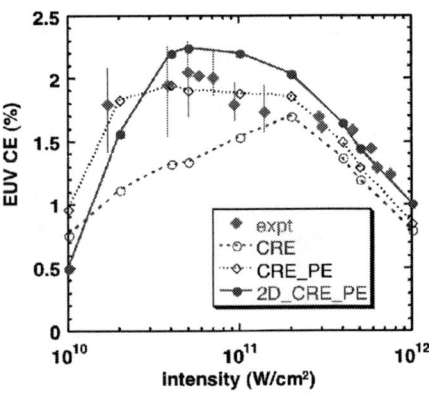

FIGURE 13. Comparison of experimental EUV CEs with simulations. Planer Sn plasma was generated with a 2.2 ns, 1.05 μm pulse [5].

Figure 13 shows comparison of the experiment with the simulations for 1-D CRE without PE process, 1-D CRE with PE, and 2-D CRE with PE. A good agreement is obtained for the cases including PE. Because of relatively short laser pulse duration, the 2-D effect on CE is not clearly seen. But, experimental results for longer duration show better agreement with simulations of 2-D with PE. Figure 14 shows comparison of calculated spectra with

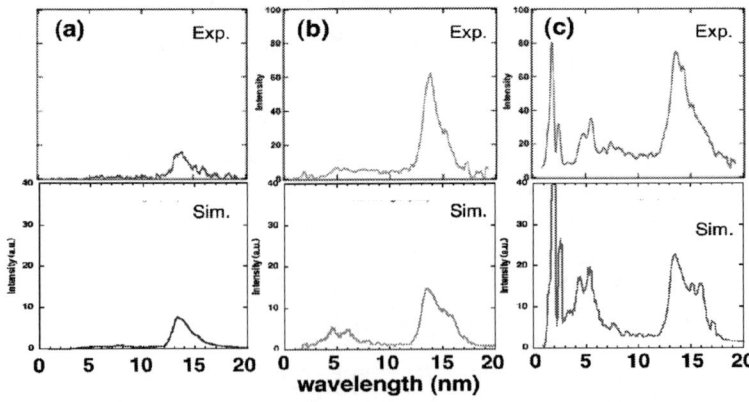

FIGURE 14. Comparison of calculated spectra with experiments at (a) 9×10^{10}W/cm^2, (b) 3×10^{11}W/cm^2, (c) 9×10^{11}W/cm^2 with a 1.2 ns, 1.06 μm laser pulse. Here, spectrum resolution is 0.42 nm [5].

experiments for a wide range of laser intensity of (a) $9x10^{10}$, (b) $3x10^{11}$, and (c) $9x10^{11}$ W/cm^2. The vertical axes of figures are relative but of the same scale. STAR simulations quantitatively well reproduce EUV CE and emission spectra. Note that 2-D plasma density profiles observed in the experiments are also well replicated with this code as discussed above with Fig. 6.

SUMMARY

Systematic studies on EUV emission from laser-produced plasma have been made. Because of high opacity, CEs are critically dependent on laser and target conditions. It has been shown that atomic parameters provided with HULLAC have been validated by a wide variety of experimental data. 2-D radiation hydrodynamic code STAR including photo-excitation effect well reproduced EUV spectra from laser produced Sn plasma. These results are now practically utilized for EUV source system development.

ACKNOWLEDGMENTS

We would like to acknowledge both Gekko XII operation and EUV target fabrication groups. This work was performed under the auspices of the Leading Project by MEXT, Japan.

REFERENCES

1. H. Nishimura, et al., *Proc. of Inertial Fusion Sci. and Applications* 2003, Monterey, CA, p.1007 (2004).
2. M. Nakai, et al., in *Proceeding of SPIE* **5196** (SPIE, Bellingham, WA, 2004) 289.
3. K. Nishihara, H. Nishimura, et al., *The Review of Laser Engineering*, **32**, 330 (2004) (in Japanese).
4. V. Banine, *Proc. of EUVL Source Workshop*, Oct. 16, 2006, Barcelona, Spain, Internat. SEMATECH.
5. A. Sunahara, et al., Radiative Properties of Hot Dense Matter, 11-15 Sep. 2006, Albufeira, Portugal; to appear in J. Plasma and Fusion Research (in Japanese) (2007).
6. K Nishihara, et al., Proc. of Inertial Fusion Science and Applications 2003, Monterey, CA, 1069 (2004).
7. A. Bar-Shalom, J. Oreg, and M. Klapisch *Phys. Rev.* E **56** R70 (1997).
8. Y. Shimada, H. Nishimura, et al., *Appl. Phys. Lett.* **86** 051501(2005).
9. R. C. Spitzer, et al., *J. Appl. Phys.* **79** 2251 (1996).
10. Y. Tao, H. Nishimura, et al., *Appl. Phys. Lett.* **86**, 201501 (2005).
11. Y. Tao, H. Nishimura, et al., *Rev. Sci. Instum.***75**, 5173 (2004).
12. M. Yamaura, et al., *Appl. Phys. Lett.* **86**, 181107(2005).
13. H. Tanaka, et al., *Appl. Phys. Lett.* **87**, 041503 (2005).
14. T. Ando, S. Fujioka, et al., *Appl. Phys. Lett.* **89**, 151501 (2006).
15. M. Murakami et al., *Phys. Plasmas* **13**, 033107 (2006).
16. T. Okuno, S Fujioka, et al., *Appl. Phys. Lett.* **88**, 161501 (2006).
17. K. Nagai, et al.,*Trans. Mater. Res. Soc. Japan* **29**, 943 (2004).
18. Q. Gu, et al., *Chem. Mater.* **17**, 1115 (2005).
19. S. Fujioka, H. Nishimura, et al., *Phys. Rev. Lett.* **95** 235004 (2005).
20. H. Takabe *et al.*, *Phys. Fluids* **31**, 2884 (1988).
21. W. Svendsen and G. O'Sullivan, *Phys. Rev.* A **50**, 3710 (1994).
22. F. Koike, Y.Azuma, et al., *J. Electron. Spec. Relat. Phenom.* **144-147,** 1227 (2005).
23. M. B. Chowdhurla, S. Morita, et al., to appear in Plasma and Fusion Research (2007).
24. C. Suzuki, H. Nishimura, et al.,*J. Plasma Fusion Res.* **81**, 480 (2005).

25. H. Ohashi, H. Tanuma, et al., to appear in *J. Phys: Conference Series* (2007).
26. H. Tanuma, J. Matsumoto *J. Chin. Chem. Soc.* **48**, 389 (2001).
27. H. Tanuma H, Ohashi, et al., *Nucl. Instr. Meth.* **B 235**, 126 (2005).
28. G. O'Sullivan and R. Faukner, *Opt. Eng.* **33** 3978 (1994).
29. R. E. Kieft, et al., *Phys. Rev.* **E 71** 026409 (2005).
30. A. F. Nikiforov, et al., *"Quantum-Statistical Models of Hot Dense Matter: Methods for Computation and Equation of State"*, Progress in Mathematical Physics, Birkhauser, (2005).

Important Atomic, Molecular and Radiative Processes in Low Pressure Discharge Lamps

Graeme G. Lister

Central Research and Service Laboratory, OSRAM Sylvania,
71 Cherry Hill Drive, Beverly, MA 01915, USA

Abstract. Low pressure discharges are used in a number of light sources, of which the most important application for general lighting is the fluorescent lamp (FL). In conventional FL, electrical energy is converted to *UV* radiation through excitation of mercury atoms; the *UV* is then converted to visible radiation using a phosphor. Other atomic radiators, such as sodium and rare gases, are used for applications but are unsuitable for general lighting. In recent years, there has been strong interest in finding alternative atomic and molecular radiators and research is continuing. The efficiency of producing light in a low pressure discharge depends on the balance between ionization processes, which sustain the plasma, and the excitation of atoms or molecules into radiating states through electron impact excitation and collisions between atoms in excited states. Since radiation emitted at one point in the discharge may be absorbed and re-emitted several times before it finally reaches the wall, radiation transport also plays a significant role in determining the fraction of electrical energy which is converted to radiation. Numerical models can help guide the development of more efficient light sources, but there is currently a lack of data for a number of important fundamental processes. This paper will describe the important physical processes in low pressure discharge light sources, and discuss the requirements for new and improved atomic and molecular data.

Keywords: lighting, radiation, gas discharges, non LTE, modeling
PACS: 31.15.-p, 33.15.-e, 33.20.-t, 52.20.-j, 52.25.Os, 52.80.Tn

INTRODUCTION

Gas discharge light sources [1,2] are produced by passing an electric current through a gaseous medium; electrons from the resulting plasma then excite constituent atoms and molecules into radiation emitting states. In low pressure discharge (LPD) lamps, electrons are far from local thermal equilibrium with the other species in the discharge. The most important commercial application of LPD light sources is the fluorescent lamp (FL), in which mercury atoms are excited to produce *UV* radiation, which is then converted to visible light by means of a phosphor. Other examples of low pressure discharge light sources are *low pressure sodium lamps* and *rare gas discharge lamps,* which have limited applications for general lighting. Current research in LPD light sources is focused on improving the efficiency of existing lamps and finding alternative atomic and molecular radiators to replace mercury in FL.

There are a number of factors which determine the suitability of a light source for a particular application [3], each of which is strongly influenced by the atomic, molecular and radiative processes in the lamp.

CP926, *Atomic Processes in Plasmas—15th International Conference on Atomic Processes in Plasmas*
edited by J. D. Gillaspy, J. J. Curry, and W. L. Wiese
© 2007 American Institute of Physics 978-0-7354-0436-6/07/$23.00

Luminous efficacy is a measure of how efficiently a lamp converts electrical energy to visible light and is measured in *lumens per watt* (*lpw*). Efficacy is defined in terms of the response of the average eye to the visible spectrum in bright viewing conditions. The eye response is a maximum for green light at 555 nm; 1 watt of radiative energy at this frequency is defined to be 683 lumens. A typical incandescent lamp operates at 12 *lpw*, a conventional florescent lamp at 80-100 *lpw* and a low pressure sodium lamp at 200 *lpw*.

Correlated Color Temperature (CCT) is the temperature of the black body whose spectrum most closely represents the spectrum of the light source and is therefore a measure of the apparent color of the light source. Incandescent lamps have a *CCT* of 2600 K, while the *CCT* for fluorescent lamps ranges from 3000 K (warm) to 6000 K (cool), depending on the choice of phosphor.

The *Color Rendering Index* (*CRI*) is a measure of how well the light source reproduces the colors of objects. In general, *CRI* is measured by comparing the apparent colors of a set of standardized pigments illuminated by a light source with those illuminated by a black body source at the same *CCT*. For incandescent lamps, *CRI*=100, which represents "perfect" color rendition. Fluorescent lamps have CRI>80, while low pressure sodium lamps have CRI<60.

Fluorescent lamps are filled with a rare gas, typically argon at around 400 Pa pressure, with 0.5-5 Pa of mercury vapor. At room temperature, mercury has the highest vapor pressure of any of the elements suitable for producing radiation. Under optimum conditions, 70-75% of electrical power is converted to *UV* radiation. The energy difference between the *UV* photons and the visible photons emitted by the phosphor (the Stokes shift) results in a total conversion efficiency of electrical power to visible light of about 25%.

Low pressure sodium (LPS) lamps [4] operate on a similar principle to FL. Sodium atoms emit principally monochromatic yellow radiation, so these lamps do not require a phosphor. They are therefore more efficient in converting electrical power to visible radiation than FL, but their poor color rendition and high operating temperature (260 °C) limits their application principally to outdoor lighting.

Rare gases atoms emit mainly in the *VUV* and the visible spectrum, but the efficiency of converting electrical energy to visible radiation is significantly less than in fluorescent and LPS lamps. Barium [5] and molybdenum [6] have been investigated as possible replacements for mercury in low pressure discharge light sources. Both atoms have green resonance lines near the center of the eye's photopic response curve (553.5 nm (Ba) and 555 nm (Mo)). Barium ions have strong blue and red spectral lines, giving the potential for a "white" light source. However, the temperature required to operate these lamps (500-700°C) is too high for practical implementation.

The potential application of molecular radiators in LPD is currently an active topic in lighting research [7]. Metal halides have been used in high intensity discharge (HID) lamps for a number of years [2], but the molecular contribution to radiation is small compared to that of the dissociated metal atoms. Excitation of molecular radiators into vibrational and rotational levels by electron impact provides an energy loss mechanism which is absent in atomic radiators. This presents a challenge to researchers to develop an efficient low pressure molecular discharge light source.

The following sections give a brief review of the physics of the fundamental processes in LPD light sources and the issues of current interest in lighting research, with particular emphasis on the important atomic, molecular and radiative processes. The physics of discharge lamps is discussed in more detail elsewhere [1,2].

THE PHYSICS OF LOW PRESSURE DISCHARGE LAMPS

Collision Processes

Collisions of electrons with atoms and molecules are the dominant mechanism in LPD lamps. Elastic collisions of electrons with atoms and ions couple the electric power to the discharge through the electrical conductivity and provide gas heating, while electron-electron collisions re-distribute the electron energy and strongly influence the electron energy distribution function (*EEDF*). Inelastic collisions between electrons and atoms are the main source of ionization and excitation into metastable and radiative states in the discharge. Inelastic collision between atoms in excited states and other atoms and molecules in the discharge, such as chemi-ionization [8], can also be important, quenching radiative states and providing an extra channel for ionization.

Radiative Processes

Atoms with a radiative transition to the ground state (resonance transition) from an energy level which is close to half the ionization energy can provide efficient radiation in low-pressure discharges. The radiative level (and neighboring metastable levels) is more readily populated by electron impact excitation than higher levels, providing efficient channels for both radiation and two-step ionization to maintain the discharge. Mercury is an efficient radiator because the energy level of the first excited state for resonance radiation is 4.89 eV, and the ionization energy is 10.4 eV. Sodium (2.1 eV and 5.1 eV) and barium (2.2 eV and 5.2 eV) have similar properties, but rare gas atoms have radiative states which are much closer to the ionization level; neon for example has levels for *VUV* resonance radiation at 16.7 and 16.9 eV and an ionization level of 21.6 eV.

Spectral lines emitted by atoms are broadened and shifted by a number of perturbing influences [2]. The most important processes in low pressure discharge lamps are Doppler broadening and resonance broadening. Radiation emitted at one point in the discharge may be absorbed and re-emitted many times before reaching the walls, broadening the radial density profiles of radiating states and consequently influencing the spectral output and power balance of the lamp.

Ambipolar Diffusion and Cataphoresis

Ion diffusion to the walls represents an important energy loss mechanism in the positive column and determines the ionization balance maintaining the discharge. The ambipolar electric field is established to maintain qausi-neutrality; ions are accelerated away from the center of the discharge, while the electron motion is retarded. In a discharge containing a minority species (such as mercury or sodium) with an ionization energy less than that of the buffer gas atoms, ambipolar diffusion causes a depletion of the minority species at the discharge axis, (*cataphoresis*). Cataphoresis is particularly important at high current densities [9]. Axial cataphoresis also occurs along the electric field of direct current discharges, so lamps operate with alternating current to prevent the minority (radiating) species from accumulating at the cathode.

Power Balance

The total electrical power W_{elec} in a gas discharge is dissipated through radiation (W_{rad}), heat conduction (W_{heat}), diffusion of particles from the discharge (W_{diff}) and acceleration of ions in the sheaths at the walls and electrodes (W_{sheath}), i.e.

$$W_{elec} = W_{rad} + W_{heat} + W_{diff} + W_{sheath} \tag{1}$$

Gas discharges for lighting are designed to maximize the fraction of electrical power emitted as visible or *UV* radiation.

FLUORESCENT LAMPS

Fluorescent lamps are arc discharges, because the electric current required to maintain the discharge is provided by thermionic emission at the cathode. The positive column is the principal source of radiation. Electrons gain energy in the axial electric field, which is dissipated by collisions with atoms and ions in the gas, establishing a steady-state *EEDF,* with an effective electron temperature ~1 eV.

The Positive Column

Typically, 60% of the electrical power in the positive column is dissipated in resonance radiation from the $6p^3P_1$ level of mercury, with a wavelength of 254 nm. A second resonance state ($6p^1P_1$) emits *UV* radiation at 185 nm. There are also a number of visible emissions (mainly in the blue and green) from higher excited states radiating to lower levels. These visible lines are particularly important at high current densities and influence the efficacy and color temperature of the lamp.

Since rare gas atoms are the dominant species in the discharge, they control the ambipolar diffusion rate. Electron momentum-transfer collisions with rare gas atoms determine the electrical conductivity and thence the lamp voltage. Rare gas ionization is negligible during normal FL operation, but is important during starting, when the mercury vapor pressure is small.

Modeling the Positive Column

Numerical models, solving the full collisonal-radiative equations, together with the power balance (cf. equation (1)), have reproduced many of the experimentally measured parameters in FL (electric field, emitted radiation, electron density) despite lack of accurate data for many fundamental processes, particularly cross sections. Details of a number of models have been summarized in [2]. Data requirements for these models are discussed in the following sections

Atomic Processes

One of the difficulties of FL modeling is the lack of reliable atomic data. Most models use the electron impact cross sections for mercury excitation and ionization obtained from swarm data by Rockwood [10]. This data is restricted to collisions with ground state atoms and must be augmented by empirical estimates of interaction with atoms in excited states. The development of theoretical methods for calculating electron impact cross sections near threshold, and computer programs to implement them, has opened the possibility that the necessary data might be accurately calculated. [11,12]. Wani [13] has noted the importance of ionization through "ladder-like" excitation into Rydberg states of Hg. Excitation of rare gas atoms is generally negligible at standard operating temperatures, but becomes important for low mercury vapor pressures, corresponding to low "cold spot" temperatures.

Chemi-ionization was suggested as an important process in FL [14] and is included in a number of models. In particular, inclusion of the Penning ionization process

$$Hg(6p^3P_2) + Hg(6p^3P_2) \rightarrow Hg^+ + Hg(6s^1S_0) + e$$

makes an important contribution to the ionization balance. However, interpretation of experimental data to determine cross sections for some of these processes is unreliable [8], and numerical computations to determine these cross sections [15] have indicated that they are much smaller than previously estimated.

Radiation Transport

The steady state rate equation for the density $n_j(\vec{x})$ of an excited atomic level j with a radiative transition $j \rightarrow k$ at a point $\vec{x}$ in a discharge is (ignoring diffusion)

$$\frac{n_j(\vec{x})}{\tau_{jk}} - \int \frac{P_{jk}(\vec{x} - \vec{x}')}{\tau_{jk}} n_j(\vec{x}') d\vec{x}' = S_j(\vec{x}) \qquad (2)$$

where τ_{jk} is the vacuum transition time, P_{jk} represents the probability of all photons emitted elsewhere in the discharge being absorbed at $\vec{x}$ and S_j represents the net rate of creation of atoms in level j by atomic and molecular collision processes.

An accurate representation of P_{jk} in numerical models is important because of the competition between radiation and collisional quenching, including ladder like

285

processes in the discharge. P_{jk} depends on the average density of absorbing atoms, the gas temperature, the dimensions of the discharge and the densities of atomic species responsible for broadening of the spectral lines.

Monte Carlo methods are the most accurate available for calculating P_{jk}, but they are too computer intensive for direct coupling to FL models. In most models, the P_{jk} integral is replaced by a suitable average quantity, the effective trapped decay rate β_{jk}, so that equation (2) is independent of the origin of photons absorbed in each cell, i.e. $P_{jk}(\vec{x} - \vec{x}') = (1 - \beta_{jk}\tau_{jk})\delta(\vec{x} - \vec{x}')$.

Resonance radiation in mercury is a special case, due to the 5 isotopes in natural mercury, resulting in a complex lumped line profile which reduces the radiation trapping of these lines [16]. Cataphoresis can also reduce radiation trapping [17]. Quenching of the 6^3P_1 level by rare gas atoms plays an important role in broadening of the major resonance line, but reliable data is currently only available for argon.

An analytic formula for β_{jk} for Hg 253.7 nm radiation, based on Monte Carlo calculations and experimental measurements was recently published [18]. The assumption in the model that the 6^3P_1 density profile is close to the "fundamental mode" is however, questionable for many ranges of FL operation, particularly for low opacity (low cold spot temperature). The importance of "non local" radiation transport on the electron kinetics was recently demonstrated using computer modeling [19] and experiment [20].

Electrodes and the Electrode Regions

The cathodes in a fluorescent lamp are multi-coiled helices of tungsten, the interstices of which are impregnated with alkaline-earth oxides to enhance electron emission. During normal operation, they are heated by passing an electric current through the tungsten wire and by ion bombardment from the plasma. The presence of excess barium dissolved in the mixed oxide crystals and at the surface makes the oxides semi-conducting at typical operating temperatures (1200-1400 K) and reduces the work function of the cathode.

The cathode region can be thought of as the "engine" of the fluorescent lamp. "Beam" electrons emerging from the cathode are accelerated through the cathode fall into a region of relatively weak electric field, the negative glow. There is an over production of ions in the negative glow, which is compensated by a dark region of low ionization, the Faraday dark space, which joins the positive column. A number of important atomic processes in this region control the performance of the lamp. These processes also reduce efficacy – in a standard FL, about 7% of the energy in the discharge is dissipated in this region.

Evaporation of barium at the cathode limits the life of the lamp. A function of the buffer gas is to prevent the liberated barium atoms from diffusing to the walls before they are ionized by beam electrons from the cathode sheath and return to that cathode.

The negative glow extends for about one tube radius on either side of the cathode and the Faraday dark space extends for a length slightly smaller than the tube diameter [1]. The cathode sheath is extremely thin (~0.1 mm) and beam electrons enter the negative glow with the full energy of the cathode fall, which depends on the operating conditions, including the amount of auxiliary heating, but it is usually not much greater than the ionization potential of the buffer gas.

In the early 1990's, there was considerable research activity to study the physics of the negative glow with the goal of eliminating the need for a ballast. During steady state operation, FLs, in common with most gas discharges, have a "negative static differential resistance" – as *rms* discharge current *increases*, the *rms* voltage required to maintain the discharge *decreases*. This situation is inherently unstable when driven with a constant voltage source and would allow the current to grow unimpeded. The ballast is the electric circuit device that supplies the necessary impedance to restore stability.

Buffer gas excitation in the negative glow of Hg-Ne discharges can be seen as a red glow from Ne lines, and can also be observed in Hg-Ar discharges using a spectrometer [21]. Absorption measurements in the negative glow of a T12 Hg-Ar fluorescent lamp operating at 400 mA [21,22] showed the absolute Hg^+ density to be a factor of 5 above that in the positive column, while Hg^+ density in the Faraday dark space was a factor of 2 lower than in the positive column. This is consistent with the view of the negative glow and Faraday dark space as "overshoot" and "undershoot" regions respectively. In contrast, the emission intensity from excited Hg was a factor of 2 lower in the negative glow than in positive column, indicating that different mechanisms are important in different regions. Measurements of Ar excited state densities indicated that Penning ionization of Hg is possibly the dominant process for producing ions in the negative glow [21].

Processes at the anode in FL are less important than at the cathode, although the anode fall during 60 Hz operation also represents an energy loss. The anode potential with respect to the plasma (the anode fall) may be positive or negative and depends on the physical processes at the anode.

ALTERNATIVE ATOMIC AND MOLECULAR RADIATORS

Rare Gas Discharges

Neon discharges are the most efficient of all rare gases in producing visible light. The bright red emission of Ne has been used for some special applications, such as the third stop light on some automobiles, but efficacy is typically 13 *lpw*, comparable to incandescent lamps. Emissions in the *VUV* (~70 nm) means the Stokes shift is much larger than in fluorescent lamps, hence they are impractical for general lighting.

Experiments on discharges in Xe [23] and He-Xe mixtures [24] have obtained radiation output ~15 Wm^{-1} with 80% conversion efficiency of electrical power to 147 nm radiation. However, the efficiency markedly decreases as electrical power is increased to levels required for a viable light source. The development of phosphors which convert each quantum of *UV* radiation to more than one photon of visible light would greatly improve the situation. Dielectric barrier discharges in Xe are used for some lighting applications [25], where efficiencies of 60% conversion from electrical power to *VUV* Xe_2^* excimer radiation (which has a broad band emission centered at 172 nm) have been obtained.

Atomic Metal Discharges

There have been a number of attempts to use alternative metallic radiators in LPD lamps, but results have been disappointing. As noted in the introduction, discharges using sodium have very high efficacies, but poor color properties. Most other metals only exist in the vapor state at temperatures which are too high for practical application in these lamps.

The efficacy of barium discharges was found to be highest for wall temperatures of 700 °C. Efficacy was limited by the strong emission in near-*IR* spectral lines. Further, barium reacts with a wide variety of standard lamp materials including glass, silica, and *PCA* and the only suitable material was yttria (Y_2O_3). Lamp operation at such high temperatures implies the need for short, high voltage lamps such as are found in HID lamps. However, the maintenance field of the positive column in these discharges was about 100 V/m, similar to FL, requiring long discharge lamps in order to provide the necessary luminance.

Investigations in atomic discharges in Mo [6], introduced in the form of MoO_3 showed that ~15% of electrical power could be converted to visible light, supplemented by ~21% in the *UV*. The optimum cold spot temperature for MoO_3 was similar to that of Ba, resulting in the same thermal issues.

Molecular Discharges

Since the possibilities for using pure metallic radiators to replace mercury in low pressure discharge lamps appear to be exhausted, interest has turned to molecular radiators. Ideal emitters radiate in the visible spectrum, but if emitters can be found which radiate efficiently in the near-UV, efficacy can be enhanced by the reduced Stokes shift compared to Hg in converting this radiation to visible light by a phosphor.

Molecular discharges may be divided into three classes, , defined by their emission spectrum [7]:

I   Pure molecular emission spectrum

II   Spectrum dominated by lines of one or more of the atomic species of the molecule

III   A mixture of I and II

Sulfur is a good example of Type I. The S_2 molecule radiates almost exclusively in the near *UV*, (280-400 nm), with a peak intensity around 305 nm. Efficiencies of 17% have been reported in low pressure sulfur discharges in argon [26].

Thallium bromide is an example of Type II. Efficiencies of 22% have been observed [7] in TlBr discharges in a rare buffer gas, with the spectrum dominated by visible (555 nm) and near *UV* lines (323-378 nm). Low pressure discharges in other metal halides such as InCl and CuBr are examples of Type III. For example, CuBr, molecules emit in the visible spectrum, while atomic lines of copper are found in the near *UV*.

The N_2 molecule has a broad band of emission lines in the range 300-400 nm, but efficiencies are typically of order 2% [27]. Water has also been investigated as a potential light source [28]. The OH radical has a strong, narrow emission line at 308 nm and a much smaller line peaking at 284 nm, but efficiencies are also less than 2% [7], principally due to energy loss by electron elastic collisions with water molecules. Experiments have also been performed using deuterium [29]. The D_2 molecule has a continuous spectrum from 250 nm to 600 nm, and this is augmented by a strong OD emission line at 308 nm, but no efficiencies have been reported.

SUMMARY AND CONCLUSIONS

Research into LPD for lighting applications has a long and distinguished history. It has only been possible to give a flavor of the subject in this paper and there are many challenges remaining for the development of new and more efficient light sources.

Development of "mercury free" low pressure discharge lamps provides a challenge for the lighting industry. There is a vast array of light emitting molecules, but methods must be found to excite these molecules into radiative sates, while minimizing loss channels due to excitation into vibrational and rotational states, which subsequently diffuse to the wall. On the other hand, great progress has been made in reducing the amount of mercury contained in fluorescent lamps – from 50 mg in 1980 to 8 mg or less today.

A better knowledge of the electron distribution function and how it can be influenced in a gas discharge can lead to new ways of generating plasmas with improved radiation output, as in the case of the dielectric barrier discharge [25]. The development of multi-photon phosphors would also revolutionize the use of LPD for lighting. Finally, the great advances in computer technology, together with parallel advances in "state of the art" diagnostics provide an opportunity for numerical modeling and experimental investigations of these discharges to make significant advances in our understanding impact on product development. Improved data for fundamental atomic, molecular and radiative processes will be of great importance to this development.

REFERENCES

1. J.F. Waymouth, *Electric Discharge Lamps*, The MIT Press: Cambridge, MA, 1971.
2. G.G. Lister, J.E. Lawler, W.P. Lapatovich and V.A. Godyak, *Rev. Mod. Phys.* **76**, 541-598 (2004).
3. J.R. Coaton and A.M. Marsden, editors, *Lamps and Lighting*, Arnold: London, 1997, pp5-69.
4. M.W. Kirby. in *Lamps and Lighting* (J .R. Coaton and A. M. Marsden, eds.), Arnold: London, 1997, pp 227-234
5. X.L. Peng,, J. J. Curry, G. G. Lister, and J. E. Lawler, *J. Appl. Phys.* **91**, 1761-1771 (2002); J. Laski, G. G. Lister, F. Palmer, P. E. Moskowitz, and J. J. Curry, *ibid,* 1772-1779.
6. G.M. Petrov, J.L. Giuliani, A. Dasgupta, K. Bartschat and R.E. Pechacek, *J. Appl. Phys.* **95**, 5284-5294 (2004)
7. R. Hilbig, A. Koerber, J. Baier and R. Scholl, in *Light Sources 2004: Proc. 10ᵗʰ Int. Symp. on the Science and Technology of Light Sources, Toulouse* (G. Zissis, ed.) , Institute of Physics Conf. Series No. 182, 2004, pp75-84.
8. V.A. Sheverev, G.G. Lister and V. Stepaniuk, *Phys. Rev. E* **71**, 056404 1-8 (2005)
9. J.J. Curry, G.G. Lister and J.E. Lawler, *J. Phys. D: Appl. Phys.* **35**, 2945-2953 (2002).
10. D. Rockwood, *Phys. Rev. A* **8**, 2348-2358 (1973).
11. K. Bartschat in *Atomic and Molecular Data and Their Applications, AIP Conference Proceedings #636,* (D. R. Schultz, P.S.Krstic, and F. Ownby Eds.), American Institute of Physics: Melville, NY, 2002, pp 192-201.
12. V. Fursa, I. Bray, and G. Lister, *J. Phys. B: At. Mol. Phys.* **36**, 4255-4271 (2003).
13. Wani, *J. Appl. Phys.* **75**, 4917-4926 (1994).
14. L. Vriens, , R. A. J. Keijser, F. A. S. Ligthart, *J. Appl. Phys.* **49**, 3807-3813 (1978).
15. J.S. Cohen, R. L. Martin, and L. A. Collins, *Phys. Rev. A* **66**, 012717 1-18 (2002)
16. J. Maya and R. Lagushenko, *Advances in Atomic, Molecular and Optical Phys.* **26**, 321-373 (1990).
17. J.J. Curry, J. E. Lawler, and G. G. Lister, *J. Appl. Phys.* **86**, 731-737 (1999).
18. M.T. Herd, J.E. Lawler and K.L. Menningen, *J. Phys. D: Appl. Phys.* **38**, 3304-3311(2005).
19. J.L. Giuliani, G.M. Petrov, J.P. Apruzese and J. Davis, *Plasma Sources Sci. Technol.* **14**, 236-249 (2005).
20. G.M. Petrov, J.L. Giuliani and R.E. Pechacek, *J. Phys. D: Appl. Phys.,* **38**, 4180-4183 (2005)
21. R.C. Wamsley, K. Mitsuhashi, and J. E. Lawler, *Phys. Rev. E* **47**, 3540-3546 (1993)
22. R.C. Wamsley, J. E. Lawler, J. H. Ingold, L. Bigio, and V. D. Roberts, *Appl. Phys. Lett.* **57**, 2416-2418 (1990).
23. T.J.Sommerer and D.A. Doughty, *J. Phys. D: Appl. Phys.* **31**, 2803-2817 (1998).
24. D. Uhrlandt, R. Bussiahn, S. Gorchakov, H. Lange, D. Loffhagen and D. Nötzold, *J. Phys. D: Appl. Phys.* **38**, 3318-3325 (2005).
25. F. Vollkommer, and L. Hitzschke, in *Proc. 8ᵗʰ Int. Symp. on the Science and Technology of Light Sources, Greifswald*, KIEBU-DRUCK GmbH, 1998, pp 138-139.
26. N.D. Gibson and J.E. Lawler, J. Appl. Phys. **79**, 86-92 (1996).; N.D. Gibson, U. Kortshagen and J.E. Lawler, *ibid*, 7523-7528.
27. M. Jinno, S. Takubo, Y. Hazata, S. Kitzinelis and H. Motomura, *J. Phys. D: Appl. Phys.* **38**, 3312-3317 (2005).
28. A.Ya.Vul,, S.V. Kidalov, V.M. Milenin, N.A. Timofeev and M.A. Khodorkovskii, *Tech. Phys. Lett.* **25**, 4-6 (1999); *ibid* 321-323.

29. S. Takubo, Y. Muguruma, H. Motomura, M. Jinno and M. Aano, in *Light Sources 2004: Proc. 10th Int. Symp. on the Science and Technology of Light Sources, Toulouse*, G. Zissis, ed.), Institute of Physics Conf. Series No. 182, 2004, pp 581-582.

The 15th International Conference on Atomic Processes in Plasmas
March 19-22, 2007
National Institute of Standards and Technology (NIST), Gaithersburg, MD

Joseph Abdallah
Los Alamos National Lab
MS B212
Los Alamos, NM 87545 USA
Phone: (505)667-7388
Fax: (505)665-6229
Email: abd@lanl.gov

Kanti Aggarwal
Queen's University
University Road
Belfast, BT7 1NN IRELAND
Phone: 44 28-9097-3239
Fax: 0044-28-9097-3110
Email: K.Aggarwal@qub.ac.uk

Mahamed Ali
NIST
100 Bureau Drive, Mail Stop 8422
Gaithersburg, MD 20899-8422 USA
Phone: (301)975-5335
Email: ali@nist.gov

Rami Ali
University of Jordan
Department of Physics
Queen Rania Street
Amman, 11942 JORDAN
Phone: 962 6 535 5000 ext. 22050
Email: ramimali@ju.edu.jo

Alexander Andriyash
RFNC VNIITF
13, Vasilyeva Street
Snezhinsk, 456770 RUSSIA
Phone: 7 351 46 56560
Fax: 7 351 46 55118
Email: a.v.andriyash@vniitf.ru

James Babb
Harvard-Smithsonian CfA
60 Garden Street
MS 14
Cambridge, MA 02138-1516 USA
Phone: (617)496-7612
Fax: (617)496-7668
Email: jbabb@cfa.harvard.edu

Christina Back
General Atomics
P.O. Box 85608
San Diego, CA 92186-5608 USA
Phone: (858)455-2025
Fax: (858)909-5949
Email: backca@fusion.gat.com

James Bailey
Sandia National Laboratories
P.O. Box 5800
MS-1196
Albuquerque, NM 87185-1196 USA
Phone: (505)845-7203
Fax: (505)845-7820
Email: Jebaile@sandia.gov

Aman Baines
Guru Nanak Dev University
 Amritsar, 143005 INDIA
Phone: (91)0183-2258802   ext. 3345
Email: bainsphysics@yahoo.co.in

Connor Ballance
Rollins College
1000 Holt Avenue
Winter Park, FL 32789 USA
Phone: (407)691-1755
Email: ballance@vanadium.rollins.edu

Mark Bannister
Oak Ridge National Laboratory
Bldg 6010, MS-6372
P.O. Box 2008
Oak Ridge, TN 37831-6372 USA
Phone: (865)574-4700
Email: bannisterme@ornl.gov

Jacques Bauche
Laboratoire Aime Cotton
Batiment 505 Campus d'Orsay
ORSAY, 91405 FRANCE
Phone: 33 1 6935 2105
Email: Jacques.Bauche@lac.u-psud.fr

Claire Bauche-Arnoult
Laboratoire Aime Cotton
Batiment 505 Campus d'Orsay
ORSAY, 91405 FRANCE
Phone: 33 1 6935 2106
Email: Claire.Bauche@lac.u-psud.fr

Djamel Benredjem
University of Paris
Avenue Georges Clemenceau
Orsay, 91405 FRANCE
Phone: 33 1 69 15 55 34
Fax: 33 1 69 15 58 11
Email: djamel.benredjem@lixam.u-psud.fr

Julian Berengut
Auburn University
206 Allison Labs
Auburn, AL 36849 USA
Phone: (334)844-4644
Email: jcb@physics.auburn.edu

Paul Bergstrom
NIST
100 Bureau Drive, Mail Stop 8460
Gaithersburg, MD 20899-8960 USA
Phone: (301)975-5567
Fax: (301)869-7682
Email: paul.bergstrom@nist.gov

Igor Bespamyatnov
Fusion Research Center
2511 Speedway Street
Austin, TX 78712 USA
Phone: (512)232-4136
Email: bespam@physics.utexas.edu

Georg Biedermann
Max-Planck-Institut fur Plasmaphysik
Wendelsteinstrae 1
Greifswald, 17491 GERMANY
Phone: 49 3834 88 10 00
Email: Biedermann@ipp.mpg.de

Jonathan Bray
1008 N. Turner Avenue, Suite 259
Ontario, CA 91764 USA
Phone: (909)483-3394
Email: jdbrayster@gmail.com

Nancy Brickhouse
Harvard-Smithsonian Center for Astrophysics
60 Garden Street, MS 15
Cambridge, MA 02138 USA
Phone: (617)495-7438
Fax: (617)495-7059
Email: nbrick@cfa.harvard.edu

Philip Bucksbaum
Stanford University
PULSE Center, MS 69
2575 Sand Hill Road
Menlo Park, CA 94025 USA
Phone: (650)926-5337
Fax: (650)926-4100
Email: phb@slac.stanford.edu

Michel Busquet
ARTEP, Inc.
2922 Excelsior Springs Court
Elicott City, MD 21042 USA
Phone: (202)404-7802
Email: busquet@this.nrl.navy.mil

Jose Castro
Rice University, Physics & Astronomy
6100 Main Street
Houston, TX 77005 USA
Phone: (713)348-4938
Fax: (713)348-2603
Email: Jose.Castro@rice.edu

Peter Celliers
Lawrence Livermore National Laboratory
M/S L-286
7000 East Avenue
Livermore, CA 94550 USA
Phone: (925)424-4531
Email: celliers1@llnl.gov

Oleg Chefonov
RFNC - VNIITF
Vasiljeva 13
Chelyabinsk region
Snezhinsk, 456770 RUSSIA
Phone: 7 35146 511 55
Fax: 7 35146 5 11 01
Email: dep5@vniitf.ru

Yu-hsin Chen
University of Maryland
IPST Department
College Park, MD 20742 USA
Phone: (301)405-4893
Email: yhchen@umd.edu

Hyunkyung Chung
LLNL
7000 East Avenue
L-399
Livermore, CA 94550 USA
Phone: (925)423-3452
Email: hchung@llnl.gov

Robert Clark
IAEA
Wagramerstrasse 5
Vienna, A-1400 AUSTRIA
Phone: 43 1 2600 21731
Fax: 43 1 26007
Email: r.e.h.clark@iaea.org

James Colgan
Los Alamos National Laboratory
Theoretical Division
MS B283
Los Alamos, NM 87545 USA
Phone: (505)665-0291
Fax: (505)667-1931
Email: jcolgan@lanl.gov

Michael Crisp
U. S. Department of Energy
1000 Independence Avenue, SW
Washington, DC 20585 USA
Phone: (301)903-4883
Fax: (301)903-1225
Email: Michael.Crisp@Science.doe.gov

John Curry
NIST
100 Bureau Drive, Mail Stop 8422
Gaithersburg, MD 20899-8422 USA
Phone: (301)975-2817
Email: jjcurry@nist.gov

Ilya Dodin
Princeton University
Astrophysics Department
Peyton Hall, Ivy Lane
Princeton, NJ 08544 USA
Phone: (609)243-2643
Email: idodin@princeton.edu

Ilija Draganic
NIST
100 Bureau Drive, Mail Stop 8422
Middletown, MD 20899-8422 USA
Phone: (301)975-3212
Email: ilija.draganic@nist.gov

Jacques Dubau
CNRS
Universite Paris-Sud
LIXAM, Bat 350
Orsay, 91405 FRANCE
Phone: 33 169157553
Fax: 33 169155811
Email: jacques.dubau@lixam.u-psud.fr

Daniel Dubin
University of California at San Diego
9500 Gilman Drive
La Jolla, CA 92093-0319 USA
Phone: (858)534-4174
Fax: (858)534-0173
Email: ddubin@physics.ucsd.edu

Tunay Durmaz
UNR
1664 N. Virginia Street
Reno, NV 89557 USA
Phone: (775)784-4815
Email: durmaz@physics.unr.edu

Steven Federman
Department of Physics and Astronomy
University of Toledo
Toledo, OH 43606 USA
Phone: (419)530-2652
Email: steven.federman@utoledo.edu

Uri Feldman
Artep Inc./Naval Research Laboratory
2922 Excelsior Spring Circle
Ellicott City, MD 21042 USA
Phone: (202)767-3246
Fax: (202)404-7997
Email: uri.feldman@nrl.navy.mil

Gary Ferland
Physics Department
University of Kentucky
Lexington, KY 40506 USA
Phone: (859)257-8795
Email: gary@pa.uky.edu

Jorge Filevich
Colorado State University
1320 Campus Delivery
Fort Collins, CO 80523 USA
Phone: (970)491-3964
Email: rage@engr.colostate.edu

Charlotte Fischer
NIST
100 Bureau Drive, Mail Stop 8422
Gaithersburg, MD 20899-8422 USA
Phone: (301)975-2099
Email: charlotte.fischer@nist.gov

Christopher Fontes
Los Alamos National Laboratory
MS F663
Los Alamos, NM 87545 USA
Phone: (505)665-7676
Email: cjf@lanl.gov

Lewis Foster
Los Alamos National Laboratory
Theoretical Division, T-4
MS B283
Los Alamos, NM 87545 USA
Phone: (505)667-0956
Email: foster@lanl.gov

Kevin Fournier
Lawrence Livermore National Laboratory
7000 East Avenue, L-072
P.O. Box 808
Livermore, CA 94551 USA
Phone: (925)423-6129
Fax: (925)422-2243
Email: fournier2@llnl.gov

Jeffrey Fuhr
NIST
100 Bureau Drive, Mail Stop 8422
Gaithersburg, MD 20899-8422 USA
Phone: (301)975-3204
Fax: (301)975-5560
Email: jeffrey.fuhr@nist.gov

John Gillaspy
NIST
100 Bureau Drive, Mail Stop 8422
Gaithersburg, MD 20899-8422 USA
Phone: (301)975-3236
Fax: (301)975-5560
Email: john.gillaspy@nist.gov

Franck Gilleron
Commissariat a l'Energie Atomique
CEA/DIF, B.P. 12
Bruyeres-Le-Chatel, 91680 FRANCE
Phone: 33 169265973
Email: franck.gilleron@cea.fr

Siegfried Glenzer
Lawrence Livermore National Laboratory
P.O. Box 808, L-399
Livermore, CA 94551-0808 USA
Phone: (925)422-7409
Fax: (925)422-0327
Email: glenzer1@llnl.gov

Tom Gorczyca
Western Michigan University
1903 West Michigan Ave
Kalamazoo, MI 49008-5252 USA
Phone: (269)387-4913
Email: gorczyca@wmich.edu

Jonathan Grava
Colorado State University
ERC Room B308
1320 Campus Delivery
Fort Collins, CO 80523 USA
Phone: (970)491-3964
Email: jgrava@engr.colostate.edu

297

Hans Griem
University of Maryland
IREAP
Energy Research Facility
College Park, MD 20742 USA
Phone: (301)405-4981
Fax: (301)314-9437
Email: griem@umd.edu

Peter Hakel
University of Nevada
Department of Physics / 220
Reno, NV 89557 USA
Phone: (775)784-6595
Fax: (775)784-1398
Email: peter_hakel@msn.com

Iain Hall
University of Nevada
1664 North Virginia Street
Reno, NV 89557 USA
Phone: (775)784-6059
Fax: (775)784-1398
Email: iain.m.hall@googlemail.com

Stephanie Hansen
Lawrence Livermore National Laboratory
P.O. Box 808
L-473
Livermore, CA 94557 USA
Phone: (925)422-6187
Email: hansen50@llnl.gov

Peter Hauschildt
Univerity of Hamburg
Hamburger Sternwarte
Gojenbergsweg 112
Hamburg, 21029 USA
Phone: 49 40 42891 4013
Email: yeti@hs.uni-hamburg.de

Scott Howard
Lawrence Livermore National Laboratory
7000 East Avenue
L-18
Livermore, CA 94551 USA
Phone: (925)423-1530
Fax: (925)423-5112
Email: hascott@llnl.gov

Lawrence Hudson
NIST
100 Bureau Drive, Mail Stop 8460
Gaithersburg, MD 20899-8460 USA
Phone: (301)975-5575
Fax: (301)869-7682
Email: lawrence.hudson@nist.gov

Una Hwang
NASA/GSFC and JHU
Code 662
Greenbelt, MD 20771 USA
Phone: (301)286-3632
Fax: (301)286-1684
Email: hwang@milkyway.gsfc.nasa.gov

Robert Ivester
NIST
100 Bureau Drive, Mail Stop 8223
Gaithersburg, MD 20899-8223 USA
Phone: (301)975-6600
Email: robert.ivester@nist.gov

Terrence Jach
NIST
100 Bureau Drive, Mail Stop 8371
Gaithersburg, MD 20899-8371 USA
Phone: (301)975-2362
Fax: (301)417-1321
Email: terrence.jach@nist.gov

Verne Jacobs
Naval Research Laboratory
Code 6390
4555 Overlook Avenue, SW
Washington, DC 20375 USA
Phone: (202)404-7147
Fax: (202)404-7546
Email: verne.jacobs@nrl.navy.mil

Heather Johns
University of Nevada
Department of Physics
900 N. Virginia Street
Reno, NV 89557 USA
Phone: (775)784-4815
Email: hmhill@physics.unr.edu

Brent Jones
Sandia National Laboratories
P.O. Box 5800
MS 1168
Albuquerque, NM 87185 USA
Phone: (505)284-9481
Email: bmjones@sandia.gov

Brian Judd
Johns Hopkins University
33rd and Charles
Baltimore, MD 21218-2686 USA
Phone: (410)516-8693
Fax: (410)516-7239
Email: juddbr@pha.jhu.edu

Henry Kapteyn
University of Colorado/JILA
JILA Tower 440 UCB
Boulder, CO 80309 USA
Phone: (303)492-6763
Email: kapteyn@jila.colorado.edu

Christopher Keane
Department of Energy
1000 Independence Avenue, SW
Washington, DC 20585 USA
Phone: (202)586-0852
Email: chris.keane@nnsa.doe.gov

Thomas Killian
Rice University
6100 Main Street, MS-61
Houston, TX 77005 USA
Phone: (713)348-2927
Fax: (713)348-4150
Email: killian@rice.edu

Mark Kinnane
100 Bureau Drive, Mail Stop 8422
Gaithersburg, MD 20899-8422
Phone: (301)975-4334
Email: mark.kinnane@nist.gov

Marcel Klapisch
Institute for Particle Physics HPK/G28
ETH Honggerberg Campus
Schafmanstrasse 20
Zurich, 8093 SWITZERLAND
Phone: 41 79 2252713
Email: klapisch@phys.ethz.ch

Michel Koenig
Laboratoire LULI
Ecole Polytechnique
Palaiseau, 91128 FRANCE
Phone: 33 1 69334799
Email: michel.koenig@polytechnique.fr

Alexander Kramida
NIST
100 Bureau Drive, Mail Stop 8422
Gaithersburg, MD 20899-8422 USA
Phone: (301)975-8074
Fax: (301)975-5560
Email: alexander.kramida@nist.gov

Thomas Kruecken
Philips Research Laboratories
Weisshausstrasse 2
Aachen, D-52066 GERMANY
Phone: 49 241 6003 ext. 474
Email: thomas.kruecken@philips.com

Martin Laming
Naval Research Laboratory
4555 Overlool Avenue, SW
WASHINGTON, DC 20375 USA
Phone: (202)767-4415
Fax: (202)404-7997
Email: laming@ nrl.navy.mil

Brian Layer
University of Maryland
4213 CSS Building
College Park, MD 20742 USA
Phone: (301)405-4850
Email: layerbd@wam.umd.edu

Richard Lee
LLNL
P.O. Box 808
Livermore, CA 94707 USA
Phone: (925)422-7209
Email: RWLee@berkeley.edu

Darrin Leonhardt
Fusion UV Systems, Inc.
910 Clopper Road
Gaithersburg, MD 20878 USA
Phone: (301)990-8700 ext. 8751
Email: dleonhardt@fusionuv.com

Graeme Lister
Osram Sylvania
71 Cherry Hill Drive
Beverly, MA 01915 USA
Phone: (978)750-1514
Email: graeme.lister@sylvania.com

Petr Loboda
RFNC VNIITF
13, Vasilyeva Street
Snezhinsk, 456770 RUSSIA
Phone: 7 351 46 54730
Fax: 7 351 46 55118
Email: p.a.loboda@vniitf.ru

Stuart Loch
Auburn University
206 Allison Lab.
Auburn, AL 36849 USA
Phone: (334)844-5154
Email: loch@physics.auburn.edu

Bachana Lomsadze
Tbilisi State University
Chavchavadze Av.3
Tbilisi, 0128 GEORGIA
Phone: 995 32 290814
Email: lombacha86@hotmail.com

Roberto Mancini
University of Nevada
Department of Physics
Reno, NV 89557 USA
Phone: (775)784-6595
Fax: (775)784-1398
Email: rcman@physics.unr.edu

Steven T. Manson
Georgia State University
Department of Physics and Astronomy
Atlanta, GA 30303 USA
Phone: (404)651-3982
Fax: (404)651-1427
Email: smanson@gsu.edu

Anna Markhotok
Maine Maritime Academy
1 Pleasant Street
Castine, ME 04420 USA
Phone: (207)326-2342
Email: anna.markhotok@mma.edu

Yitzhak Maron
Weizmann Institute of science
Hertzel Street
Rehovot, 76100 ISRAEL
Phone: 972 8 934 4055
Email: Yitzhak.Maron@weizmann.ac.il

William Martin
NIST
100 Bureau Drive, Mail Stop 8422
Gaithersburg, MD 20899-8422 USA
Phone: (301)975-3213
Email: william.martin@nist.gov

Reinke Matthew
MIT-PSFC
77 Massachusetts Avenue
NW17-225
Cambridge, MA 02139 USA
Phone: (617)253-5395
Email: mlreinke@mit.edu

Stephane Mazevet
CEA-DAM-DIF
BP12
Bruyeres le Chatel, 91680 FRANCE
Phone: 33 1 69266701
Email: stephane.mazevet@cea.fr

Ronald McKnight
8 Michele Court
Gaithersburg, MD 20878 USA
Phone: (301)869-7212
Email: rmcknight@starpower.net

Brendan McLaughlin
Queen's University Belfast
Deparment Applied Maths & Theo. Phys.
David Bates Bldg.
Belfast, BT7 1NN UNITED KINGDOM
Phone: 44 28 90973305
Email: b.mclaughlin@qub.ac.uk

Howard Milchberg
University of Maryland
IPST Building, Room 2123
College Park, MD 20742 USA
Phone: (301)405-4816
Fax: (301)314-9404
Email: milch@ipst.umd.edu

Warren Mori
UCLA
4-913 PAB
UCLA
Los Angeles, CA 90095 USA
Phone: (310)206-0372
Fax: (310)825-4057
Email: mori@physics.ucla.edu

Richard Mushotzky
NASA/GSFC
Code 662
Greenbelt, MD 20771 USA
Phone: (301)286-7579
Fax: (301)286-1684
Email: richard@milkyway.gsfc.nasa.gov

Taisuke Nagayama
Dr. Roberto Mancini
1664 N. Virginia Street
Reno, NV 89557 USA
Phone: (775)784-4815
Email: taisuke@physics.unr.edu

Gillian Nave
NIST
100 Bureau Drive, Mail Stop 8422
Gaithersburg, MD 20899-8422 USA
Phone: (301)975-4311
Email: gillian.nave@nist.gov

Hiroaki Nishimura
Osaka University
Institute of Laser Engineering
2-6 Yamada-oka, Suita
Osaka, 565-0871 JAPAN
Phone: 81 6879 8772
Fax: 81 6877 4799
Email: nishimu@ile.osaka-u.ac.jp

Justin Oelgoetz
Los Alamos National Laboratory
X-1-NAD, MS F663
P.O. Box 1663
Los Alamos, NM 87544 USA
Phone: (505)606-0763
Fax: (505)665-2879
Email: oelgoetz@lanl.gov

Jeffrey Okamitsu
Fusion UV Systems, Inc.
910 Clopper Road
Gaithersburg, MD 20878 USA
Phone: (301)527-2660
Fax: (301)527-2661
Email: jokamitsu@fusionuv.com

Jean-Christophe Pain
Commissariat a l'Energie Atomique (CEA)
CEA/DIF, B.P. 12
Bruyeres-le-Chatel, 91680 FRANCE
Phone: 33 1 69 26 41 85
Fax: 33 1 69 26 70 94
Email: jean-christophe.pain@cea.fr

Olivier Peyrusse
CELIA, Universite Bordeaux
Universite Bordeaux 1
351 Cours de la Liberation
Talence, 33405 FRANCE
Phone: (33)540003767
Email: peyrusse@celia.u-bordeaux1.fr

Michael Pindzola
Auburn University
Auburn, AL 36849 USA
Phone: (334)844-4127
Email: pindzola@physics.auburn.edu

Larissa Podobedova
NIST
100 Bureau Drive, Mail Stop 8422
Middletown, MD 20899-8422 USA
Phone: (301)975-5832
Email: larmitage@nist.gov

Josh Pomeroy
NIST
100 Bureau Drive, Mail Stop 8423
Gaithersburg, MD 20899-8423 USA
Phone: (301)975-5508
Fax: (301)975-5485
Email: joshua.pomeroy@nist.gov

Ryan Porter
University of Kentucky
177 Chem-Phys Bldg.
Lexington, KY 40506 USA
Phone: (859)257-6737
Email: rporter@pa.uky.edu

Cara Rakowski E
Naval Research Laboratory
4555 Overlook Avenue, SW
Washington, DC 20375 USA
Phone: (202)767-6202
Email: Crakowski@ssd5.nrl.navy.mil

Yuri Ralchenko
NIST
100 Bureau Drive, Mail Stop 8422
Gaithersburg, MD 20899-8422 USA
Phone: (301)975-3210
Email: yuri.ralchenko@nist.gov

Joseph Reader
NIST
100 Bureau Drive, Mail Stop 8422
Gaithersburg, MD 20899-8422 USA
Phone: (301)975-3222
Fax: (301)975-5560
Email: joseph.reader@nist.gov

Patrick Renaudin
CEA
DPTA, CEA-DAM Ile-de-France
Bruyeres-le-Chatel, 91680 FRANCE
Phone: (33) 1 69 26 74 18
Fax: (33) 1 69 26 70 77
Email: patrick.renaudin@cea.fr

John Rice
MIT
175 Albany Street
Cambridge, MA 02139 USA
Phone: (617)253-5395
Email: rice@psfc.mit.edu

Jorge Rocca
Colorado State University
1320 Campus Delivery
ERC B329
Fort COllins, CO 80523 USA
Phone: (970)491-8847
Email: matilda@engr.colostate.edu

Gregory Rochau
135 Elite Drive
Tijeras, NM 87059 USA
Phone: (505)845-7540
Email: garocha@sandia.gov

William Rowan
Fusion Research Center
The University of Texas
1 University Station, C1510
Austin, TX 78750 USA
Phone: (512)471-4559
Fax: (512)471-8865
Email: w.l.rowan@mail.utexas.edu

Edward Saloman
NIST
100 Bureau Drive, Mail Stop 8422
Gaithersburg, MD 20899-8422 USA
Phone: (301)975-5554
Email: ebs@nist.gov

Craig Sansonetti
NIST
100 Bureau Drive, Mail Stop 8422
Gaithersburg, MD 20899-8422 USA
Phone: (301)975-3223
Email: craig.sansonetti@nist.gov

Akira Sasaki
Japan Atomic Energy Agency
8-1 Umemidai
Kizu-cho, Soraku-gun
Kyoto, 619-0215 JAPAN
Phone: 81-774-71-3398
Email: sasaki.akira@jaea.go.jp

Daniel Wolf Savin
Columbia Astrophysics Laboratory
550 West 120th Street
MC 5247
New York, NY 10027 USA
Phone: (212)854-4124
Fax: (212)854-8121
Email: savin@astro.columbia.edu

Dieter Schneider
LLNL
7000 East Avenue, L-466
Livermore, CA 94550 USA
Phone: (925)423-5940
Email: schneider2@llnl.gov

Ralf Schneider
Max-Planck-Institut fur Plasmaphysik
Wendelsteinstrae
Greifswald, D - 17491 GERMANY
Phone: 49 3834 882400
Email: ralf.schneider@ipp.mpg.de

David Schultz
Oak Ridge National Laboratory
Physics Division, Bldg. 6010
P.O. Box 2008
Oak Ridge, TN 37831-6372 USA
Phone: (865)576-9461
Fax: (865)574-1118
Email: schultzd@ornl.gov

John Seely
Naval Research Laboratory
Code 7670
4555 Overlook Avenue, SW
Washington, DC 20375 USA
Phone: (202)767-3529
Email: john.seely@nri.navy.mil

Manolo Sherrill
Los Alamos National Laboratory
P.O. Box 1663
Los Alamos, NV 87545 USA
Phone: (505)665-8559
Fax: (505)665-6229
Email: manolo@lanl.gov

Eric Silver
Harvard-Smithsonian Center for Astrophysics
60 Garden Street, MS 15
Cambridge, MA 02138 USA
Phone: (617)496-7858
Email: esilver@cfa.harvard.edu

Randall Smith
JHU
Code 662
NASA/GSFC
Greenbelt, MD 20771 USA
Phone: (301)286-1155
Fax: (301)286-1684
Email: rsmith@milkyway.gsfc.nasa.gov

Vsevolod Soukhanovskii
Lawrence Livermore National Laboratory
7000 East Avenue, L-637
Livermore, CA 94550 USA
Phone: (609)243-2064
Fax: (609)243-2874
Email: vlad@llnl.gov

Peter Stoltz
Tech-X Corp.
5621 Arapahoe Avenue
Boulder, CO 80303 USA
Phone: (720)563-0336
Email: pstoltz@txcorp.com

Phil Stone
NIST
100 Bureau Drive, Mail Stop 8422
Gaithersburg, MD 20899-8422 USA
Phone: (301)975-3211
Email: philip.stone@nist.gov

Dan Stutman
Johns Hopkins University
3400 N. Charles
Baltimore, MD 21218 USA
Phone: (410)516-7929
Email: stutman@pha.jhu.edu

Joseph Tan
NIST
100 Bureau Drive, Mail Stop 8422
Gaithersburg, MD 20899-8422 USA
Phone: (301)975-8985
Email: joseph.tan@nist.gov

Lydia Tchang-Brillet
Universite Pierre et Marie Curie and Observatoire
LERMA Observatoire de Meudon
Meudon, 92195 FRANCE
Phone: 33-1-45 07 75 76
Fax: 33-1-45 07 71 00
Email: lydia.tchang-brillet@obspm.fr

Yong Chia Thio
U.S. Department of Energy
19901 Germantown Road
Office of Fusion Energy Sciences, SC-24.2
Germantown, MD 20874 USA
Phone: (301)903-4678
Fax: (301)903-1225
Email: francis.thio@science.doe.gov

Dan Thomas
General Atomiics
3550 General Atomics Court
San Diego, CA 92121-1122 USA
Phone: (858)455-2403
Fax: (858)455-4156
Email: dan.thomas@gat.com

Todd Ditmire
University of Texas at Austin
Physics Department
2511 Speedway RLM 12.202
Austin, TX 78712 USA
Phone: (512)471-3296
Fax: (512)471-8865
Email: tditmire@physics.utexas.edu

Sanjay Varma
Univ. of Maryland
4213 CSS Building
University of Maryland
College Park, MD 20742 USA
Phone: (301)405-4847
Email: srv@umd.edu

Daniel Vrinceanu Inceanu
Los Alamos National Laboratory
MS B283
Los Alamos, NM 87545 USA
Phone: (505)606-0079
Email: vrinceanu@lanl.gov

Leslie Welser-Sherrill
Los Alamos National Laboratory
4920 Sombra
Los Alamos, NM 87544 USA
Phone: (505)665-3651
Email: lwelser@lanl.gov

Brian Wilson
Lawrence Livermore National Laboratory
Mail Stop L-473
P.O. Box 808
Livermore, CA 94551-0808 USA
Phone: (925)423-4636
Fax: (925)423-7228
Email: wilson9@llnl.gov

Riccardo Tommasini
LLNL
7000 East Avenue
Livermore, CA 94550 USA
Phone: (925)423-8943
Email: tommasini2@llnl.gov

Seth Veitzer
Tech-X Corp.
5621 Arapahoe Avenue
Suite A
Boulder, CO 80303 USA
Phone: (720)974-1848
Fax: (303)448-7756
Email: veitzer@txcorp.com

Glenn M. Wahlgren
Catholic University of America
Department of Physics, 200 Hannan Hall
620 Michigan Avenue, NE
Washington, DC 20064 USA
Phone: (301)286-4531
Fax: (301)286-1752
Email: wahlgren@milkyway.gsfc.nasa.gov

Wolfgang Wiese
NIST
100 Bureau Drive, Mail Stop 8422
Gaithersburg, MD 20899-8422 USA
Phone: (301)975-3201
Email: wolfgang.wiese@nist.gov

Jean Wyart
CNRS
Bat. 505, Univ Paris-Sud
Orsay, 91405 FRANCE
Phone: 33 1 69 35 20 12
Email: jean-francois.wyart@lac.u-psud.fr

Andy York
University of Maryland
4213 CSS Building
College Park, MD 20742 USA
Phone: (301)405-4850
Email: york@physics.umd.edu

Oleg Zabaydullin
CNRS
Universite Paris-Sud
LIXAM, Bat 350
Orsay, 91405 FRANCE
Phone: 33 169157558
Email: oleg.zabaydullin@lixam.u-psud.fr

Sadia Zaheer
Govt. College University, Lahore, Pk
Physics Department
Lahore, 54000 PAKISTAN
Phone: 00923334355201
Email: s4sadi@hotmail.com

Honglin Zhang
Los Alamos National Laboratory
M.S. 663, Group X-1
P.O. Box 1663
Los Alamos, NM 87545 USA
Phone: (505)665-3676
Fax: (505)665-2879
Email: zhang@lanl.gov

AUTHOR INDEX

A

Abdallah, Jr., J., 180
Aglitskiy, Y., 110
Alessi, D., 135
Alexiou, S., 4
Ali, R., 216
Aliabadi, H., 197
Ando, T., 270
Antonsen, T. M., 152
Aota, T., 270
Apruzese, J. P., 229
Atzeni, S., 110
Audbert, P., 19
Audebert, P., 24

B

Back, C. A., 19
Bahati, E. M., 197
Bannister, M. E., 197
Barbrel, B., 19
Baroso, P., 110
Bar-Shalom, A., 206
Bastiani-Ceccotti, S., 19, 24
Baton, S. D., 19
Benuzzi-Mounaix, A., 110
Berrill, M., 135
Blancard, C., 24
Boehly, T. R., 18
Bouquet, S., 110
Bradley, D. K., 18
Brickhouse, N. S., 79, 102
Brygoo, S., 18
Bucksbaum, P. H., 3
Busquet, M., 206

C

Castro, J., 69
Celliers, P. M., 18
Chen, G. X., 79
Chen, Y.-H., 152
Clark, R. W., 229
Cohen, O., 145

Colgan, J., 180
Collins, G. W., 18
Cooley, J. H., 238
Cossé, P., 24
Courtois, C., 110
Coverdale, C. A., 229

D

Davara, G., 4
Davis, J., 229
Deeney, C., 229
Dodin, I. Y., 149
Dubin, D. H. E., 66

E

Eggert, J. H., 18
Ehlerding, A., 197

F

Faenov, A. YA., 110
Falize, E., 110
Faussurier, G., 24
Federman, S. R., 190
Fisch, N. J., 149
Fogle, M. R., 197
Fontes, C. J., 180
Fujima, K., 270
Fujioka, S., 270

G

Gao, H., 69
Gauci, E., 19
Gaudiosi, D. M., 145
Geindre, J.-P., 24
Geppert, W., 197
Gillaspy, J. D., 79, 80
Giraldez, E., 19
Glenzer, S. H., 8
Golovkin, I. E., 238

309